Industry 4.0, Smart Manufacturing, and Industrial Engineering

Industry 4.0 is a revolutionary concept that aims to enhance productivity and profitability in various industries through the implementation of smart manufacturing techniques. This book discusses the profound impact of Industry 4.0, which involves the seamless integration of digital technologies into manufacturing processes within the realm of industrial engineering.

Industry 4.0, Smart Manufacturing, and Industrial Engineering: Challenges and Opportunities thoroughly examines the intricate facets of Industry 4.0 and Smart Manufacturing, offering a comprehensive overview of the challenges and opportunities that this paradigm shift presents to industrial engineers. It provides practical insights and strategies to help professionals navigate the complexities of this evolving landscape. Fundamental components of Industry 4.0 and Smart Manufacturing, ranging from the incorporation of sensors and data analytics to the deployment of cyber-physical systems and the promotion of sustainable practices are covered in detail. The book addresses the obstacles and prospects brought about by Industry 4.0 in the digital age and offers solutions to issues such as data security, interoperability, and workforce preparedness.

The book sheds light on how Industry 4.0 combines various disciplines, including engineering technology, data science, and management. It serves as a valuable resource for researchers, undergraduate and postgraduate students, as well as professionals operating in the field of industrial engineering and related domains.

Advances in Intelligent Decision-Making, Systems Engineering, and Project Management

This new book series will report the latest research and developments in the field of information technology, engineering and manufacturing, construction, consulting, healthcare, military applications, production, networks, traffic management, crisis response, human interfaces, and other related and applied fields. It will cover all project types, such as organizational development, strategy, product development, engineer-to-order manufacturing, infrastructure and systems delivery, and industries and industry-sectors where projects take place, such as professional services, and the public sector including international development and cooperation etc. This new series will publish research on all fields of information technology, engineering, and manufacturing including the growth and testing of new computational methods, the management and analysis of different types of data, and the implementation of novel engineering applications in all areas of information technology and engineering. It will also publish on inventive treatment methodologies, diagnosis tools and techniques, and the best practices for managers, practitioners, and consultants in a wide range of organizations and fields including police, defense, procurement, communications, transport, management, electrical, electronic, aerospace, requirements.

Smart Technologies for Improved Performance of Manufacturing Systems and Services
Edited by Bikash Chandra Behera, Bikash Ranjan Moharana, Kamalakanta Muduli, and Sardar M. N. Islam

Wireless Communication Technologies
Roles, Responsibilities and Impact of IoT, 6G, and Blockchain Practices
Edited by Vandana Sharma, Balamurugan Balusamy, Gianluigi Ferrari, and Prerna Ajmani

Digital Technology Enabled Circular Economy
Models for Environmental and Resource Sustainability
Edited by Bikash Moharana, Bikash Behera, and Kamalakanta Muduli

Industry 4.0, Smart Manufacturing, and Industrial Engineering
Challenges and Opportunities
Edited by Amit Kumar Tyagi, Shrikant Tiwari, and Sayed Sayeed Ahmad

For more information about this series, please visit: www.routledge.com/Advances-in-Intelligent-Decision-Making-Systems-Engineering-and-Project-Management/book-series/CRCAIDMSEPM

Industry 4.0, Smart Manufacturing, and Industrial Engineering

Challenges and Opportunities

Edited by Amit Kumar Tyagi, Shrikant Tiwari, and Sayed Sayeed Ahmad

CRC Press
Taylor & Francis Group
Boca Raton London New York

CRC Press is an imprint of the
Taylor & Francis Group, an **informa** business

Designed cover image: Shutterstock—Jackie Niam

First edition published 2025
by CRC Press
2385 NW Executive Center Drive, Suite 320, Boca Raton FL 33431

and by CRC Press
4 Park Square, Milton Park, Abingdon, Oxon, OX14 4RN

CRC Press is an imprint of Taylor & Francis Group, LLC

ISBN: 978-1-032-75327-0 (hbk)
ISBN: 978-1-032-75411-6 (pbk)
ISBN: 978-1-003-47388-6 (ebk)

DOI: 10.1201/9781003473886

Typeset in Times
by Apex CoVantage, LLC

Contents

Preface

Today, Industry 4.0 and the rise of Smart Manufacturing have changed smart era for industrial engineering and manufacturing processes. In an increasingly interconnected and data-driven world, traditional manufacturing paradigms are being reshaped, and new horizons of innovation are emerging. This book discusses the interesting changes and exciting prospects that this new industrial revolution (i.e., Industry 1.0 to 4.0) brings to the forefront.

Industry 4.0 represents a convergence of cutting-edge technologies, including the Internet of Things (IoT), artificial intelligence (AI), big data analytics, and advanced automation, among others. These technologies are reshaping the manufacturing landscape, enabling companies to achieve unprecedented levels of efficiency, flexibility, and productivity. Smart Manufacturing leverages these technologies to create intelligent and adaptable production systems that respond swiftly to market demands, reduce waste, and enhance product quality.

As the readers go through this book, they will be on a journey through the multi-faceted facets of Industry 4.0 and Smart Manufacturing. Our aim is to provide a comprehensive overview of the challenges and opportunities that these trends present to industrial engineers, as well as to offer practical insights and strategies for navigating this complex terrain.

Hence, this book will discuss the key components of Industry 4.0 and Smart Manufacturing, from the integration of sensors and data analytics to the implementation of cyber-physical systems and the adoption of sustainable practices. We will explain few real-world case studies, showcasing how leading companies have embraced these concepts to gain a competitive edge in their respective industries.

This book will also address the challenges and difficulties that must be overcome to fully realize the potential of Industry 4.0 and Smart Manufacturing. We hope that you will enjoy a lot in reading this book.

With regards,
Amit Kumar Tyagi
Shrikant Tiwari
Sayed Sayeed Ahmad

About the Editors

Amit Kumar Tyagi is an assistant professor at the National Institute of Fashion Technology, 110016, New Delhi, India. Previously he worked as an assistant professor (Senior Grade 2) and senior researcher at Vellore Institute of Technology (VIT), Chennai Campus, 600127, Chennai, Tamil Nadu, India for the period of 2019–2022. He received his Ph.D. in 2018 from Pondicherry Central University, 605014, Puducherry, India. He joined the Lord Krishna College of Engineering, Ghaziabad (LKCE) from 2009–2010 and 2012–2013. He was assistant professor and head researcher at Lingaya's Vidyapeeth (formerly known as Lingaya's University), Faridabad, Haryana, India for the period of 2018–2019. His supervision experience includes more than 10 master's theses and one Ph.D. dissertation. He has contributed to several projects such as "AARIN" and "P3-Block" to address some of the open issues related to the privacy breaches in Vehicular Applications (such as Parking) and Medical Cyber Physical Systems (MCPS). He has published over 200 papers in refereed, high-impact journals, conferences and books, and some of his articles have been awarded best paper awards. Also, he has filed more than 25 patents (nationally and internationally) in the area of deep learning, Internet of Things, cyber physical systems, and computer vision. He has edited several books for CRC Press. Also, he has authored books on intelligent transportation systems, vehicular ad-hoc network, machine learning, and Internet of Things. He is a winner of a Faculty Research Award for the years of 2020, 2021, and 2022 (consecutive three years) given by Vellore Institute of Technology, Chennai, India. Recently, he has been awarded the best paper award for a paper titled "A Novel Feature Extractor Based on the Modified Approach of Histogram of Oriented Gradient", in ICCSA 2020, Italy (Europe). His current research focuses on next-generation machine-based communications, blockchain technology, smart and secure computing, and privacy. He is a regular member of the ACM, IEEE, MIRLabs, Ramanujan Mathematical Society, Cryptology Research Society, and Universal Scientific Education and Research Network, CSI and ISTE.

Shrikant Tiwari (Senior Member, IEEE) received his Ph.D. in Computer Science & Engineering (CSE) from the Indian Institute of Technology (Banaras Hindu University), Varanasi (India) in 2012 and M.Tech. in Computer Science and Technology from University of Mysore (India) in 2009. Currently, he is working as associate professor in the Department of Computer Science & Engineering (CSE), School of Computing Science and Engineering (SCSE), at Galgotias University, Greater Noida, Uttar Pradesh (India). He has authored or co-authored more than 75 national and international journal publications, book chapters, and conference articles. He has five patents filed to his credit. His research interests include machine learning, deep learning, computer vision, medical image analysis, pattern recognition, and biometrics. Dr. Tiwari is a member of ACM, IET, FIETE, CSI, ISTE, IAENG, and SCIEI. He is also a guest editorial board member and a reviewer for many international journals of repute.

Sayed Sayeed Ahmad is a seasoned academician with nearly 20 years of experience in the educational sector across the UAE. He earned his Ph.D. in management from Banasthali Vidyapith, India, and another Ph.D. in computer science and engineering from Integral University, India. Dr. Ahmad has served at prestigious institutions like De Montfort University Dubai, Rochester Institute of Technology Dubai, University of Dubai, and Al Ghurair University Dubai, showcasing his expertise in a wide array of subjects from machine learning to computer engineering. His contributions extend beyond teaching to include curriculum development, quality assurance, and research with publications and patents in advanced technology fields.

Contributors

O. Abioye
Nigerian Defence Academy
Kaduna, Nigeria

R. Olayinka Adelegan
Nigerian Defence Academy
Kaduna, Nigeria

Arivazhagan N.
Department of Computational Intelligence
SRM Institute of Science and Technology
Kattankulathur, India
Chennai, Tamil Nadu, India

J. Olalekan Awujoola
Nigerian Defence Academy
Kaduna, Nigeria

Vishalsinh V. Bais
Department of Computer Science and Engineering
Prof Ram Meghe College of Engineering & Management
Badnera, Amravati, Maharashtra, India

Kiran Bellam
Prairie View A&M University
Prairie View, Texas

Amol P. Bhagat
Department of Information Technology
Prof Ram Meghe College of Engineering & Management
Badnera, Amravati, Maharashtra, India

Aswani Kumar Cherukuri
Vellore Institute of Technology
Vellore, Tamil Nadu, India

Yugen Chokshi
Illinois Institute of Technology
Chicago, Illinois

Dilliraj Ekambaram
SRM Institute of Science and Technology, Kattankulathur
Chennai, Tamil Nadu, India

T. Aniemeka Enem
Airforce Institute of Technology
Kaduna, Nigeria

Gouthaman P.
Department of Networking and Communications
SRM Institute of Science and Technology
Kattankulathur, India
Chennai, Tamil Nadu, India

Shrikant M. Harle
Department of Civil Engineering
Prof Ram Meghe College of Engineering & Management
Badnera, Amravati, Maharashtra, India

Ida Hector
Vellore Institute of Technology (VIT)
Chennai, Tamil Nadu, India

Jayaprada S. Hiremath
Dept. of Computer Science and Engineering
Sai Vidya Institute of Technology
Bangalore, Karnataka, India

Mrutyunjaya S. Hiremath
Math Technology, Pvt. Ltd
Bengaluru, Karnataka, India

Shantala S. Hiremath
Tata Elxsi Limited
Bengaluru, Karnataka, India

Chetan R. Ingole
Department of Computer Science and Engineering
Prof Ram Meghe College of Engineering & Management
Badnera, Amravati, Maharashtra, India

Archana Jadhav
Dr. D.Y. Patil Institute of Engineering Management & Research
Dr. D.Y. Patil International University
Pune, Maharashtra, India

Sharath Kumar Jagannathan
Saint Peter's University
Jersey City, New Jersey

Rekha Kashyap
KIET Group of Institutions
Ghaziabad, UP, India

Thong Chee Ling
UCSI University
Kuala Lumpur, Malaysia

R. Maheswari
Vellore Institute of Technology
Chennai, Tamil Nadu, India

Bijoy Kumar Mandal
NSHM Knowledge Campus
Durgapur-Group of Institutions
Durgapur, West Bengal, India

Bireshwar Dass Mazumdar
Bennett University
Greater Noida, Uttar Pradesh, India

Dibyendu Mukherjee
Bengal College of Engineering and Technology
Durgapur, West Bengal, India

Krishnaraj Nagappan
SRM Institute of Science and Technology
Chengalpattu, Tamil Nadu, India

Meghna Manoj Nair
Computer Science and Engineering Department
New York University Tandon School of Engineering
New York, New York

Nallarasan V.
Department of Networking and Communications
SRM Institute of Science and Technology
Chennai, Tamil Nadu, India

F. N. Ogwueleka
University of Abuja
Abuja, Nigeria

Priya Parate
Department of Computer Engineering
Rajiv Gandhi Institute of Technology
Mumbai, Maharashtra, India

Nikunj R. Patel
UW—Madison
Madison, Wisconsin,

Rahul K. Patel
Illinois Institute of Technology
Chicago, Illinois

Vijayakumar Ponnusamy
Department of Electronics and Communications
SRM Institute of Science and Technology
Chennai, Tamil Nadu, India

Poornima
Sri Sairam Engineering College, Chennai
Chennai, Tamil Nadu, India

Praveen N.
University of Technology and Applied Sciences, Ibra
Al-Sharqiyah North Governorate, Sultanate of Oman

Benson Edwin Raj
Higher Colleges of Technology, Fujairah Women's Campus
Fujairah, Emirate of Fujairah, United Arab Emirates

Nilesh Rathod
Mumbai, Maharashtra, India

Sharmila Rathod
Department of Computer Engineering
Rajiv Gandhi Institute of Technology
Mumbai, Maharashtra, India

R. Ravinder Reddy
Chaitanya Bharathi Institute of Technology
Hyderabad, Telangana, India

Rukmani P.
Vellore Institute of Technology (VIT)
Chennai, Tamil Nadu, India

Rajasegar Rajendhiran Shanthi
Cyber Security, IT Industry
Dundalk, County Louth, Ireland

Arti Singh
Dr. D.Y. Patil Institute of Engineering Management & Research
Dr. D.Y. Patil International University
Pune, Maharashtra, India

Prachi Singh
University of Glasgow (Scotland)
Glasgow, Scotland, United Kingdom

Richa Singh
KIET Group of Institutions
Ghaziabad, UP, India

Sujith Kumar Sivanandan
Electronics for Imaging India Pvt. Ltd.
Karnataka, India
Bengaluru, Karnataka, India

T. Sridevi
Chaitanya Bharathi Institute of Technology
Hyderabad, Telangana, India

Priyanga Subbiah
SRM Institute of Science and Technology
Chengalpattu, Tamil Nadu, India

Lipsa Subhadarshini
Kalinga Institute of Technology
Bhubaneswar, India

Shirly Sudhakaran
Vellore Institute of Technology
Chennai, Tamil Nadu, India

Tamil Selvi A.
SRM Institute of Science and Technology
Chengalpattu, Tamil Nadu, India

Lakshmi Sravani Thummapudi
Vellore Institute of Technology
Vellore, Tamil Nadu, India

Amit Kumar Tyagi
National Institute of Fashion Technology
New Delhi, Delhi, India

Ravi Uyyala
Chaitanya Bharathi Institute of Technology
Hyderabad, Telangana, India

Samrudh Wagh
Cloud Engineer
Dublin, Ireland

Nemanja Zdravkovic
Metropolitan University, Serbia
NIS, Serbia

1 Introduction to Industry 4.0

Dibyendu Mukherjee and Bijoy Kumar Mandal

1.1 INTRODUCTION: INDUSTRY 4.0—WHAT IT IS AND WHY IT MATTERS

"Industry 4.0" is another name for the current technology growth, which is sometimes called the "Fourth Industrial Revolution". Combining digital and physical systems into smart ones that are all linked together constitutes the First and Second Industrial Revolutions. In 2011, Germany's "High-Tech Strategy 2020" launched "Industry 4.0". The concept was originally used to how digital businesses will automate, improve efficiency, and increase productivity. The successes of Industry 4.0 were built on those of earlier industrial revolutions:

1.2 THE PROGRESSION OF THE NUMEROUS INDUSTRIAL REVOLUTIONS

Technology has gradually transformed manufacturing and production processes into Industry 4.0 [1]. The four industrial revolutions and Industry 4.0 are summarized as follows:

1. **Industrial Revolution I:** "Industry 4.0", often referred to as the "Fourth Industrial Revolution", signifies a notable advance in fusing digital and physical systems, forming connected bodies. The principal groundwork of this evolution is that introducing digitization into corporations will automate workflows, enhance effectiveness, and escalate manufacturing capacities.
2. **Industrial Revolution II:** Industry underwent a remarkable shift during the closing decades of the 19th century and was well underway into its transition by the early 20th century, signifying what can only be described as an industrial metamorphosis. Innovations in large-scale production methods drove this change, along with breakthroughs in assembly line systems. Signature inventions like the internal combustion engine reshaped our mobility cultures, while telegraphy and telephony redefined how we communicate over distances. This time frame also brought about accelerated developments within rail transportation technologies alongside energy creation through electricity which pushed forward progress globally at breakneck speed, fuelling international commerce to greater heights.
3. **Industrial Revolution III:** Leaping more modern eras, we notice significant progress brought on by internet-based information technology changes, the use of microprocessors, and the broad acceptance of personal computers. The

DOI: 10.1201/9781003473886-1

end years of the 20th century ushered in what is often referred to as both the "digital revolution" and "information era", entwining seamlessly with today's Third Industrial Revolution definition. During this period, various economic landscapes experienced growth led primarily by IT investments reminiscent of automation solutions coupled with continually advancing electrical components. This progression resulted in unparalleled advancements, like digital communication strategies and effective computer-regulated systems, allowing for deeper integration across different social fields within society.

4. **Industrial Revolution IV:** Industry 4.0 is a time of rapid industrial technology advancement. Artificial intelligence, data analytics, virtual-physical amalgamations, and the Internet of Things are discussed at this time. Industry 4.0's underpinnings include cloud computing, robots, additive manufacturing, 3D printing, Big Data processing, simulations, and AR/VR technologies, all of which work together to create an interconnected, smart production infrastructure that makes fast decisions for bespoke customer experiences and supports supply chains. Key improvements include synchronized digital and physical processes driving enterprise-wide automation based on actionable insights.

1.3 KEY OBJECTIVE OF INDUSTRY 4.0

Increasing functional efficiency, enabling real-time data-driven decisions, product sophistication and customization, supply chain optimization, and innovation are the main aims. Industry 4.0 promotes "intelligent factories", where equipment works together to boost production, robustness, and speed. This transition doesn't simply influence industrial settings; healthcare, transportation, energy, and farming may all alter [2]. Industry 4.0 has many benefits for enterprises, but it may also pose economic and societal issues. Novel business models are needed to address data solvent protection and labour force dynamics. If these organizations want to develop, they must find ways to overcome current challenges while exploiting the many advantages of industry four-point zero evolution.

1.4 KEY FEATURES OF INDUSTRY 4.0

a) **Linkage:** The fusion of multiple technologies and tools is instrumental in enabling machinery to relay and examine data instantaneously.
b) **Mechanization:** Industry 4.0's innovative tech houses the capability to fully mechanize many factory processes, thus minimizing human interference while augmenting productivity rates and effectiveness.
c) **Information Evaluation:** Predictive sustenance and enhancement are achievable through accumulation, preservation, and scrutiny of sizeable quantities of information procured from industrial activities which can then be channelled towards cost savings as well as quality improvement.
d) **Diffusion of Power:** Independent decision-making capabilities brought forth by Industry 4.0 imply that robots, along with other systems, maintain discretion for autonomous choices, thereby reducing dependence on human

intervention significantly. This results in not just heightened convenience but also augmentation in personalized production, once again contributing positively to both terms customer satisfaction and revenues.

1.5 TECHNOLOGIES FOUNDATIONAL TO INDUSTRIES 4.0

Digitalization and system unification drive Industry 4.0, which redefines industrial practices. Through the Internet of Things, common goods with sensors, motors, and network connections exchange data, forming CPS (Cyber-Physical Systems) that combine computation and communication to link physical and digital space [3]. IoT and CPS provide immediate communication between machines, devices, and integrated systems, improving decision-making processes.

Big data analytics extracts useful insights and recurring trends from massive amounts of data, while artificial intelligence uses cognitive computing and deep learning to mimic human thought patterns [4]. By merging AI and big-data analysis, organizations may extrapolate functional insight from massive datasets to make wiser choices about maintenance and process improvements.

Cloud-based solutions allow customers to store and retrieve important data online and provide flexible processing with adaptable scalability possibilities at low cost. However, Edge Computing moves these storage capabilities closer to the dataset's origin, reducing latency and enabling real-time important decisions in the workplace. Combining Cloud and Peripheral corporate computing designs guide Industry 4.0 deliverables, simplifying bulk analysis/processing/storage procedures are crucial operational metrics.

Robotic automation technologies operating in smartly mechanized environments using complex advanced sensor tech and advanced actuators slapped on measures made possible by potentially limitless, unimagined manifestations of AI capabilities empowering these new-age worker-bots to tackle numerous tasks with incredible efficiency and certainty without compromising their collaborative modus operandi. Additive manufacturing, also known as 3D printing, lets users magically convert mundane, everyday objects into magical, three-dimensional, manipulated contours to capture unfettered, Ipsilateral, original creation designers' thoughts and intentions. Additive printing allows flexibility to explore unexplored design zones and offers new creative alternatives. Curious experts may accelerate product development while exploring on-demand customized manufacturing for specialized needs.

AR and VR technology creates domain-specific environments and generates, monitors, and stores sensory input. Virtual Reality creates a complex artificial environment, whereas AR adds digital information to the actual world, making cognition simpler. These technical advances in AR/VR may help coordinate specialized applications like training programmes and routine maintenance, and they may enable effective remote collaborations for seeing physical prototypes' restricted eyesight.

Multiple cutting-edge technologies surrounding the Industry 4.0 platform enable smart factories to integrate immediate monitoring protocols, control mechanisms, predictive maintenance scheduling, automated operations, and efficiency optimization. Much noise has been made in related forums about how mainstreaming

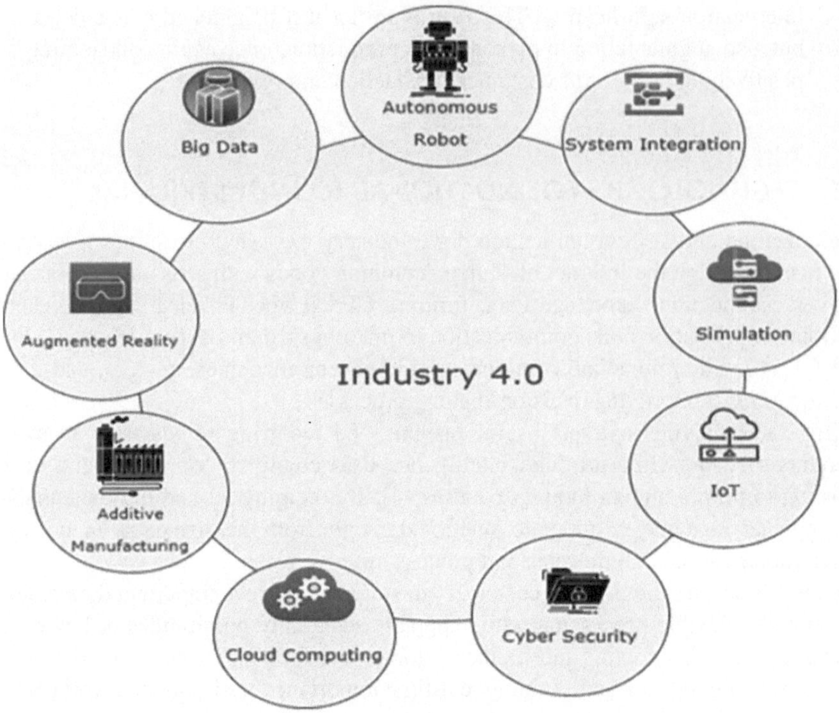

FIGURE 1.1 Technologies Crucial to the Fourth Industrial Revolution.

such new-age pioneering techniques is essential to leveraging positive outcomes to help corporations succeed in rapidly evolving markets with tight competition. Figure 1.1 shows some of the technologies that are crucial to the Fourth Industrial Revolution.

1.6 INDUSTRY 4.0 SIGNIFICANT IMPLICATIONS FOR VARIOUS INDUSTRIES

Industry 4.0 impacts an array of sectors, from healthcare and transportation to communications, alongside manufacturing lines. It paves the path for custom-flexible production facilities in line with consumer preferences. Its footprint manifests within the health sector through affordable yet superior patient-focused solutions whilst also empowering autonomous transport systems and sophisticated delivery infrastructure that consequently reduce costs.

The fourth wave has launched interconnected gadgets, creating a platform where people can converse live-time with automated processes. However, leveraging such pioneering progressions associated with Industry 4.0 poses risks, including cyber vulnerabilities, necessitates substantial investment in state-of-the-art technologies plus overall infrastructural upliftment, and demands upskilling workforce techniques along with establishing new regulations or legislations. To summarize: The Fourth

Industrial Revolution harbours immense potentiality towards restructuring organizational dynamics and unearthing untapped growth avenues albeit it warrants thorough groundwork, significant economic input, as well as involved negotiations amongst pivotal stakeholders.

1.7 RESPONSE OF MANUFACTURING TO INDUSTRY 4.0

Industry 4.0, also termed the Fourth Industrial Revolution, transitions from conventional factory set-ups to "intelligent facilities". These institutes use state-of-the-art tech in a synchronized space thriving on data for refinement and efficiency of manufacturing procedures via digital means. Industry 4.0 allows companies to gather and dissect data promptly enabling improved methods while reducing expenses and enhancing quality assurance processes across their functions [5]. The noteworthy result of this revolution is that it brings automation into play superseding traditional production operations—roles like assembly line monitoring, packaging, or maintaining consistent product quality can be effectively managed by robotics combined with artificial intelligence mechanisms.

Automated systems are instrumental in uplifting product consistency besides trimming down operational costs significantly. Another core aspect that amplifies manufacturing strategies under Industry 4.0 revolves around predictive maintenance powered by sensors along with precise analytics that monitor industrial machinery in real life [6], letting manufacturers steer clear of costly impediments and thereby curtailing regular upkeep expenses rendering them dynamic, productive, and responsive due involvement digital tools and automated mechanics facilitating control regulation tracking over operation phases [7].

Decoration markings distinguish smart factories and represent interlinked networks where machines devices equipment communicate unceasingly, concurrently preparing action, plans data, sourced deputy instruments, technological installations, and fabric lines assessed imparting intelligent ability. These institutions exploiting collected knowledge processing, it using algorithms, contributing greatly, cautious election early invasive decisions. Automation robots pivotal fast forwarded work routine, decrease manual interventions they bring about and handling materials, assembling checks balances, and packages great precision effectiveness.

- **Digital Twin:** A digital double stands as a digitally created doppelganger of an actual product, service, or system. Through this mechanism, one can simulate and visualize real-life manufacturing methods for productivity enhancement and robust preventive maintenance.
- **Imitated Modelling & Simulation:** Digital transformation paves the path for making virtual models/simulations usable in refining production mechanisms. Virtual enactments assist in identifying bottlenecks to streamline layouts and testing schemas, thus elevating performance efficiency even before the physical application becomes necessary.
- **System Connectivity & Integration:** One key aspect of digitization is linking numerous systems/mechanisms/devices within a singular production domain. Such systematic unification simplifies data sharing among

stakeholders throughout the entire value chain spectrum, hence dealing meticulously with comprehensive information.

- **Data Analysis through Digitization Process:** Digitized protocols enable end-to-end collection/analysis cycle at every step of developing products/services. This allows improvement in overall delivery by embracing new-age technologies like machine learning, which makes it possible to identify patterns and inefficacies, thereby opening up scope for optimizing processes
- **Connection across Supply Chain Network:** The emergence of digitization encourages meticulous amalgamation along supply chains stretching from raw material providers right up to consumers. Absolute transparency over communication channels coupled with stock management optimization are some takeaways from such integrations that prove beneficial towards improved business functions leading up to a satisfied clientèle base. This results in a better-connected ecosystem paving the way-out smoother manufacturing measures.

1.8 INTELLIGENT SUPPLY CHAIN AND LOGISTICS

Industry 4.0 requires smart logistic systems and supply chains to optimize product flow, operational effectiveness, and customer happiness. These clever networks monitor items live using RFID, GPS, and IoT sensors. Without shipment tracking, logistics monitoring and coordination are almost impossible. The deployment analytics enabled by AI evaluates historical data, current market trends, and consumer behaviour patterns to estimate demand accurately. Leading organizations to smarter inventory management strategies involves managing manufacturing schedules and building robust transportation setups to project and control costs, improving productivity output plans, and reducing expenses.

AI algorithms used in supply chain processes help maintain ideal stock levels across all nodes by taking inherent factors like varying demands, procurement time frames, potential production capabilities, and transport-related limitations into account, reducing transportation costs and preventing future shortages or surplus stock availability issues.

An intelligent delivery network with real-time data and powerful, AI-aided tactics significantly improves shipping efficiencies. Smart, dynamic routing modules consider fuel efficiency, window period timings, and traffic conditions to cut unnecessary spending and schedule times and avoid compromising global environmental protections.

Digital platforms between company partners helping achieve similar objectives enhance embedded collaboration qualities. Reducing lead times, ensuring quality, standardizing procedures, and adopting effective need replacement techniques sooner is a benefit of this programme.

Blockchain technology supports major add-ons where transaction recording maintenance is hassle-free, avoiding fraudulent activities, if any builder's confidence level among sectors involved mandates, smoother operations digitally, tamper-proof, decentralized ledger trains record of both parties transactions agreements certifications, proven product source, and displaying authenticity.

Critical in logistics, "last-mile delivery" is the last chapter in a flawless chain-to-customer delivery. Intelligent Link uses modern technology like route optimization, drones, automated self-driving vehicles, and crowd shipping nodes to ensure fast, efficient distribution schedules, improve end-user experience, and reduce operational costs. Smart supply chains incorporate sustainability and environmental effect. They employ data analysis and optimization algorithms to control transportation routes, fuel use, emissions, and eco-friendly packaging. This meets sustainability objectives and reduces environmental damage.

Supply chain analytics may boost efficiency and customer satisfaction, which are valuable in Industry 4.0. By leveraging digital technology, big-data analysis tools, and collaborative platforms, they optimize process-flow operations, inventory management solutions, cost expenditure, and risk reduction, boosting supply line efficiency.

1.9 CUSTOMIZATION AND PERSONALIZATION OF PRODUCTS

Industry 4.0 relies on product modification and customization due to changing technology and customer preferences [8]. These ideas focus on personalizing things to match client needs, creating extraordinary experiences, and using modern tech for mass customization. Product adaptability in Industry 4.0 is examined here.

In Industry 4.0, corporations discover client needs and meet them. Online surveys and social media monitoring assist them in understanding their customers' habits, personalities, and buying patterns. These pieces of information provide companies an advantage by allowing them to personalize products and services to today's discriminating customers.

When customizing procedures, data inspection tools and AI are crucial. Businesses may identify recurrent patterns, new trends, and individual preferences by digging into massive consumer data sets. Implementing AI systems provides important forecasts and suggestions that aid in design tuning, improving the shopping experience.

Customers have more opportunities to customize products to meet their demands. Companies provide diverse colours, textures, and functions. "Thumbnail Url": Industrial-size, perfect manifestation-vision gives consumer ownership and a unique identity. Persistence via customizable commodities makes platform voice expressiveness normal.

Newer elements like additive manufacturing, agile digital modelling platforms, and flexible generating methods have enabled model transition beyond laboratories for large-scale applications. Mass customized productivity levers help both economic scales, enabling businesses to generate considerable tailored catering during peak demand periods with rising expenses and delayed delivery. Suitable interface interaction tools provide new engagement to create a dream, produce virtual attribute, pick modifiable forms, and visualize future changes by previewing output. These outlets boost consumer interaction and ease procedure customisation.

In Industry 4.0, organizations may gather feedback-inclusive analytics to see how consumers respond to personalized items before creating an outstanding feedback loop. This helps firms understand customer happiness and identify areas for development in customized tactics and account-specific improvement journeys. By incorporating personalization and customization into a product line, a company enables its customers to generate growth, boost happiness, and get the competitive edge they deserve. Conglomerates may create full experiences that fulfil one-step symbol uniqueness by combining big-data analytics, AI, and digital technologies. This optimizes production connections, which are vital in group assembly tasks like line identification.

1.10 WORKFORCE TRANSFORMATION AND HUMAN-MACHINE COLLABORATION

Change within the workforce and synergizing humans with machines are vital facets of the Fourth Industrial Revolution, or Industry 4.0, a phase marked by advanced technology and automation which cause drastic shifts in roles performed by flesh-and-blood workers.

Industry 4.0-related elements like artificial intelligence and mechanized robots, along with other forms of automation, contribute to an evolving demand for novel areas of skill expertise at workstations. The increasing tendency towards automating rule-based, monotonous tasks has spiked up demands for skills such as data interpretation and software coding abilities, alongside problem-solving prowess and robust analytical thinking capabilities [9]. Both individuals as well enterprises need to invest heavily into learning ventures that align seamlessly with fluctuating requisites introduced due to advancements in digital era technologies.

Imperative is understanding how the collaboration between man and machines could maximize both parties' strengths rather than becoming substitute entities; while human beings offer cognitive excellence coupled with creativity besides possessing empathy and highly sophisticated decision-making capacities, state-of-the-art systems excel in repetitive performances, data extrapolation techniques, and robotic precision, aiding base level processes [10, 11]. Humans leveraging machinery should ideally involve clear demarcations about collective responsibilities and delineated duties, paving the way for seamless interaction interfaces, besides promoting mutual, respect-infused, collaborative spirits.

Human operators can also be enhanced via integrating them through multifaceted, digitized tools, inclusive but not parqueted exclusively wearable apparatuses, alternate reality (AR) experiences supplemented by immersive Virtual Reality (VR), improving floor shop worker efficiency, largely keeping hands free, adding onto workspace safety precautions, too. Moreover, this presents real-time analytics and strategic guidance, helping improve productivity metrics significantly one step further because operations remain spearheaded, taking context-driven perspectives far off from superfluous noise constraints

Moreover, Co-bots hybrid mechanical helpers—operating side-by-side with fellow human employers—get built-in hi-tech sensors, serving dual purposes, first initializing productive alliterations reinforcing higher trust-building strategies plus mitigating risks effectively with humans. On the contrary, they can rotate jobs by attending repetitive work, instead of a human freeing up time spaces, helping devote attention towards cerebral chores and improving their happiness quotient overall, hence leading to higher quality and output.

Positively aiding these set tasks within the Industry 4.0 context are data analysis AI algorithms and overseen machine learning capabilities, enabling minute examination of greater details, extracting insights/trends/patterns for decision-making processes, promoting enhanced workflow productivity.

Lastly, defined Industry 4.0 wouldn't be possible without instilling perennially growing cultural change, where workers constantly stay updated and stay ahead, making agile embraces, mastering new stride areas supported organizations' fully by curating tailored training programs, growing digital campuses, or creating mindset strengths, driving resource progressive experimental innovations, incorporating learnings, failure stories alike. "Industry 4.0" sparks a reimagining of professional roles and duties. With the rise in automation handling everyday tasks, employee engagement shifts towards more intricate, innovative, and value-driven actions. Occupational profiles gradually adapt to centre on troubleshooting issues, stimulating innovation, interacting with customers effectively, or making decisions intelligently. Awareness should be there within corporate bodies that they need to reassess job descriptions while nurturing opportunities for talent augmentation. All this is geared towards facilitating an effortless shift for workers as their traditional role gets disrupted.

As digital solutions gain robust footing inside our workplaces ethical considerations paired with worker welfare demand the utmost attention. Our corporations must guarantee that implementing automated systems along with AI technologies shouldn't put jobs under threat or displace employees from work. They are required to maximize focus on staff welfare by offering formidable support networks addressing relevant concerns as well as creating conducive environments promoting trust inclusion coupled with healthy balancing between office-life balance.

1.11 PRODUCTION BENEFITS FROM INDUSTRY 4.0

Industry 4.0 enables the accumulation and analysis of data in real time, allowing factories to become more efficient. As a result, manufacturing has become more efficient and productive. Numerous benefits for manufacturers are available in Industry 4.0:

> Automation and predictive maintenance let firms make reliable products without costly recalls. Robotics cut labor costs due to Industry 4.0. This helps companies make trustworthy products, satisfy customers, and prevent product withdrawals due to faults, lowering labor costs [12]. This innovative approach gives factories a lot of flexibility; they can react quickly to consumer needs and production requirements, reducing potential issues through proactive preservation methods and increasing productivity and competitiveness.

1.12 THE STRUGGLES OF MANUFACTURING WITH INDUSTRY 4.0

Industry 4.0 has many positive effects, but it also presents factories with certain difficulties. These challenges include:

1. **High Implementation Costs:** In the present age, infrastructural and implementation costs are increasing, so, Industry 4.0 technology comes with several challenges [13].

2. **Need for Skilled Workers:** With the new technology, there is a need for skilled workers, which are hard to find or train.
3. **Cybersecurity Risks:** Cybersecurity risks due to increased use of connected devices, leading manufacturers towards securing their networks and data.
4. **Integration Challenges**: Integration difficulties with existing manufacturing procedures demand major changes in workflows.

1.13 RESPONSE OF COMMUNICATION TO INDUSTRY 4.0

A lot has changed in the communication structure within organizations and across the entire value chain. Industry 4.0 has had profound impacts on communication structures. Some of these include:

Enhanced Interaction: In this regard, Industry 4.0 depends on broad-based interconnectivity involving different tools, machines, sensors, and systems to facilitate seamless communication and information sharing. Consequently, communication channels become more extensive, enabling effective real-time information flow and collaboration.

Real-Time Communication: The emphasis on real-time communication through Industry 4.0 enhances dynamic decision-making as well as adaptable operations. For instance, IoT devices and sensors collect and transmit data in real-time, thus allowing instant analysis and response. This increases reaction time, which enables proactive decision making as well as problem solving.

Collaborative Platforms: Industry 4.0 has gifted us with collaborative platforms, which are digital forums that enable interactions between employees and stakeholders such as suppliers, partners or customers. These online spaces foster virtual collaborations and facilitate the sharing of information on a vast range of work-related topics or any other subject matter for discussion through text-based communication.

Data-Driven Communication: With Industry 4.0 comes an era anchored in data-driven communication, too; devices like machines and sensors now collect immense quantities of data. The evolution in our communicative ways is a response to this influx—we're shifting towards more data-oriented decision-making processes. Implementing tools designed for visibility into analytics empowers robust expression by visualizing critical metrics-signals along with providing trending recommendations.

Vertical Melding: There's an encouragement of vertical melding in every stage of manufacturing powered by Industry 4.0; it links the ground-level factory workings with tech-savvy enterprise-wide elements. This amalgamation requires reliable conduits for transferring wisdom and directives through diverse organizational tiers, boosting communication coherence, visibility, and accord.

Organized Making of Alternatives: Industry 4.0 reallocates strength in decision-making throughout the organization, paving the way for decentralized idea generation. Advanced communication networks empower collaborative effort amongst hierarchical participants by sharing information which allows distributed choices.

Customer-Centric Communication: With a focus on consumer-centric interaction, businesses leverage digital platforms along with analytical data methods to get insights into customer needs, predilections, and patterns of behaviour enabling them to craft personalized client engagements that augment targeting strategies while refining service offerings and user experience.

Cybersecurity Considerations: Given Industry 4.0's inclination towards amplified connection and data exchange through its networking channels, necessitates addressing cybersecurity concerns, diligently safeguarding such connectivity premises from cyber threats, that becomes crucial using protocols, bestowing secure communications alongside, imperative elements like encryption methodologies and authentication systems.

Fundamentally, the Fourth Industrial Revolution transforms communicative routes by promoting boosted connectivity and immediate conversation. It also advocates for collaborative-based frameworks accompanied by judgements rooted in data scrutiny. Additionally, it promotes vertical-horizontal integration along with decentralized decision-making methodologies followed by cyber protection measures that give precedence to customer-centric dialogue. These modifications enable enterprises to enhance operational efficiency, swift adaptability, as well as team cohesiveness, paired with elevated client satisfaction within our digital era [14].

1.14 POSITIVE COMMUNICATION EFFECTS OF INDUSTRY 4.0

Industry 4.0 has numerous benefits for communication providers. These benefits include:

Boosted Conversational Effectiveness: Real-time data tracking and analysis offered by Industry 4.0 helps communication providers detect inefficiencies and refine their workflows, leading to heightened creativity as well as enhanced conversation.

Better Communication Quality: By streamlining regular tasks via automation and initiating preemptive maintenance, providers can bolster the calibre of consistent dialogue, enhancing customer happiness while decreasing costly re-dos or fixes requirement.

Slashed Work Costs: The use of robots alongside process mechanization significantly cuts workforce charges, allowing service operators to allot resources towards other important aspects financial potential projects.

Augmented Adaptability: A fast response from communication agents in addressing demands or conversing needs is made possible due to Industry 4.0, which further speeds up consumer satisfaction through prompt servicing capabilities.

Decreased Non-Operating Periods: Potential enhancements in network performance, along with reduced glitches for players, providing services, are some benefits that come with predictive conservation.

1.15 THE STRUGGLES OF COMMUNICATION WITH INDUSTRY 4.0

Industry 4.0 has many benefits, but it also faces issues for the communication industry. Implementing Industry 4.0 technology can be expensive, requiring significant investment in new equipment and infrastructure. In Industry 4.0, automation, AI, and the Internet of Things require qualified labour. It could be complicated to locate and train ample individuals. With the increased use of connected devices and data sharing, cybersecurity risks are also a concern. Communication providers need to take steps to secure their networks and data. Integrating new technology with existing communication processes can be challenging, requiring significant changes to workflows and procedures.

The Fourth Industrial Revolution is distinguished by the pervasive adoption of digital technologies across all industries. Advanced robots, AI, ML, and the IoT are included in this category. Industry 4.0 is defined by intelligent transportation, which includes vehicle-infrastructure connectivity in real time.

1.16 TRANSPORTATION'S APPROACH TO INDUSTRY 4.0

By transforming the conveyance, management, and monitoring of commodities, Industry 4.0 arrives. Principal transport repercussions of Industry 4.0:

Self-Operating Vehicles: The advent of Industry 4.0 has brought forth autonomous vehicles, such as drones and self-driving trucks, purposed for transportation activities. These sophisticated machines employ cutting-edge technologies like artificial intelligence, sensors, and interconnectivity to function without human aid [15]. Autonomous vehicles promote efficiency by reducing errors caused by humans while offering the flexibility of uninterrupted operations around the clock; besides optimizing routes, they also minimize fuel usage enhancing delivery speed.

Integrated Logistics Approach: IoT devices underpin connected logistics in Industry 4.0 that create a seamless link among distribution centres, warehouses, and transport-sector operators, allowing smooth coordination between them [16]. This live data-relay places asset tracking and improved inventory management at our fingertips, making supply chain visibility better than ever before.

Proactive Maintenance Strategies: Incorporating predictive maintenance machinery health checks can now be carried out regularly, thanks to Industry 4.0 enabling infrastructural integrity preservation alongside vehicle performance optimization [17]. Propelled on real-time information captured via omnipresent sensor technology along with assisting advanced analytic capabilities are optimized diagnostics combined with preventative intervention strategies which enhance both operational efficiency coupled with financial savings primarily derived from reduced downtime.

Advancements in Final Stretch Delivery: The crucial last leg of delivery, involving the transportation from warehouse to end user, referred to as final stretch or last-mile delivery, is at the forefront for sectors implementing

Industry 4.0 solutions. Tools such as navigation-enhancing algorithms, drone services, and self-navigating transport robots are part of strategies aimed at enhancing this phase of operations. These innovations help diminish the time taken for deliveries while improving accuracy—ultimately uplifting customer experience.

Emphasizing Sustainable Green Transportation: An important agenda being pushed forward by advocates promoting Industry 4.0 includes sustainable green transportation alternatives that reduce environmental footprint. Options ranging from electrically powered vehicles to hybrid cars/vehicles through alternative fuel emitters are gaining traction within industrial communities seeking climate-friendly steps [18]. Supported with smart charging outlets and energy regulation mechanisms implemented via technologies introduced under industry trends known commonly today as "Industry 4.0", these options have become more viable than ever before.

Promoting Cooperation & Transparency across Sectors: When considering how advancement-rich fields, including but leaving out any like industries revolving around manufacturing technological components, improve daily task completion using current technology-driven tools equally applies when logistical planning becomes a fully digitized system where everyone involved can gain live information allowing sector-wise transparency made possible due sharing data actively [7].

1.17 INDUSTRY 4.0'S BENEFICIAL EFFECTS ON TRANSPORTATION

Industry 4.0 has numerous benefits for transportation providers. These benefits include:

Boosted Strategic Operations in Transportation: Backed by Industry 4.0's real-time data collection and analysis, transportation providers spot operational gaps to maximize efficiency and elevate productivity, essentially making transit smoother.

Honed Transport Safety Precautions: The blend of automation with regular tasks featuring predictive maintenance not only boosts safety measures but also secures the reliability of transport services. Happy clientele with lowered accident risks is a direct offshoot.

Shaved Labour Expenses: By employing robotics for routine responsibilities, labour costs can be critically reduced letting transport firms invest their core energies into other critical business aspects.

Upgraded Responsiveness Levels: With the rise of Industry 4.0, revolutions allow quicker adaptability towards reflecting customers' needs and commuting stipulations at commercial vehicle company's margins.

Minimized Unproductive Hours: To clip disruption rates that could hinder vehicular efficacy or infrastructure quality, pre-emptive maintenance championed by these outfits is a significant means embraced.

1.18 THE STRUGGLES OF TRANSPORTATION WITH INDUSTRY 4.0

Despite its many benefits, Industry 4.0 also faces some challenges for transportation providers. These challenges include:

Expensive Uptake: Adopting the advancements of Industry 4.0 is costly, necessitating substantial funds for new vehicles and foundational systems.

Demand for Apt Experts: The Internet of Things, robots, and AI—all components of Industry 4.0—need proficient manpower; recruiting new workers and training them could be a daunting task.

Security Threats in Cyberspace: With enhanced usage of interconnected devices and data-sharing practices come increased cybersecurity perils, as well. Transportation companies have to take measures to safeguard their networks along with their data.

Assimilation Roadblocks: Harmonizing newer technological developments into existing transportation procedures can pose challenges, due to its requirement for profound adjustments in workflows and protocols.

1.19 THE RESPONSE OF HEALTHCARE IN INDUSTRY 4.0

Industry 4.0 ushers in a new era of transformation for the healthcare sector, reshaping how health services are provided, overseen, and managed. Let's discuss some major contributions from Industry 4.0 to healthcare:

Remote Health Supervision: The propagation of universal digital health observation devices is made possible by Industry 4.0 techniques; sensors along with interconnected gadgets carry out real-time monitoring tasks like tracking glucose levels or observing heart rate and blood pressure fluctuations continuously. Using data gathered remotely can assist medical providers in taking proactive steps towards prevention tactics and developing unique patient-centric treatment strategies, leading to more positive outcomes.

Telemedicine & Remote Medical Consultations: With the introduction of Industry 4.0 technology platforms comes telemedicine plus remote consultations, allowing individuals to access quality care without stepping outside their own homes.

Patients consult physicians over internet-powered platforms through video calls, utilizing instant communication technologies, making high-quality medicine affordable whilst assisting them living in far-off places where else it would be hard to access such facilities.

Electronic Health Records (EHRs): A crucial development facilitated by the industry revolution "Industry-4.0" involves digitization coupled with incorporation of electronic patients' records, permitting secure storage, besides efficient exchange; moreover, prompt retrieval possibilities. Real-time availability results in healthier choices, giving rise to streamlined workflow procedures and improving coordination processes "Healthcare works". It also opens doors even wider when dealing with advanced analytics and managing market trends and population-based statistics extensively.

Precision Medicine Techniques: Advancement of science techniques, personalized medicinal treatments offering interventions, matching an individual's particular genetic composition, utilize the adoption of state-of-the-art, multimedia capability, holding immense prospective transforming therapeutic patterns employing large-scale genome, resulting highly efficacious treatments, minimizing unwanted reactions, and enhancing better experiences.

Robotics and Automation: The utilization of automation and robotics from Industry 4.0 significantly improves the healthcare sector's efficiency, providing invaluable assistance in surgeries, carrying out repetitive tasks, and offering support for rehab and elderly care patients while also streamlining administrative duties efficiently.

Predictive Analytics and AI: With AI-based, predictive analytics at its backbone, massive amounts of health-related data can be processed by Industry 4.0 to find patterns or predict accurate future outcomes that may assist with early disease detection as well as personalized treatment methods. Precise diagnostic practices have improved considerably due to image recognition technologies empowered through Artificial Intelligence.

Remote Patient Monitoring: Monitoring patients remotely is now a reality because IoT devices supported by this revolution allow real-time monitoring, which helps nurses or physicians detect problems earlier thereby promoting preventive patient care. Remotely monitored treatments are connected closely with enhanced medication compliance, reduced hospital readmissions, etc.

Supply Chain Management: Industry 4.0 has managed even supply chain matters effectively, ensuring the availability and traceability of medicines and medical equipment. IoT sensors provide network connectivity facilities, helping businesses to run inventory checks more competently. This results in a streamlined workload and improved safety protocols, reducing waste generation and hence proving beneficial for both—establishing firms as well as potential clients.

Data Security and Privacy: Ensuring information security along with maintaining privacy stands is crucial within evolving days, where digitalization continues exceeding limits, multi fold, since the industry practices many protective algorithms including but not restricted to multi-factor authentication services, and encryption methodologies, preventing possible cyberattacks, breaching, and stored content.

The emergence of basic Institute Industrial Revolution came tagged alongside, numerous benefits across multiple niches, especially terms shifting the gaze towards better served customer reliability. Includes telehealth consultations, electronic records precision-oriented cures mark the beginning of new age amidst, widespread change being observed in the current methodology that is available rendering satisfactory care giving jobs [19].

1.20 POSITIVE EFFECTS OF INDUSTRY 4.0 ON HEALTHCARE

Healthcare providers stand to gain a lot from the adoption of Industry 4.0 in their operations. These advantages are as follows:

Boosted Healthcare Efficiency: An important advantage that comes with Industry 4.0 is its ability for real-time data collecting and analysis, helping medical professionals identify inefficiencies resulting in enhanced working methods—making healthcare establishments more productive.

Elevated Quality Assurance and Patient Safety Levels: By automating different processes within healthcare and incorporating predictive analytics, there can be an overall improvement of quality assurance levels while also enhancing patient safety parameters—leading to fewer mistakes made medically, causing better treatment outcomes at lowered costs involved.

Lower Employment Expenditures: Automation put into place by these health service providers allows them efficient deployment labour resources where they matter most, therefore effectively reducing associated staffing expenses considerably through automation of routine tasks performed involving robotics, etc.

Enhanced Engagement with Patients on a Real-Time Basis: Result of using selected technologies from the fourth industrial revolution (industry & seemingly), doctor-patient interactions are becoming more immediate that enabling increased participation, patients in their care which leads to better outcomes overall health wise.

More Effective Resource Utilization: This is achieved through data collection analysis techniques adopted that allow locating areas where typically there might either be underused or overuse occurring within operations.

1.21 THE STRUGGLES OF HEALTHCARE WITH INDUSTRY 4.0

Industry 4.0 possesses a wealth of advantages, yet concurrently, it imposes certain predicaments on medical professionals.

Adopting Industry 4.0 technology entails high initial costs due to the expenditure required for fresh software applications, devices, and related frameworks. Expertise in the Internet of Things (IoT), robotics, as well as artificial intelligence stands crucial within the realm of the Industry 4.0 jobs market, which can be tedious when identifying and training such personnel. Escalating utilization of interconnected gadgets, combined with data interchange, heightens cybersecurity threats, thereby compelling healthcare providers to fortify their networks, along with storage data protection measures [11]. An incorporation process encompassing state-of-the-art tech amongst extant healthcare procedures could pose significant difficulty necessitating a notable shift in workflow strategies plus prevailing operations.

1.22 INDUSTRY 4.0: THE HURDLES AND PROMISES

In the realm of Industry 4.0, data safety and personal privacy emerge as central elements, due to an intensified interconnection between devices, leading to an augmented exchange of information through systems. Here are important points worth noting:

Potential for Infiltration: The cross-linked nature inherent in Industry 4.0 presents possibilities for breaches into secured virtual territories where critical data is stored. Companies need strong security protocols against potential unauthorized intrusions, cyber-attacks, or theft.

Respecting Privacy Guidelines: Given the numerous sources from which they accumulate relevant details, the onus lies with businesses to abide by existing legal standards about user-data confidentiality, such as those elucidated under the European Union's General Data Protection Regulation (GDPR). Transparency should be maintained regarding how collected individual information would be utilized or kept secure.

Risk from Within: Internal threats serve a significant risk warranting attention towards securing any crucial records. Organizations must tighten access entry limits, maintain regular checks along system pathway tracing capability, and spread awareness.

Protective Shield Through Encryption: Packaging sensitive content before proceeding with its storage, transmission processes, etc., may help reinforce safeguards.

Securing Communications across Networks: Systems intertwine together to form the resultant "industry scale web", characterizing industry quadrant progression. Robust protective mechanisms like network sections allocate specific firewalls, and intrusion alert platforms, alongside adopting TLS (Transport Layer Security) protocols, aid pre-emptively staving off suspicious infiltrative efforts and uninterrupted communications flow.

Data Lifecycle Management: Companies call for guidelines and sequential procedures to handle data all through its lifespan. Discuss your strategies to gather, store, process, and eliminate data. Data eradication on devices that have been replaced or discarded safeguards against cyberattacks.

Secure Supply Chain: With Industrial Revolution 4.0 comes enhanced interconnectedness in the supply chain, involving numerous parties involved.

Incident Response and Recuperation: Despite having preventive safety measures employed, unforeseen breaches sometimes occur, leading to potentially extreme damage. The plan drafted must include detection efforts and resource dedication towards response efforts along recovery processes from containments; regular testing plus updates will cater towards more recent threats.

Continual Surveillance and Examination: A proactive approach consists of monitoring potential hazards continuously; this helps get an insight into the vulnerability of a system, preventing future incidents, thereby confirming systems are secure.

Building trust among consumers, partners, and stakeholders within Industry 4.0 requires careful implementation of digital technology, focusing particularly on areas like securing individual private details and eliminating possible risks.

1.23 SKILLS GAP AND WORKFORCE READINESS

The advent of the Fourth Industrial Revolution heralded substantial alterations in job demands, calling for a workforce that possesses both adaptability and high-level skills. Yet, there are certain hurdles regarding readiness and skill gaps within this area.

Incorporating AI technologies such as robotics, big data analytics, and cybersecurity along with the Internet of Things (IoT) into Industry 4.0 calls for technical professionals possessing exceptional expertise. This leads to an observed disparity, or "skills gap", between business requisites and what's currently accessible; it's quite rare to locate individuals adept at these domains. As digital resources persistently gain momentum amidst rising dominance within our lives today, having strong-ensured familiarity towards anything digital becomes indispensable on every level—entry positions included! Essential tasks include: operating computers successfully alongside effectively utilizing software applications plus other virtual communication tools necessary whilst engaging proficiently inside an Industry 4.0 environment.

Employer-affiliated organizations must provide appropriate training to prepare workers for job standards changes in "Industry 4.0". In reaction to fast technological developments, workers are under continual pressure to acquire new skills and enhance their present ones. Industry 4.0 personnel must adapt to new technologies in their areas. Promoting research and development, incorporating practical experiences into the classroom, and aligning academic curriculum with industrial needs may help businesses and educational institutions close the skills gap.

Diverse teams and technological knowledge enable innovative problem-solving in Industry 4.0. Diversity in recruiting and inclusive settings boost innovation and readiness. Government programmes that complement technology advances in the "industry-education" linked with concentrated training may reduce the skills gap. Public-private partnership, finance ideas, and group skill development help jobless people into the workforce.

By connecting academics with sectors, governments may provide multidimensional digital-divide solutions to the industry. Innovative investment tactics focused on nutritional scalability provide prospects with a persistent cultural education to meet the abrupt and asynchronous needs of the digital age.

1.24 ETHICAL AND SOCIAL IMPLICATIONS

Industry 4.0 raises social and ethical issues that must be addressed. The Industry 4.0 automation revolution may change job duties or replace workers. Transitionally difficult jobs may need reskilling. Promoting a fair and beneficial transition is crucial. Industry 4.0 may increase the "digital divide". To address this gap, underprivileged communities need universal technology and moral digital literacy courses.

Industry 4.0 relies on massive data gathering, processing, and use. This raises privacy concerns and the need for safe data storage. It is difficult to protect user privacy when concluding massive data sets. Algorithms that provide biased results and perpetuate prejudice must also be addressed.

Misuse and mistreatment of ethical assessments powered by AI authorities must use prudence while designing and implementing machine learning systems to protect human authority. Limiting AI armament and creating "deep fakes", or false information, during use case investigations is necessary. Cyberattacks have increased, owing to interconnectedness. Because they use networked technology, fronting industries are more vulnerable to security threats. Cyberattacks on critical infrastructure may benefit society. Strengthening cybersecurity principles, raising awareness, and defining standard practices may mitigate these threats. The Fourth Industrial Revolution enables ethically sourced items via open and verifiable global trade. These digital technologies must be used more to improve global supply networks' social and environmental implications.

Strong governance and adaptable legal frameworks are needed as Industry 4.0 spreads internationally. The foregoing systems should promote innovation while addressing global ethical and social issues. Academics, corporations, and government officials from many cultures collaborate to reduce the socio-ethical effects of the second industrial revolution. Technology must be ethical from conception to implementation to promote equality, inclusion, environmental tolerance, and human well-being.

1.25 NEW BUSINESS MODELS AND MARKET DISRUPTIONS

In Industry 4.0, companies are using digital technology to create new business models and gain a competitive advantage. Significant variables drive this change:

> Fourth Industrial Revolution capabilities enable companies to customise offerings for individual customers, gain insights from real-time data analysis, implement predictive maintenance strategies, or integrate components holistically to increase customer engagement.

Industry 4.0 prioritises subscription-based or results-driven services above physical goods. By transitioning from product sales to outcome-based pricing, companies may improve client relationships and sustain income. Leading companies should take advantage of these possibilities, since industrial preferences might provide a wealth of useful data with high growth potential. The insights and analytical tools from these reviewed standard processes may revitalize consumer-supplier portals, providing favourable circumstances. Industry innovations encourage enterprises to cut intermediaries and communicate directly with customers, saving money, building loyalty, and improving maintenance.

In conclusion, cooperative channels help corporate structures maximize platform utilization while underserved production capacity storage places—notably, skilled manpower—provide undeveloped markets unique trading prospects. Professional patterns that foster resource cooperation and information exchange help collaboration.

Industry 4.0 creates dynamic platforms and networks with many key participants. Companies may use their capabilities and cooperate to provide full solutions as intermediates. These ecosystems enable innovation, scalability, and complicated client needs. As technologies synchronise, Industry 4.0 blurs industrial borders and creates breakthrough notions that create new rivalry venues. Unless they change, established firms risk slipping behind.

Industry 4.0 requires constant change, pushing long-standing company practices to survive. It also drives companies to surf the wave of emerging technology developments that are changing customer tastes and forcing them to adapt quickly to market changes. Change is increasingly related to successful, creative, and knowledge-acquiring cultures in forward-thinking companies.

1.26 CASE STUDIES AND SUCCESS STORIES OF INDUSTRY 4.0

There are some case studies and success stories in the context of Industry 4.0. They are as follows—

i) **Siemens:** Siemens has fundamentally integrated the theories of Industry 4.0 into their functioning mannerisms, crafting digital doppelgangers during production to test and refine both products and manufacturing systems before they take physical form, not only enhancing productivity but also cutting costs through early maintenance—which monumentally escalates overall profit margins.
ii) **Bosch:** Bosch, a reputable international engineering/tech entity, thrives on imbibing the principles from Industry 4.0 throughout its framework—a commendable feat being an Internet of Things-driven intelligent factory at their semiconductor facility in Germany, augmented with real-time monitoring data interpretation tools [17]. The upshot was amplified process transparency, coupled with supreme efficiency, resulting in marked quality enhancements as well enhanced effectiveness across all domains.
iii) **General Electric (GE):** Through General Electric (GE)'s strategies, "Brilliant Factory" emerges, incorporating key industrial modules, cleverly combined with state-of-the-art machine learning statistics, innovatively merged using IoT assets meshing together, conventional methods causing mass scale shifts altering quantity and produce scenarios including superior asset utilization levels worth emulating by other large-scale corporations exploring innovative techniques while intersecting traditional practices.
iv) **BMW:** This renowned global automobile connoisseur applies modernized industry tactics such as deploying robotic protocols working simultaneously alongside human counterparts, enabling considerable accuracy advancements, progressively maintaining basic safety standards within workplace environs, showcasing promising possibilities between anthropomorphic mechanical cooperation, and reflecting future amalgamation perspectives.
v) **Procter & Gamble (P&G):** A celebrated global consumer goods giant, shrewdly strengthening supply chain architectures through inventive application abilities, unveiled from groundbreaking analytical models, bypassing old-fashioned device restrictions, consequently revolutionizing classic logistics networks, thus triggering cascading results, saving incremented overhead expenditure, thereby reducing customer dissatisfaction events, whilst amplifying operational flexibility score cards.

This underlines how firms from varied sectors are harnessing the tactics and methodologies of Industry 4.0 to ignite transformation and bolster their competitive edge. They've deployed ground-breaking steps like the Internet of Things (IoT), data interpretation, and artificial intelligence (AI)—which include automation—fuelling increase in efficiency, productivity rates, and inventive resolutions.

1.27 FUTURE TRENDS AND OUTLOOK OF INDUSTRY 4.0

As we look into the future, several trends and outlooks are expected to shape the evolution of Industry 4.0. As advancements in AI and machine learning algorithms continue, computers will be increasingly proficient at making informed decisions by interpreting vast clusters of data. This progress promises enhanced automation, predictive maintenance, and a streamlined manufacturing process. The integral use of the Internet of Things (IoT) is expected to grow more pronounced—existing not only as interconnected devices but also as sensors embedded within industrial products, allowing for real-time, calculable readings, thus augmenting remote oversight abilities alongside fostering superior communication between machines, contributing towards heightened productivity efficiently.

Edge computing stands out, further asserting its significant role; it allows processing closer to its genesis, thereby ensuring speedier decision-making coupled with heightening security protocols whilst maintaining privacy—an essential feature, considering Industry 4.0 prospects. Undoubtedly, robust safety regulations are a set-up of paramount importance as networks become progressively digitized globally and draw attention towards safeguarding sensitive industry-based systems from cyber threats while concurrently securing integrity-inclusive, availability-related intricacies regarding indispensable infrastructures serving magnified emphasis shifting forward. Another transformative technology, known widely as additive manufacturing, or simply 3-D printing, continues evolving, unabatedly facilitating production featuring intricacy combined with adequate detailing, providing increased swiftness plus precision, possibly revolutionizing supply chain paradigms and eventually aiding drastic cost reduction, additionally enabling flexible, on-demand productions whenever necessary, effortlessly inspiring admiration universally [20].

Anticipate witnessing collaborative robots, fondly referred Co-bots, assuming popularity, gradually industrially functioning maximising their potential roles, collaborating uninterruptedly along humans geared, assisting repetitive even physically taxing tasks, yet relentlessly endorsing overall optimisation, enhancing general efficiency, remarkably established furthermore through extended applications associated digital twin technological practices, producing virtual counterparts, detailed physical assets, beneficial revisualisations processes, simulating leading noticeable difference, product qualities reducing downside risks, presenting extensive additional possibilities, similarly emphasise rise relating focus that amidst environmentally cautious sustainable forms, adopting better regulatory measures and consequently responding actively raising concerns, naturally integrating renewable energy sources wherever viable, analyzing profoundly available Industrial principles, encapsulated conveniently under Industry 4.0.

Practically envision manufacturers adapting newer business roles progressively involving manageable solutions apart from routine providing for products caused by significant strategic shift, adding greater value-oriented services, inbuilt advantages, predictive maintenance model added benefits, outcomes-based business advancement that heavily drawing influences analytics including IoT connectivity. Naturally, such an evolution will reshape future job prospects, demanding the necessity of developing novel skills within the workforce, digital technology proficiency cyber security methods, robotics alongside data analysis, all refreshingly up-skilled and re-skilled effectively that preparing individuals better suited to accommodating futuristic jobs.

1.28 CONCLUSION

Industry 4.0 has transformed manufacturing, communications, transportation, and healthcare. This led to "smart factories". Industry 4.0 improves productivity, quality, labour costs, flexibility, and equipment idle time for firms. This new industry paradigm replaces inefficient traditional ways of interaction with intelligent communication, which benefits communication service providers by increasing efficiency, quality, low operational costs, adaptability, and downtime. Intelligent transport improves traditional transportation. Industry 4.0 boosts transit productivity, lowers costs, improves safety, increases adjustment capacity, and reduces stale times. Linked health care revolutionizes medicine. Hospitals and clinics might speed up daily operations with industrial advances.

REFERENCES

[1] S. Vaidya, P. Ambad and S. Bhosle, *Industry 4.0–A Glimpse*, Procedia Manufacturing 20 (2018), pp. 233–238.

[2] J. Qin, Y. Liu and R. Grosvenor, *A Categorical Framework of Manufacturing for Industry 4.0 and Beyond*, Procedia CIRP 52 (2016), pp. 173–178.

[3] B. Bagheri, S. Yang, H.-A. Kao and J. Lee, *Cyber-Physical Systems Architecture for Self-Aware Machines in Industry 4.0 Environment*, IFAC-PapersOnLine 48 (2015), pp. 1622–1627.

[4] J. Lee, H.-A. Kao and S. Yang, *Service Innovation and Smart Analytics for Industry 4.0 and Big Data Environment*, Procedia CIRP 16 (2014), pp. 3–8.

[5] M.B. Raval and H. Joshi, *Categorical Framework for Implementation of Industry 4.0 Techniques in Medium-Scale Bearing Manufacturing Industries*, Materials Today: Proceedings 65 (2022), pp. 3531–3537.

[6] K.-D. Thoben, S. Wiesner and T. Wuest, *"Industrie 4.0" and Smart Manufacturing–A Review of Research Issues and Application Examples*, International Journal of Automation Technology 11 (2017), pp. 4–16.

[7] G. Schuh, T. Potente, C. Wesch-Potente, A.R. Weber and J.-P. Prote, *Collaboration Mechanisms to Increase Productivity in the Context of Industrie 4.0*, Procedia CIRP 19 (2014), pp. 51–56.

[8] A. Schumacher, S. Erol and W. Sihn, *A Maturity Model for Assessing Industry 4.0 Readiness and Maturity of Manufacturing Enterprises*, Procedia CIRP 52 (2016), pp. 161–166.

[9] D. Kolberg and D. Zühlke, *Lean Automation Enabled by Industry 4.0 Technologies*, IFAC-PapersOnLine 48 (2015), pp. 1870–1875.

[10] A.K. Tyagi, T.F. Fernandez, S. Mishra and S. Kumari S, Intelligent Automation Systems at the Core of Industry 4.0. In: Abraham A., Piuri V., Gandhi N., Siarry P., Kaklauskas A., Madureira A. (eds.), Intelligent Systems Design and Applications. ISDA 2020. Advances in Intelligent Systems and Computing, vol 1351. Cham: Springer, 2021.

[11] L. Gomathi, A.K. Mishra and A.K. Tyagi, *Industry 5.0 for Healthcare 5.0: Opportunities, Challenges and Future Research Possibilities*, 2023 7th International Conference on Trends in Electronics and Informatics (ICOEI) (2023).

[12] K. Sipsas, K. Alexopoulos, V. Xanthakis and G. Chryssolouris, *Collaborative Maintenance in Flow-Line Manufacturing Environments: An Industry 4.0 Approach*, Procedia CIRP 55 (2016), pp. 236–241.

[13] M. Landherr, U. Schneider and T. Bauernhansl, *The Application Center Industrie 4.0—Industry-Driven Manufacturing, Research and Development*, Procedia CIRP 57 (2016), pp. 26–31.

[14] Z. Wan, Z. Gao, M. Di Renzo and L. Hanzo, *The Road to Industry 4.0 and Beyond: A Communications-, Information-, and Operation Technology Collaboration Perspective*, IEEE Network 36 (2022), pp. 157–164.

[15] M.A. Kamarul Bahrin, M.F. Othman, N.H. Nor Azli and M.F. Talib, *Industry 4.0: A Review on Industrial Automation and Robotic*, Jurnal Teknologi 78 (2016).

[16] K. Witkowski, *Internet of Things, Big Data, Industry 4.0–Innovative Solutions in Logistics and Supply Chains Management*, Procedia Engineering 182 (2017), pp. 763–769.

[17] S. Wang, J. Wan, D. Li and C. Zhang, *Implementing Smart Factory of Industrie 4.0: An Outlook*, International Journal of Distributed Sensor Networks 12 (2016), p. 3159805.

[18] T. Stock and G. Seliger, *Opportunities of Sustainable Manufacturing in Industry 4.0*, Procedia CIRP 40 (2016), pp. 536–541.

[19] V.V. Popov, E.V. Kudryavtseva, N. Kumar Katiyar, A. Shishkin, S.I. Stepanov and S. Goel, *Industry 4.0 and Digitalisation in Healthcare*, Materials 15 (2022), p. 2140.

[20] A. Behl, R. Singh, V. Pereira and B. Laker, *Analysis of Industry 4.0 and Circular Economy Enablers: A Step Towards Resilient Sustainable Operations Management*, Technological Forecasting and Social Change 189 (2023), p. 122363.

2 Security Concerns and Controls of Intelligent Cobots of Industry 4.0

Lakshmi Sravani Thummapudi, Aswani Kumar Cherukuri, and Thong Chee Ling

2.1 INTRODUCTION: BACKGROUND AND DRIVING FORCES

In recent years, the integration of collaborative robots (cobots) into the industry has received a lot of attention, with a focus on improving productivity, efficiency, and safety. The potential of cobots to collaborate with humans in shared workspaces has paved the way for innovative solutions in Industry 4.0. Like all new and existing technologies, security concerns and drawbacks must be addressed for successful implementation and adoption. Cobots are a leading technique that is evolving with Industry 4.0. A major concern is safety and techniques to assure safety are needed to be identified. Industries can put in place a variety of security measures, such as industrial controls, security standards, and security policies, to lessen these worries. Industrial controls and security standards can offer guidance for the creation, implementation, and administration of secure devices, respectively. This work discusses intelligent cobots and industrial robots, their security issues and concerns, and different controls to mitigate the security issues.

The first industrial revolution introduced steam power; the second, electricity; and the third, the use of miniature electronics to increase industrial production. Cloud, automation, and big data are all part of Industry 4.0. Via robots, this has had a significant impact on industrial production. The development of cobots as a new technology happened recently. Its primary characteristic is their ability to interact and collaborate with people and other technologies, as they are fitted with sensors that enable them to monitor their surroundings and take appropriate action [1]. These cobots eliminate the constraints that are inherent in conventional industrial robots. These robots are completely isolated from people by security barriers and are programmed to work autonomously while functioning at maximum speed. As cobots share a workspace with human operators and work together to complete tasks, they have quite different applications [2]. Smart robots are being used in the most recent digital revolution and strong dependence on artificial intelligence (AI) technology to speed up the transformation of digital operations [3]. In fact, robots play an important part in modern civilizations, as they offer a great number of options to assist in a variety of professions. Yet, there are a few problems associated with the utilization of robots. The key concerns at hand are those of safety, accuracy, trust, and protection [4].

DOI: 10.1201/9781003473886-2

The platform provided by Industry 4.0 is suitable for promoting direct interaction between human and robots as they adjust to changes and uncertainty in system operations. Yet, security and safety must be guaranteed during a physical human-machine connection [5]. Industry 4.0 has an impact on the organization and efficiency of work in industrial systems [6]. Because they are significantly lower in weight than the massive industrial robots, these collaborative robots, also known as cobots, are significantly more moveable and easier to move around the factory or industry in which they are installed. Cobots have various benefits over industrial robots, including their versatility, which enables them to be employed in a variety of industries and use only one robot for several different jobs. One of the primary advantages of cobots is their ability to perform multiple tasks simultaneously [7].

2.2 INDUSTRY 4.0 AND INDUSTRY 5.0

Discrete and process manufacturing, as well as the oil and gas, mining, and other industrial sectors, can all benefit from the concepts and technologies of Industry 4.0. The technologies that are driving Industry 4.0 are Internet of Things, cloud computing, AI, machine learning, edge computing, cyber security, digital twin, and many more, while the focus of Industry 5.0 is human-machine collaboration. Figure 2.1 illustrates different industrial revolutions and corresponding technologies. In the future, collaboration between humans and machines will combine the greatest aspects of both worlds. As a result, integration is strengthened, enabling quicker, better automation combined with the power of human brains. Industry 5.0 advances the idea of personalization by involving people more in the design process than in the automated manufacturing process.

Manufacturing is being revolutionised by cobots, which offer safer and more effective ways for humans and robots to collaborate. As a result of the Fourth Industrial Revolution, or Industry 4.0, the manufacturing process is becoming more automated and digitalized. Without cobots, which are programmable machines designed to work alongside people, Industry 4.0 would not be complete. The work of Weiss et al. [8]

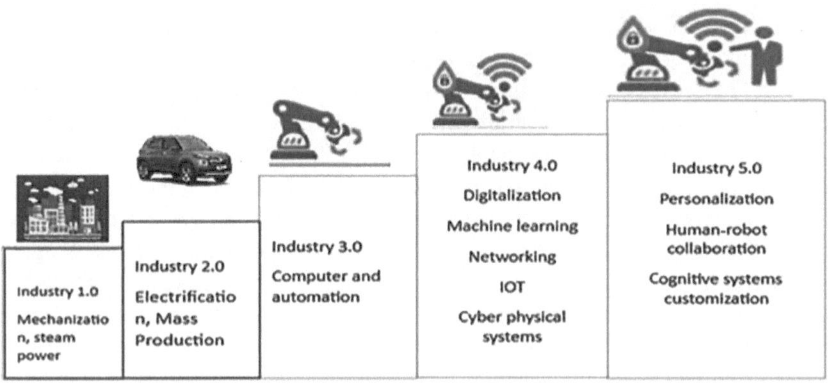

FIGURE 2.1 Evolution of Industrial Revolution.

emphasize the need for future field research on human-robot collaboration to ensure successful integration of cobots in Industry 4.0. The authors propose a roadmap for these studies and highlight the importance of understanding cobot implementations, human-robot interactions, and the role of context in shaping these interactions.

Current trends in cobots for Industry 4.0 include the development of cobots equipped with advanced sensors and machine learning capabilities to improve their ability to work safely and efficiently with humans. increase. The use of cobots in collaborative assembly lines is also increasing. This allows greater flexibility and customization of the manufacturing process. Additionally, integrating cobots with cloud computing and the Internet of Things (IoT) will enable manufacturers to collect and analyze real-time data, enabling predictive maintenance and improving overall efficiency.

Beyond Industry 4.0, collaborative robots might be crucial to the future of manufacturing and other sectors like logistics and healthcare. Collaborative robots can perform labour-intensive or dangerous tasks, freeing people to focus on more creative and challenging work. Cobots will also collaborate with people to promote accuracy and efficiency, leading in more productivity and lower costs. Cobots could influence the future of manufacturing and other industries, and they are a key component of Industry 4.0. Cobots are becoming more and more popular, thanks to technological advancements in advanced sensors and machine learning as well as their integration with cloud computing and the Internet of Things (IoT). This is helping these sectors develop and succeed.

2.3 COBOTS AND INTELLIGENT ROBOTS

Cobots are made to operate securely and productively alongside people in open workspaces. Cobots have the ability to engage with human employees in a safe, effective, and user-friendly manner, in contrast to typical industrial robots that are frequently caged or isolated from them for safety reasons. The role and importance of cobots in industry is very important, as they offer several advantages over traditional industrial robots. First and foremost, cobots are designed to work cooperatively with human workers and can perform tasks that are difficult or impossible for traditional industrial robots. For example, cobots can be used to assemble complex products or perform delicate tasks that require human-like dexterity and precision. Because of their flexibility and adaptability, cobots are perfect for use in production settings where duties and procedures change regularly [9]. Cobots may be easily reprogrammed and reconfigured to execute a range of jobs and processes, in contrast to typical industrial robots, which are frequently designed to perform a specific task or group of tasks.

The ability of cobots to increase worker safety is another significant benefit. Cobots have several safety measures to assist them detect and avoid accidents with people and other objects in the environment because they are made to work alongside human workers. This decreases the possibility of harm coming to human workers as well as downtime and delays, boosting total productivity. Overall, as businesses explore for the ways to increase the effectiveness, productivity, and safety of their manufacturing processes, the usage of cobots in industry is expanding in popularity. Cobots are positioned to play a significant role in the future of manufacturing, thanks to their capacity to work collaboratively with human workers, adapt to changing tasks and procedures, and enhance workplace safety [10]. Following are some of the examples of industrial robots. Automated machines called industrial robots are made to carry out a range of duties in production settings. As businesses look for ways to increase the effectiveness, productivity, and safety of their production operations, these robots have grown more and more in popularity in recent years. Here are some of the current industrial robot trends.

a) Collaborative robots (cobots):
   Cobots are made to function securely with humans in shared workspaces. These robots are equipped with various sensors and safety features that can detect and avoid collisions with people or other objects in the environment, so they are used in production environments where tasks and processes change frequently.
b) Mobile robots:
   Mobile robots are designed to move autonomously within production facilities to perform tasks such as transporting materials, inspecting products, and maintaining facilities. These robots are often equipped with advanced sensors and navigation systems that allow them to navigate complex environments and avoid obstacles.
c) Articulated robots:
   The articulating robot is one of the types of industrial robots that receives the most attention. The mechanical design of it makes it look like it may be a human arm. The arm is attached to the base by means of a twisting joint. The number of rotary joints that connect the arm's links can range anywhere from two to ten, and each joint adds an additional degree of flexibility to the overall structure. It's possible for the joints to be perpendicular to one another or parallel to one another.
d) SCARA robots:
   These robots are efficient in terms of cost, and they can carry out operations in between two planes that are parallel to one another. Because they are lightweight, they are perfect for applications that take place in confined areas. Due to the fact that SCARA robots have fixed swing arms, they are unable to perform jobs that need them to operate around or reach inside objects within a work cell. Examples of such objects include fixtures, jigs, and machine tools. This design choice can be advantageous in certain contexts.
e) Delta robots:
   Delta robots are sometimes referred to as parallel link robots because their construction consists of parallel joint links that are attached to a single base. Delta robots, which are often commonly referred to as "spider robots," utilize three motors that are mounted to the base in order to operate control arms that position the wrist. Although there are also 4- and 6-axis variants available, the most popular type of delta robot is the 3-axis variety.
f) Cartesian robots:
   Cartesian robots, often referred to as gantry robots, are able to travel in straight lines along the X, Y, and Z axes thanks to their three-dimensional

coordinate system. Pick-and-place operations, as well as jobs including material handling, assembly, and inspection, are frequently carried out using these robots. Robots built with Cartesian principles are extremely accurate and capable of learning intricate movements.

g) Autonomous guided vehicles (AGVs):
Self-driving cars, known as autonomous guided vehicles, are utilized for transportation and material handling in industrial and production settings. They are frequently used to move goods or materials between phases of the manufacturing process. AGVs can be configured to travel a certain path and have sensors to detect potential hazards. They are frequently employed in settings where effectiveness and automation are crucial, like sizable production facilities or storage facilities.

h) Modular robots:
Modular robots can be quickly changed and reprogrammed to carry out various activities and processes because they are very versatile and adaptable. These robots frequently consist of modular parts that may be assembled in a variety of ways to produce a unique solution for your particular production requirements.

Overall, the field of industrial robotics is developing rapidly, with new technologies and trends constantly emerging. Industrial robots may continue to play an important role in manufacturing and industry as companies look for ways to improve the efficiency, productivity and safety of their manufacturing processes.

Cobots are capable of more complicated duties than just basic material manipulation. Additionally, they are capable of handling a variety of machine tending tasks, such as computer numerical control (CNC), injection moulding, press brakes, stamping presses, and manual loading and unloading parts or components. Both employees and businesses can profit from cobot welding in various ways. It guarantees flawless uniformity on straightforward welding jobs, which would be challenging for a human employee to uphold for hours on end, therefore minimizing material waste. Cobots can perform Arc, TIG, laser, MIG, ultrasonic, plasma, and spot welding, although they won't be appropriate for all welding jobs. Therefore, allowing cobots to conduct welding does not result in the loss of employment for qualified welders. In fact, since these staff members will have the opportunity to concentrate on more complex, low-volume projects that call for their expertise, it may increase employee happiness.

Due to their ability to collaborate with humans effectively and efficiently, cobots are becoming more and more well-known in businesses throughout the world. Cobots are created to interact with human workers in a shared workspace, fostering a cooperative atmosphere that enhances efficiency, productivity, and safety. These robots differ from typical business robots in that they operate in remote locations without interacting with people. Industrial robots have been in use for many years, especially for dangerous and repetitive tasks that demand high precision and accuracy. However, conventional business robots are bulky, heavy, and hard to program, making them mistaken for collaborative workspaces. On the other hand, cobots are lightweight, flexible, and smooth to program, making them best for small and medium-sized enterprises (SMEs) that can't manage to pay for steeply-priced automation systems. Cobots may be programmed to carry out a huge variety of obligations, which include assembly, packaging, first-rate control, and inspection, amongst others.

The potential use cases of cobots are as follows:

Healthcare: Cobots can help doctors and nurses with activities including moving patients, cleaning surfaces, and assisting with operations.

Agriculture: Cobots can be used to harvest crops, apply fertilizer and pesticides, and carry out other operations that are challenging or hazardous for people.

Education: Cobots can help teachers and students in the classroom, resulting in a distinctive and engaging learning environment.

Cobots can be utilized in amusement parks and other places of entertainment to handle jobs including ticketing, crowd management, and ride operation.

Hospitality: At hotels and restaurants, cobots can be employed to complete activities like cleaning, preparing food, and serving.

Manufacturing: The employment of cobots in manufacturing can free up human workers to concentrate on more complicated jobs by automating repetitive processes like packing, assembly, and quality control.

In general, cobots have the potential to revolutionize a variety of industries by enhancing productivity, efficiency, and safety while also presenting fresh chances for growth and innovation. Several industries can benefit from the introduction of industrial automation. Even if every machine is capable of doing a variety of tasks, from sophisticated labour to palletizing, some models are more useful in some applications than others.

Cobots have numerous benefits over conventional business robots, inclusive of elevated protection, decreased costs, and progressed productiveness. The protection functions of cobots, which include pressure and torque sensors, permit them to come across and reply to human presence, decreasing the chance of injuries within the workplace. Cobots also are cost-powerful, as they require much less protection and may be programmed fast and easily. Cobots also can enhance productiveness with the aid of using automating repetitive obligations, permitting human employees to consciousness on greater complicated and innovative obligations. Some of the examples of the cobots are presented in the following subsections.

2.3.1 Fanuc CRs

In 2015, the CR series of cobots was introduced by Fanuc. The versatility, dependability, and customizability of the CR series are its major selling points. High motion repeatability and durable materials are features of Fanuc cobots. They also support a wide range of Fanuc accessories, including as force control, vision, and screwdrivers. This, along with their exceptional performance, enables them to be effective in many industrial applications.

2.3.2 TM-Omron Cobots

The Techman Robot TM cobot series was introduced in 2016 and has since established itself as one of the most cutting-edge products available. Modern sensory technology is crucial to the success of TM cobots. Accurate force and torque sensors are another feature of the TM series that reduce the risk of worker collisions. The TMS cobots efficiently handle complex applications when combined with an intuitive UI.

2.3.3 Panda

Panda offers an exceptional user experience and technological performance. It can particularly benefit from an ecosystem of open-source software. Users can share and discover new customized applications on the Franka World cloud platform. Students can test their cobot on applications right away and instantly access programming materials.

Despite the benefits of cobots, there are, nevertheless, demanding situations that want to be addressed to make sure a hit implementation and adoption. These demanding situations consist of protection and protection concerns, programming complexities, regulatory issues, and moral considerations. Addressing those demanding situations calls for collaboration among researchers, policymakers, and enterprise stakeholders to expand powerful answers that sell the secure and green use of cobots in industries.

2.4 SECURITY CONCERNS AND THREAT LANDSCAPE

Robots play a key role in contemporary civilizations, providing numerous opportunities to aid in a variety of fields, including the military and the agricultural as well as the civilian and military sectors. Yet, there are several issues with the use of robots. The key concerns that need to be addressed are trust, safety, accuracy, and security. The level of these robots' resistance to different kinds of cyberattacks is the primary factor determining their level of security. The level of contentment and the capacity of these robots to accurately perform and take the place of people in specific domains and activities are the two primary factors that determine trust. Accuracy is based on performing the intended task without any flaws or mistakes. Robots are subject to a variety of attacks, including those on their hardware, firmware, and communications. Worms, Ransomware, Trojan horses, rootkits, botnets, spyware, and other forms of malicious software are examples of software attacks.

Robot-to-robot and robot-to-human communication is disrupted by jamming attacks, which causes further suspension. Traffic analysis attacks affect the bots and data's confidentiality and privacy. Attacks involving eavesdropping also target confidentiality by obtaining sensitive information. Data accessibility is affected by replay attacks. Accuracy in Masquerading attacks has an impact on the robot's availability and integrity. Confidentiality, integrity, and authentication are impacted when an attacker may listen in between two entities, alter the information, and inject it without being noticed. These various attacks are based on the targeted layer [11]. The threat landscape refers to the current and potential security threats that an organization or system may face. This includes a wide range of possible attacks, vulnerabilities, and risks that could compromise the confidentiality, integrity, and availability of systems and data. The threat landscape of industrial robots refers to the potential security threats that robots can face, including both physical and cyber threats. Robots can be physically attacked and they are connected devices making them vulnerable to cyberattacks. Attackers gain access to control systems and they can remotely control the robot and use it for malicious purposes. Phishing or spear phishing are used to gain access to robot's control system or to trick human operators into performing activities that could threaten the safety of the robot or the data that it manages.

Threats that are part of threat landscape of cobots are malware, unauthorized access, human error, and physical safety. The powerful motors movements may be dangerous to humans if they are not programmed correctly. Malware can be introduced through malicious mails, USB drives that reduce the functionality of cobots. If they are not properly secured, the sensitive information may be leaked and it can be used in unintended ways. To reduce the impact of these dangers, it is essential to implement strong passwords, carry out routine security audits, keep an eye out for unusual activities, and make certain that robots are physically protected from unauthorized access.

2.5 ATTACK SURFACE AND ATTACK TREE

In this section, the attack surface and tree for few robots and cobots are discussed. The attack surface could be either digital or physical. When it comes to the Panda robot, these attack surfaces can be broken down into two categories: digital and physical. The digital surfaces are the points of entry via which an attack can be performed against a cobot without the need for direct physical access. An attack can be performed against a cobot in several different ways. In the case of the Panda, these include the web application, services, Internet Protocol, and Ethernet. Interfaces on the cobot's arm and on its controller are two instances of physical attack surfaces [12].

The attack surface of cobots refers to the potential entry points that attackers can use to compromise the security of cobots. The attack surface of a particular cobot would be determined by its design, its configuration, and the environment in which it is deployed. Some of the attack surfaces are sensors and cameras, operating system, external interfaces such as USB ports, and wireless interfaces. Cobots can be vulnerable to attacks such as malware and DOS if the control software is not secured. Protocols such as Ethernet/IP, Modbus TCP, and PROFINET are utilized in the communication process between cobots and other types of systems. Attackers can intercept and manipulate data that is being communicated between the cobot and other systems if these protocols are not protected by security measures. Figure 2.2 provides attack tree of Panda robots. Each of these attacks can be further broken down into sub-attacks, and appropriate mitigations can be put in place to address the different attack vectors. By considering the different ways that an attacker may try to compromise the cobot's security, we can implement measures to reduce the attack surface and protect the cobot from potential attacks.

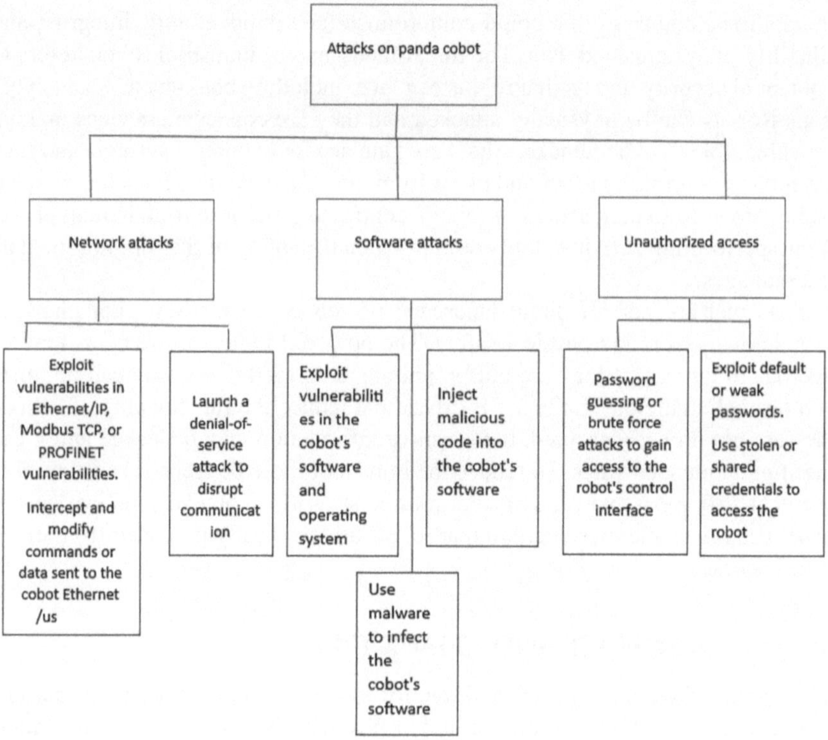

FIGURE 2.2 Attack Tree of Panda Robots.

2.6 SECURITY CONTROLS AND MECHANISMS

The process of managing and monitoring the functioning of the cobot, as well as its level of safety and performance within an industrial setting, is an important concern. One of the industrial controls, known as dynamic braking, can immediately put a stop to the cobot's movement in the event of an emergency. Cobots are programmed to always remain inside their allotted workspace in order to prevent them from colliding with human employees or other pieces of machinery. Various control systems such as programmable logic controllers (PLCs), which are digital computers used to monitor operations of industrial robots. They monitor sensors, actuators, etc. Human machine interfaces allow human operators to interact with cobots; they are used to control performance of robot and to set parameters for its operation. Vision systems are used to guide movement. They use cameras to track object position and to provide feedback to control system. Monitor controllers are used to ensure that the cobots are moving accurately and precisely.

Cobots are equipped with a wide variety of safety features, such as the following, to protect the human employees that use them:

Safety Monitored Stop: The motion of a robot is put on hold by a safety-monitored halt when an operator is present in the collaborative workspace. Power is still being supplied to the robot, but it is motionless. This workspace is the shared area in which a human and a robot carry out their respective responsibilities. Even though there are no humans present in the workspace, the cobot is still operating at a rapid rate.

Speed and Separation Monitoring: Fenceless robot systems are frequently referred to as having speed and separation monitoring capabilities. As long as the operator and collaborative robot stay at a certain distance apart from one another, the cobot is able to move in sync with the operator. This type of cobot application is typically monitored by a safety-rated laser area scanner.

Limitations on Power and Force: In order to reduce the likelihood that human employees may sustain an accident, cobots are built to function with only a limited amount of power and force.

Hand Guiding: Human hand controls the robot motion. Cobots can be guided by hand, enabling operators to provide movement instructions directly to the machine. Cobots that are guided by hand can also function on their own when their operators are not there.

Many security rules and guidelines have been established in order to guarantee the safety of both human workers and cobots in the workplace. The following is a list of some of the most important cobot security standards [12–14]:

NIST SP 800–82: This is a guideline for protecting industrial control systems, such as the control systems used for cobots. It addresses issues such as risk management, access control, and incident response, among other things.

IEC 62443: This is a set of standards that, together, constitute a set of rules for the safety of industrial control systems, such as the systems that are used to command cobots. Network security, access control, and incident response are some of the subjects that are discussed in this course.

ANSI/RIA R15.06–2012: Guidelines for the design, building, and operation of industrial robots and cobots are provided by this standard. It discusses things like hazard analysis, risk assessment, and robot safety, among other things.

ISO/TS 15066: This standard offers direction for the cooperative operation of industrial robots and cobots. It discusses things like risk assessment, safety regulations, and several ways for determining whether the joint activity is safe to do. Figure 2.3 provides hierarch of these standards.

In general, security policies for robots are an essential element of a comprehensive security plan for a variety of different situations, including the workplace and other settings. By adhering to these principles, firms can lessen the likelihood of security mishaps and make certain that their robotic systems will continue to function in a risk-free and protected manner [15].

Network security policy lays out the principles for protecting the network infrastructure that the robot uses to communicate with other devices and systems. It may be necessary to deploy virtual private networks (VPNs), firewalls, or other security measures in order to guard against unauthorized access and data breaches.

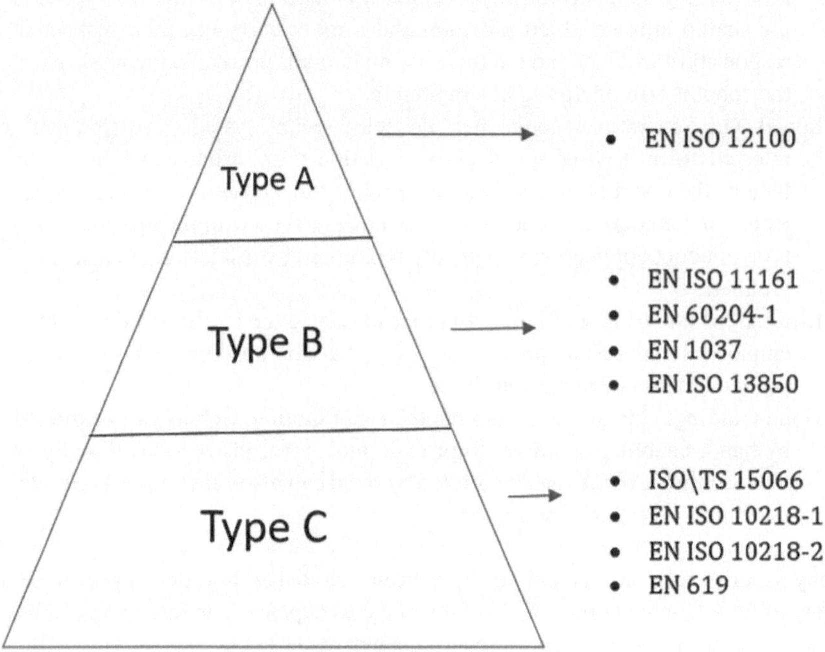

FIGURE 2.3 Standards Hierarchy.

Data privacy policy outlines the procedures to follow when dealing with sensitive data that the robot may gather or process. To defend against data breaches and privacy violations, it may be necessary to encrypt the data, store the data in a secure location, or anonymize the data. Incident response policy lays out the processes for dealing with data breaches and security events that involve robots. It creates reporting protocols, outlines standards for reducing the effects of security events, and defines roles and duties for all parties involved.

2.7 CONCLUSIONS

As a result of Industry 4.0, there has been a rapid increase in the use of cobots as well as industrial robots in the quest of manufacturers to enhance the effectiveness, productivity, and safety of their operations. Robots are becoming even more powerful and clever as a result of the integration of cutting-edge technology like as artificial intelligence, machine learning, and the Internet of Things (IoT). In addition, they are now able to operate more independently than they ever have before. To summarize, industrial robots and cobots are essential elements of the Fourth Industrial Revolution (Industry 4.0), and they are a driving force behind the transformation of the manufacturing and industrial sectors. Robots and collaborative robots are expected to play an increasingly essential role in the future of industry and manufacturing as technology continues to improve. These robots are expected to be much more advanced and powerful than they are today.

REFERENCES

[1] Cohen, Yuval, Shraga Shoval, and Maurizio Faccio. "Strategic view on cobot deployment in assembly 4.0 systems." *IFAC-PapersOnLine* 52, no. 13 (2019): 1519–1524.

[2] Cohen, Yuval, Shraga Shoval, Maurizio Faccio, and Riccardo Minto. "Deploying cobots in collaborative systems: Major considerations and productivity analysis." *International Journal of Production Research* 60, no. 6 (2022): 1815–1831.

[3] Rüßmann, Michael, Markus Lorenz, Philipp Gerbert, Manuela Waldner, Jan Justus, Pascal Engel, and Michael Harnisch. "Industry 4.0: The future of productivity and growth in manufacturing industries." *Boston Consulting Group* 9, no. 1 (2015): 54–89.

[4] Chui, Michael, James Manyika, and Mehdi Miremadi. "Where machines could replace humans-and where they can't (yet)." *The McKinsey Quarterly* (2016): 1–2.

[5] Bi, Z. M., Bin Chen, Lida Xu, Chong Wu, C. Malott, M. Chamberlin, and T. Enterline. "Security and safety assurance of collaborative manufacturing in Industry 4.0." *Enterprise Information Systems* 16, no. 12 (2022): 2008512.

[6] Müller, Sarah L., Mohammad A. Shehadeh, Stefan Schröder, Anja Richert, and Sabina Jeschke. "An overview of work analysis instruments for hybrid production workplaces." *AI & Society* 33 (2018): 425–432.

[7] Castillo, Javier F., Jesús Hamilton Ortiz, María Fernanda Díaz Velásquez, and Diego Fernando Saavedra. "Cobots in Industry 4.0: Safe and efficient interaction." *Collaborative and Humanoid Robots* (2021): 13.

[8] Weiss, Astrid, Ann-Kathrin Wortmeier, and Bettina Kubicek. "Cobots in Industry 4.0: A roadmap for future practice studies on human–robot collaboration." *IEEE Transactions on Human-Machine Systems* 51, no. 4 (2021): 335–345.

[9] Rossi, Filippo, Fabio Pini, Andrea Carlesimo, Enrico Dalpadulo, Francesco Blumetti, Francesco Gherardini, and Francesco Leali. "Effective integration of cobots and additive manufacturing for reconfigurable assembly solutions of biomedical products." *International Journal on Interactive Design and Manufacturing (IJIDeM)* 14 (2020): 1085–1089.

[10] Castillo, Javier F., Jesús Hamilton Ortiz, María Fernanda Díaz Velásquez, and Diego Fernando Saavedra. "Cobots in Industry 4.0: Safe and efficient interaction." *Collaborative and Humanoid Robots* (2021): 13.

[11] Yaacoub, Jean-Paul A., Hassan N. Noura, Ola Salman, and Ali Chehab. "Robotics cyber security: Vulnerabilities, attacks, countermeasures, and recommendations." *International Journal of Information Security* (2022): 1–44.

[12] Hollerer, Siegfried, Clara Fischer, Bernhard Brenner, Maximilian Papa, Sebastian Schlund, Wolfgang Kastner, Joachim Fabini, and Tanja Zseby. "Cobot attack: A security assessment exemplified by a specific collaborative robot." *Procedia Manufacturing* 54 (2021): 191–196.

[13] Castillo, Javier F., Jesús Hamilton Ortiz, María Fernanda Díaz Velásquez, and Diego Fernando Saavedra. "Cobots in Industry 4.0: Safe and efficient interaction." *Collaborative and Humanoid Robots* (2021): 13.

[14] Pauliková, Alena, Zdenka Gyurák Babeľová, and Monika Ubárová. "Analysis of the impact of human–cobot collaborative manufacturing implementation on the occupational health and safety and the quality requirements." *International Journal of Environmental Research and Public Health* 18, no. 4 (2021): 1927.

[15] Kóczi, Dávid, and József Sárosi. "The safety of collaborative robotics-A review." *International Journal of Engineering* 20, no. 2 (2022): 73–76.

3 Big Data Analytics (BDA) for Industry 5.0

Sharmila Rathod, Priya Parate, Nilesh Rathod, and Samrudh Wagh

3.1 INTRODUCTION

3.1.1 Role of Big Data and Analytics

Today's world is incomplete without online shopping sites that results in huge data, which is coming by daily transactions done by customers. The daily transactional data is big data that cannot be managed by relational DBMS as data that can be of any type like structured, semi structured and unstructured data. Data that cannot be handled by a traditional approach needs separate analysis by using tools or by using different techniques. After storing huge data on a daily basis, online transactions should be analyzed to understand the trends of buying patterns as well as to profit the business and help to take quick decisions to company stakeholders. This can help them compete in a market with many companies following the same trend. So, the analysis is very important and should be correctly carried out to get correct insights of the data and make sure we are not missing the important parts of the data. Analysis depends on the data size and the uniformity of data in various ways. To maintain healthy competition in world analysis is the key factor, and it can cause the increase of revenue generation in business.

3.1.1.1 Objectives

The objectives of the chapter are to understand and embed the big data techniques with industry 5.0 to gain maximum profit in the business with an increase in the intelligence of the business by combining features of big data and Industry 5.0. It's cost effective to integrate the techniques of big data and Industry 5.0 that will result in automation of the business reducing the manpower required to handle the business. Automation will increase the productivity of the business in all ways.

3.2 BIG DATA AND ANALYTICS: CONCEPTS AND PRINCIPLES

3.2.1 What is Big Data?

Data that is huge in size cannot be managed by traditional database systems, and it can have any format as structured, semi-structured and unstructured data with the size in petabytes. If you think about the example, then every e-commerce will come under the category of big data having volume with velocity, which gives the nature of big data. This type of data needs special care as its treasure for the company that

DOI: 10.1201/9781003473886-3

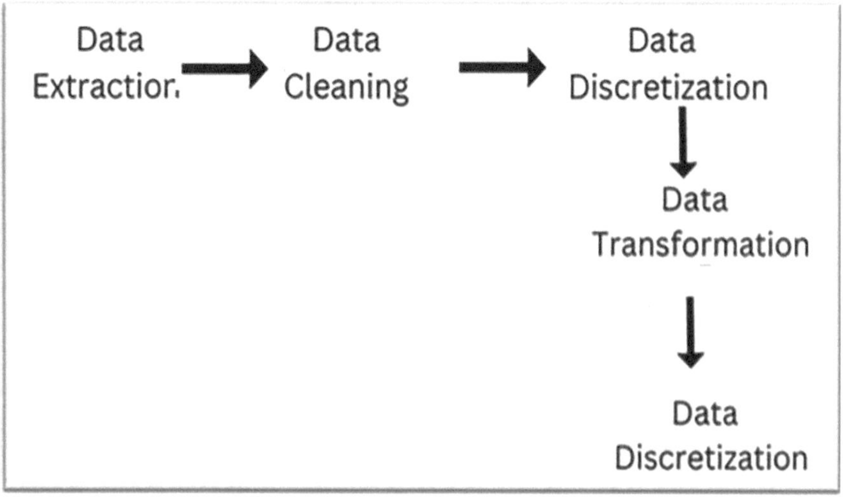

FIGURE 3.1 Big Data Preprocessing Steps.

will find the summary of the transactions affecting the decision making by the stakeholders. Big data should follow data preprocessing steps, which is mandatory to get correct results in order to get correct data analysis from tremendous amount of data and are listed as follows.

As shown in the Figure 3.1, it starts with data extraction to data discretization steps. All should be carried with care and none of the steps should be missed. Data cleaning is done after extraction, as data can have noise which is collected from different sources of data, and only valuable data will be transformed in one format for analysis for business decision-making. The next step is data discretization to segment data for proper analysis.

3.2.2 Key Features of Big Data

The features of big data are most commonly called as 5 V's that are given as:

Volume

The size of the data, we can call the volume of data. If data is available and if large amount the variations of the data can be handled properly as with small amounts of data, we cannot withdraw any conclusion for business intelligence.

The size of the data is called the volume of data.

Value

Volume will generate huge data with different trends and with the different combinations of the data generated by customers. So, it must capture the important data that need to be analyzed, which is important from business point of the view or for the expansion of the business by comparing the profits. Big data typically has value to the data of the customers for keeping good customer relationships and other benefits that gives reason to find the changing requirements of the customers.

The importance that is given to data is called as value of data.

Variety

Every e-commerce site has structured data like customers data, semi-structured data that will be excel-sheet stored from transactions done by the customers and unstructured data that are images of the products or advertisements offers displayed on the screen.

The different types of data present on e-commerce is called the variety of data.

Velocity

The speed at which data comes is very important, depending on the speed, variety of tools are used to manage data, for example, social networking sites that have high speed of data on a daily basis with categories of data like writing posts or releasing different pictures or statuses.

The rate at which businesses gain and manage data is known as velocity.

Veracity

The accuracy of data is important for the analysis for business intelligence. The decision is dependent on the quality and correctness of data. The noise in big data can affect the quality of the data, and it's difficult to maintain the quality of the data due to heterogeneity of data from different sources of data.

3.2.3 Analytics and Understandings from Big Data

Big data analytics has advantages for business whether it is used in health care, government or some other industry, as it uses advanced analytics on structured, semi-structured as well as unstructured data to find valuable insights for businesses for decision making. It is used across industries such as insurance, education, stock market artificial intelligence and manufacturing to understand if its working or not and if there is a need to improve the process or design of the products as per the changing requirements of the customers.

3.2.4 Analytics in Industry 5.0 with Big Data

Big data is an inseparable part of daily life for an e-commerce site, and a shopping mall can't run without the study of big data; otherwise, they would have to shut down, as no consumers will get to know about the offers or discounts that are available. Big data helps to take a wide view of the customers and look into their frequent choice of the product or the prior products, depending on the category of needs, like daily needs or the monthly needs to live a life.

By integrating big data with Industry 5.0, maximum automation in various shopping malls is gained, and it will update the status of the human lifestyle.

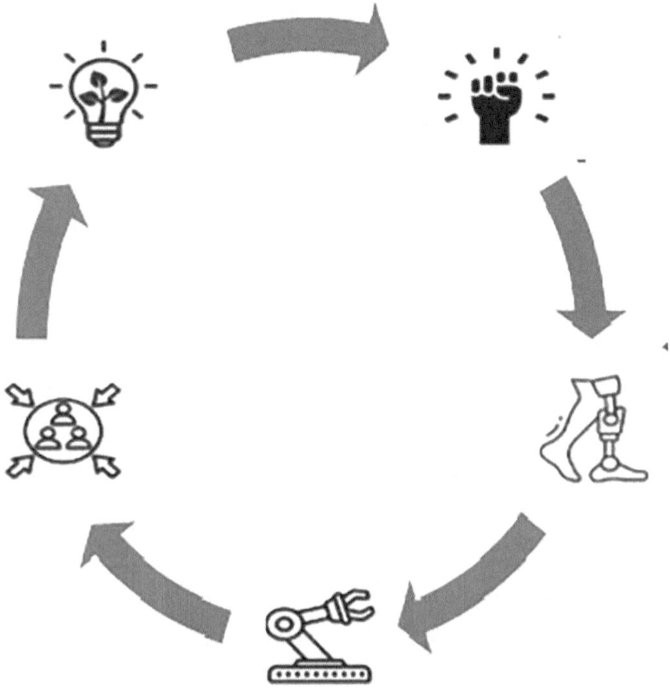

FIGURE 3.2 Industry 5.0.

To understand the necessity of the big data and Industry 5.0, consider the example of shopping. Nowadays, we all do online shopping for our comfort and follow all the protocols for doing transactions. But if you go for physical shopping, there can be a kiosk at the payment desk. Business intelligence is very important if you want to gain the maximum number of the customers, and it's possible with the help of integrating big data with Industry 5.0 technology that will cope with the drawbacks of the shopping mall regarding transactions as well as products customers are searching for in the mall.

If we combine the features of big data and Industry 5.0, we can bring evolution in automation and life will become easy and problem-free to some extent when there is a point of doing some work manually in case of physical shopping [1], as shown in Figure 3.2.

3.3 DATA COLLECTION AND MANAGEMENT IN INDUSTRY 5.0

Industry 5.0 is based on principles such as social benefits and human-centricity. The current state of Industry 5.0 gets human interaction back in the manufacturing industry, which means Industry 5.0 products provide customers with a way of realizing their requirements to express easily. We can say that Industry 5.0 is based on the concept that tries to make industry more sustainable, human centric, and resilient [2].

The emergence of Industry 5.0 is, indeed, an interesting evolution in the realm of industrial automation and smart manufacturing. Here are some key points of industry 5.0:

Human Centric Approach

- Industry 5.0 concentrates on the collaboration between humans and technology, particularly robots and IoT devices.
- The objective is to integrate advanced technologies to enhance human capabilities, creativity and decision-making rather than replacing human workers.

Integration of Big Data and Industry 5.0

- Industry 5.0 seeks to integrate the Internet of Things (IoT) and big data to empower human work which involves harnessing the big amounts of data generated by connected devices to make informed decisions to improve efficiency of the system.
- Industry 4.0 originated in Germany and was initially associated with the European business environment. Industry 5.0 has a more global and international fame from its inception.

Impact of COVID-19

- The global pandemic highlighted the need for resilient and flexible industrial systems. Industry 4.0 practices faced challenges during the COVID-19 crisis, prompting a revaluation of existing concepts.
- The observation gained from companies implementing the concept Industry 4.0 solutions during the pandemic revealed areas requiring transformation, contributing to the rising popularity of Industry 5.0.

Business Transformation

- The drawback of Industry 4.0 is that industry is overcome by 5.0 with the target to deliver business value to support unforeseen challenges, improving supply chain resilience and ensuring the continuity of operations in case of crisis.

3.3.1 Data Sources and Types

In Industry 5.0, data is an important source of decision-making by optimizing processes and enhancing human-machine collaboration. Some of the key data sources in the context of Industry 5.0 are listed as follows:

IOT Devices

IoT devices are a major source of data in Industry 5.0. These devices are embedded in machinery and equipment that collect real-time information in the form live streaming of data.

Sensors and Actuators

The applications with advanced sensors and actuators provide data on the physical state and performance of machines. This includes information on wear and tear, maintenance needs and overall equipment health.

Machine Learning and AI

Machine learning algorithms and AI systems analyze large datasets to derive insights, predict outcomes and optimize processes. These technologies enhance the decision-making process and identify patterns that may be challenging for humans to detect.

Human-Machine Interaction

Data related to human-machine interactions is important in Industry 5.0, which includes information on how workers interact with machines to make various decisions and contribute to the overall efficiency and productivity of industrial processes.

Supply Chain Data

Data related to supply chain and supports inventory levels, demand forecasts and logistics is essential for optimizing the flow of materials and ensuring timely production without affecting the quality of the product.

Digital Twins

These are virtual representations of physical assets or systems. They continuously collect and update data about the real-world counterpart that shows real-time monitoring, analysis, and representation of industrial processes.

Collaborative Platforms

Platforms that facilitate collaboration and communication among humans and machines contribute valuable data. These platforms may include project management tools, communication apps and collaborative workspaces that capture user interactions and workflows.

Cloud Services

Cloud computing services play a crucial role in storing, processing and managing vast amounts of data. Cloud platforms enable scalable and flexible data storage and analysis, supporting Industry 5.0 initiatives.

In Industry 5.0, the integration and analysis of data from diverse sources gives clear and brief understanding of industrial processes that motivates human workers to work with the machine effectively and make informed decisions that optimize overall system performance.

3.3.2 Data Collection Methods and Technologies

There are various data collection methods and technologies in Industry 5.0 which are miscellaneous to get data from various sources. Here are some key data collection methods and technologies:

IOT Sensors

Sensors are inseparable part of the IoT and are widely used to collect real-time data from physical assets such as machinery, equipment and production lines.

RFID

RFID technology is used for tracking and identifying objects in the industrial environment. RFID tags attached to products, components or equipment enable automated data collection as they move through different stages of the production process.

Machine Vision System

The subcategory of AI which is machine vision makes the use of cameras and image processing algorithms to capture and analyze visual information, and this technology is used for quality control, defect detection and monitoring of production processes.

Smart Actuators

Actuators with built-in sensors can provide information about the physical state and performance of machinery. These smart actuators contribute to data collection related to equipment health and usage.

Wearable Devices

Nowadays, advanced technologies used in wearable devices re aequipped with sensors that can monitor the health and activities of workers. This includes devices that track vital signs, motion and location, providing insights into worker well-being and optimizing task allocation.

Augmented and Virtual Reality

The combination of the digital and real world which are AR and VR technologies are useful for training, maintenance and visualization purposes in Industry 5.0. Data is collected through powerful sensors in AR/VR devices, providing insights into user interactions and improving training outcomes.

Advanced Analytics and Machine-Learning Algorithms

The increase in the speed of data that is coming in daily life applications demands to process and analyze large datasets and to extract valuable insights. These techniques and algorithms contribute to predictive maintenance, quality control and optimization of industrial processes.

3.3.3 Data Storage and Integration

In Industry 5.0, effective data storage and integration are main components for gaining knowledge and optimizing processes. Here are some key considerations and technologies in data storage and integration related with the context of Industry 5.0:

Cloud Storage

Cloud storage is used by many organizations to keep balance in work when there is a large space requirement and gives the solutions to provide scalable and flexible options for storing large volumes of data generated by diverse sources in Industry 5.0. Cloud platforms offer accessibility and the ability to scale storage capacity as needed.

Distributed Databases

Distributed databases allow data to be stored across multiple nodes or locations that are connected through the network. This ensures fault tolerance and efficient retrieval of data. An example includes NoSQL databases.

Blockchain Technology

Security is a crucial point to be considered when data storage is playing a role, so blockchain offers a decentralized and secure way to store and manage data. Blockchain offers immutability, transparency and trust in data transactions.

Data Integration Platform

Also very important is data integration platforms that facilitate the continuous flow of data between different applications and devices that ensures aggregated, transformed and made available for analysis.

APIs

APIs make it possible for different software applications to communicate and share data with each other. They play a key role in integrating various components of Industry 5.0 systems to permit interoperability between different technologies.

Middleware Solutions

Middleware acts as a bridge between different software applications and systems. It facilitates communication and data exchange between disparate components of the industrial ecosystem, promoting interoperability.

Data Virtualization

Data virtualization makes possible for users to access and update data without concern about its physical location or format that enables real-time data access and integration across different databases and platforms.

Master Data Management

MDM is managing the core data entities of an organization, ensuring consistency and accuracy across the enterprise. It is essential for maintaining a single, accurate version of sensitive data.

Data Analytics Platforms

The various platforms that analyze data with machine learning and artificial intelligence are essential for deriving actionable insights from the stored data.

3.3.4 Data Quality and Governance

The key considerations for data quality and governance in the context of Industry 5.0:

Data Quality Standards

Establishing and adhering to data quality standards is fundamental. These standards define criteria for accuracy, completeness, consistency and timeliness of data. Ensuring high-quality data is essential for reliable insights and decision-making.

Data Governance Framework

A comprehensive data governance framework provides guidelines, policies and procedures for managing data across the organization. This includes roles and responsibilities, data ownership and protocols for data access and usage.

Data Ownership

Clearly defining data ownership ensures accountability for data quality. Assigning responsibility to individuals or teams helps maintain the accuracy and integrity of data throughout its lifecycle.

Data Stewardship

Data stewards are responsible for overseeing and managing specific sets of data. They play a crucial role in implementing data governance policies, monitoring data quality and addressing issues as they arise.

Data Catalogues

Implementing data catalogues helps organize and document available datasets. This includes metadata such as data lineage, definitions and usage, facilitating better understanding and management of data assets.

Data Privacy and Security

Given the sensitivity of industrial data, robust measures for data privacy and security are essential. This involves implementing encryption, access controls and compliance with relevant data protection regulations to safeguard sensitive information.

Data Quality

Continuous monitoring of data quality ensures that issues are identified and addressed promptly. Automated tools and processes can be employed to monitor data for anomalies, inconsistencies and deviations from established standards.

Data Quality Metrics

Establishing key performance indicators (KPIs) and metrics related to data quality helps quantify and measure the effectiveness of data governance practices. Regularly assessing these metrics enables continuous improvement.

Training and Awareness

Educating employees about the importance of data quality and governance is essential. Training programmes and awareness initiatives help foster a data-centric culture within the organization.

Agile Data Governance

Adopting agile data governance practices allows organizations to adapt quickly to changing data requirements. This flexibility is crucial in dynamic Industry 5.0 environments.

Effective data quality and governance practices in Industry 5.0 ensure that the data used for decision-making is accurate, reliable.

3.4 BIG DATA ANALYTICS TECHNIQUES AND TECHNOLOGIES

Big data analytics involves processing and analyzing large and complex datasets to extract meaningful insights, patterns and trends. In Industry 5.0, big data analytics plays a pivotal role in optimizing processes and facilitating data-driven decision-making. Figure 3.3 shows big data analysis types.

Here are some techniques and technologies associated with big data analytics:

3.4.1 Hadoop

Hadoop is an open-source framework that permits distributed processing of large data across clusters of computers. It consists of a distributed file system (HDFS) and a processing framework (MapReduce) and is well-suited for parallel processing and storing vast amounts of data.

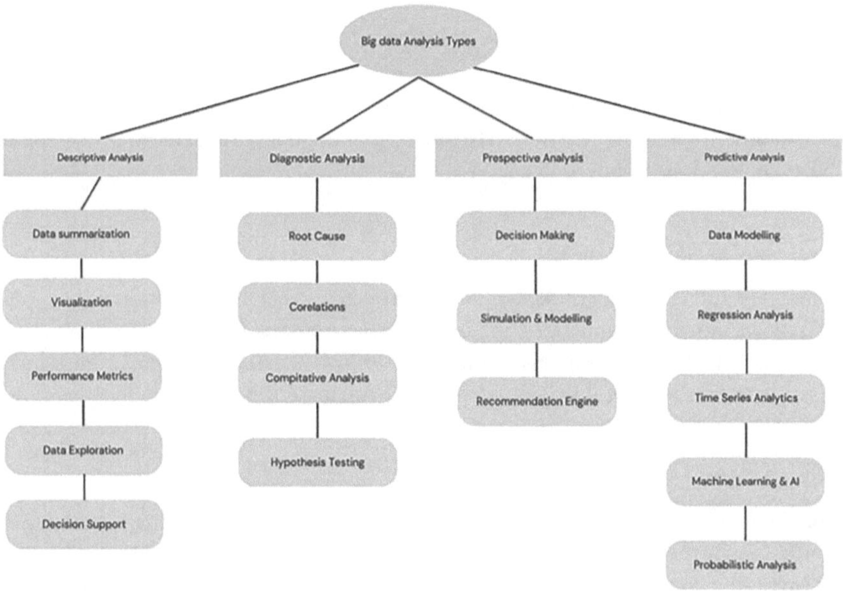

FIGURE 3.3 Big Data Analysis Types.

3.4.2 Apache Spark

Apache Spark is a fast and general-purpose, distributed computing system. It provides in-memory data processing, making it well-suited for iterative algorithms and interactive data analysis. Spark supports various programming languages, including Scala, Java and Python.

3.4.3 NoSql Databases

NoSQL databases, such as MongoDB, Cassandra and Couchbase, are designed to handle unstructured and semi-structured data with flexibility in storing and retrieving diverse data types, making them suitable for big data applications.

3.4.4 Machine-Learning Algorithms

Machine-learning algorithms that can be supervised and unsupervised learning are applied to big data for predictive analytics and pattern recognition. Algorithms such as decision trees, random forests and neural networks are commonly used.

3.4.5 Data Mining

Data-mining techniques are employed to discover patterns and relationships in large datasets. This includes clustering, association rule mining and anomaly detection, which help uncover valuable insights from complex data.

3.4.6 NLP (Natural Language Processing)

NLP techniques are applied to analyze and understand human language in textual data. Sentiment analysis, text summarization and language translation are examples of NLP applications in big data analytics.

3.4.7 Data Visualization

The most useful and easily understandable part is data visualization and various tools like Tableau, Power BI and D3.js are used to create interactive and informative visualizations. These tools help communicate complex findings and make data more accessible to non-technical stakeholders of the organization.

3.4.8 Graph Database

Graph databases, such as Neo4j and Amazon Neptune, are used to analyze and traverse relationships in interconnected data. They are particularly valuable for applications involving social networks, fraud detection and network analysis.

3.4.9 Time Series Analysis

Time series analysis gives us the brief study of historical patterns in data and analysis of data is important for understanding the trends over time.

3.4.10 Descriptive Analytics

Descriptive analytics is the study of data in brief that concentrates on summarizing historical data to describe and understand nature of data stored on the server that consists of the examination of data to identify trends and key information that we cannot find manually.

Key characteristics and components of descriptive analytics include:

Data Summarization

Data summarization is very important for every organization to get status of a product. The organization's overall summary gives insights to the stakeholders, based on which the decision making can be done. Descriptive analytics involves summarizing huge datasets into meaningful and easily understandable metrics. The measures such as averages, totals, counts, percentages and other statistical summaries are used for summary report.

Visualization

Visualization tools are commonly used in descriptive analytics to present data in a visually appealing and accessible manner. Charts, graphs and dashboards help convey trends and patterns effectively to stakeholders.

Key Performance Indicators

Descriptive analytics often revolves around the identification and tracking of key performance indicators. These are specific metrics that provide insights into the performance of a business or a specific process.

Reports and Dashboards

Reports and dashboards are generated to elaborate the findings of descriptive analytics. Reports describes detailed information, while dashboards provide a visual snapshot of key metrics for quick decision-making.

Data Exploration

Analysts explore the data for deeper understanding of its features and value. This can involve identifying outliers and examining data distributions.

Tools and Techniques

Tools and techniques in descriptive analytics consist of basic statistical measures (mean, median and mode), frequency distributions, pie charts, bar charts, etc.

Decision Support

Descriptive analytics provides decision-makers with the necessary information to assess historical performance and make informed decisions based on past trends.

Examples of descriptive analytics applications include:

- Summarizing sales data to understand monthly, quarterly or annual trends.
- Analyzing website metrics to identify popular pages, user demographics and traffic sources.
- Creating financial statements and reports to understand revenue, expenses and profitability.
- Categorizing customers based on purchasing behaviour, demographics or other criteria.

3.4.11 Diagnostic Analytics

Diagnostic analytics is the branch of data analysis that gives clear understanding cause of events or outcomes occurred in the past. It goes beyond the descriptive analytics that merely describes what happened and aims to identify the components contributing to specific situations or trends. Diagnostic analytics is precious for gaining depth insights into historical data and unseen patterns.

Key characteristics and components of diagnostic analytics include:

Root Cause Analysis

Diagnostic analytics consists of research on the root causes of observed trends or outcomes. Analysts study the root cause of specific events and its impact on performance metrics.

Correlations

To find correlations and causal relationships between different variables is a key aspect of diagnostic analytics. This helps in understanding how changes in one variable may influence others.

Comparative Analysis

Comparative analysis involves comparing different groups, time periods or segments of data to identify variations and differences. This aids in understanding why certain outcomes differ across various conditions.

Drill-Down Analysis

Analysts perform drill-down analysis to delve deeper into specific data points or segments. This helps in uncovering details and nuances that may not be apparent in high-level summaries.

Identification of Anomalies

Detecting anomalies or outliers in the data is important for understanding unusual patterns or events that might have influenced outcomes.

Examples of diagnostic analytics applications include:

- Research for a decline in sales to determine whether it is needed to do changes in marketing strategies, economic conditions or product quality issues.
- Understanding the reasons behind customer churn by examining factors such as customer service interactions, product satisfaction or pricing changes.
- Identifying the components contributing to employee turnover, such as workplace conditions, management practices or compensation issues.

3.4.12 Predictive Analytics

Predictive analytics is the brief study of data that utilizes statistical algorithms, machine learning and modelling techniques to identify the likelihood of future outcomes based on historical data. This analytical approach involves predicting trends, patterns and future events to support decision-making at strategic level.

Key characteristics and components of predictive analytics include:

Data Modelling

Building predictive models involves selecting and applying appropriate algorithms to historical data. These models learn patterns from the data and use this knowledge to make predictions about future events.

Feature Selection

Identifying relevant features or variables that contribute to predictive accuracy is crucial. Feature selection helps improve model performance by focusing on the most influential factors.

Training and Testing Data

Predictive models are trained on historical datasets, and their accuracy is evaluated using separate testing datasets. This process helps ensure the model's ability to generalize and make accurate predictions on new, unseen data.

Regression Analysis

Regression models are commonly used in predictive analytics to establish relationships between independent and dependent variables. Linear regression, logistic regression and other regression techniques are applied depending on the nature of the prediction task.

Time-Series Analysis

Predictive analytics often involves analyzing time-dependent data, such as stock prices or sales figures. Time series analysis helps forecast future values based on historical patterns.

Classification and Clustering

Classification models are used to categorize data into predefined classes, while clustering identifies natural groupings within data. These techniques are applied for tasks like customer segmentation or fraud detection.

Probability

Predictive models often produce a probability or score associated with a predicted outcome. This score indicates the likelihood of a particular event occurring, allowing decision-makers to assess and prioritize risks or opportunities.

The accuracy and performance of predictive models are evaluated using metrics such as accuracy, precision, recall, etc.

Examples of predictive analytics applications include:

- Predicting future demand for products or services based on historical sales data.
- Assessing the creditworthiness of individuals or businesses to predict the likelihood of default on loans.

3.4.13 Prescriptive Analytics

Prescriptive analytics is the advanced stage of data analysis that focuses on providing actionable recommendations to optimize outcomes.

Key characteristics and components of prescriptive analytics includes:

Decision Optimization

Prescriptive analytics involves optimizing decisions based on specified objectives, constraints and available resources. Optimization algorithms identify the most favourable actions to find the desired outcomes.

Simulation Modelling

Simulation models are used to assess the potential impact of different decisions under various scenarios. Decision-makers can explore different options and their consequences before making choices.

What-If Analysis

What-if analysis allows users to explore the impact of changes in variables or parameters on outcomes. It helps in understanding how adjustments might influence results and supports decision-making under uncertainty.

Recommendation Engine

Prescriptive analytics often incorporates recommendation engines to suggest optimal decisions. These engines analyze data and user preferences to provide personalized recommendations, such as product recommendations in e-commerce.

Rule-Based System

Rule-based systems apply predefined rules to guide decision-making. These rules are based on domain expertise and business rules, ensuring that decisions align with organizational objectives.

Machine Learning for Optimization

Machine learning techniques are employed to optimize decision-making processes. This may involve reinforcement learning, genetic algorithms or other optimization-focused machine learning approaches.

Prescriptive analytics helps in efficiently allocating resources, whether it be manpower, funds or other assets, to achieve maximum efficiency and effectiveness in a given scenario.

3.4.14 Machine Learning and Artificial Intelligence in Analytics

Machine Learning (ML) and Artificial Intelligence (AI) play significant roles in advancing analytics by enhancing the ability to derive insights, make predictions and automate decision-making processes.

The types of analysis that can be done with the help of ML and AI are listed as follows:

Predictive Analysis

ML algorithms are used in predictive analytics to build models that can predict future outcomes based on historical data.

Classification and Clustering

ML algorithms are used to classify data into different categories or clusters. Classification models help in identifying patterns and making decisions, while clustering algorithms group similar data points with common features together [3].

Natural Language Processing

NLP, a subset of AI, is used to analyze and understand human language in textual data. It enables sentiment analysis, text summarization and language translation, enhancing the analysis of unstructured data.

Recommendation System

AI-driven recommendation systems use ML algorithms to analyze user behaviour and preferences, providing personalized suggestions. These systems are widely used in e-commerce, streaming services and content platforms.

Anomaly Detection

ML algorithms are used in finding anomalies or outliers in datasets. This is important for identifying unusual patterns or events that may indicate fraud, system failures or other irregularities.

Image and Video Analysis

ML and AI technologies, including computer vision, are employed for image and video analysis. This includes image recognition, object detection and facial recognition, contributing to applications in security, healthcare and retail.

Speech Recognition

ML algorithms are utilized in speech-recognition systems to convert spoken language into text. This technology is integrated into virtual assistants, voice-controlled devices and transcription services.

Predictive Analytics

AI, along with optimization techniques, is used in prescriptive analytics to recommend optimal actions based on predicted outcomes. This involves decision-making under specified constraints to achieve the best results.

Reinforcement Learning

Reinforcement learning, a type of machine learning, involves training models to make sequential decisions through trial and error. This is applied in dynamic decision-making environments, such as robotics and gaming.

3.5 APPLICATIONS OF BIG DATA AND ANALYTICS IN INDUSTRY 5.0

There are various application of big data and analytics in industry 5.0 like healthcare, education field, entertainment, retail, marketing and so on, as shown in Figure 3.4.

- **Predictive Maintenance**
  Big data analytics makes it possible for predictive maintenance by analyzing large datasets from sources like sensors and other devices, which helps

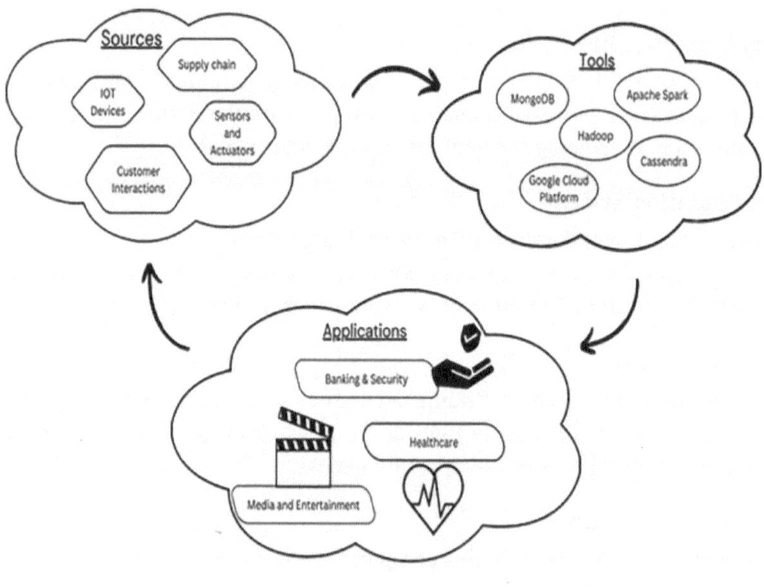

FIGURE 3.4 Big Data Applications, Tools and Sources of Data.

in predicting potential equipment failures, optimizing maintenance schedules and minimizing downtime [4].

- **Quality Control and Assurance**
  Analyzing data from production processes allows for real-time quality monitoring. Big data analytics can identify patterns and anomalies in product quality, facilitating early detection of defects and ensuring consistent product quality.
- **Human Machine Collaboration**
  Analytics facilitates human-machine collaboration by analyzing data related to human performance and machine operations. This helps in creating a safer and more productive working environment.
- **Real-time Monitoring of Manufacturing Process**
  Streaming data analytics enables real-time monitoring of manufacturing processes. This includes analyzing data from sensors and IoT devices to ensure that production processes are operating within specified parameters.

3.5.1 Smart Cities and Urban Planning

Smart cities leverage big data and Industry 5.0 principles to transform urban areas into more efficient, sustainable and liveable spaces.

Big data analytics is used to monitor and analyze data from various urban infrastructure components, including transportation systems, utilities and public services. This enables real-time monitoring of infrastructure performance and identifies areas for improvement. Big data is important in optimizing transportation systems. Data from sensors, GPS devices and traffic cameras are analyzed to manage traffic flow, optimize public transportation routes and improve overall mobility. Big data analytics enables intelligent traffic management systems that analyze real-time traffic data to optimize signal timings, manage congestion and improve overall traffic flow. This contributes to reduced travel times and enhanced safety.

Data from various sources, including surveillance cameras, sensors and social media, is analyzed to enhance public safety and security. Predictive analytics and real-time monitoring help in identifying potential security threats. Industry 5.0 principles, such as the integration of humans and machines, are applied in urban planning. Advanced technologies, including 3D modelling, simulations and collaborative design tools, support the planning and development of smart cities. Industry 5.0 principles influence the design and management of smart buildings. Data from building automation systems, IoT sensors and energy management systems are analyzed to optimize building operations [5].

3.5.2 Healthcare and Personalized Medicine

There will be innovation by using big data and Industry 5.0 principles in healthcare for transforming the industry, leading to personalized medicine and causing to efficient healthcare delivery.

Big data analytics enables the analysis of large volumes of patient data, including electronic health records (EHRs), genomic data and real-time monitoring from wearable devices. This data is used to gain insights into patient health, identify patterns and tailor treatments.

Big data analytics is applied to predict and prevent diseases by analyzing health trends, risk factors and genetic predispositions. This supports proactive interventions and personalized preventive care.

Big data and analytics accelerate drug discovery by analyzing vast datasets related to molecular biology, genomics and clinical trials. Predictive models help identify potential drug candidates. Industry 5.0 principles are applied to create personalized treatment plans that consider individual patient characteristics, preferences and responses to treatments. This human-centric approach enhances treatment effectiveness and patient satisfaction. Industry 5.0 principles contribute to optimizing the healthcare supply-chain. Big data analytics is used to manage inventory, track pharmaceuticals and ensure the availability of medical supplies, contributing to efficient healthcare operations.

3.5.3 Transportation and Traffic Management

Big data analytics analyze the data from various sources such as cameras, sensors, and GPS devices to provide real-time insights into traffic conditions. This information is crucial for managing congestion, improving traffic flow, and reducing travel times. Predictive analytics manages historical and real-time data to predict future traffic conditions. By analyzing patterns and trends, transportation authorities can proactively address potential congestion points and optimize traffic management strategies.

3.5.4 Energy Management and Sustainability

Big data and Industry 5.0 principles in energy management and sustainability is used in optimizing energy use, increasing efficiency and support environmentally friendly practices.

Big data analytics enables predictive maintenance for energy infrastructure, such as power plants and renewable energy facilities. By analyzing sensor data and historical performance, maintenance activities can be scheduled proactively, minimizing downtime and optimizing asset performance. Big data supports the integration of renewable energy sources, such as solar and wind, into the power grid. Analytics helps manage the variability of renewable energy production. Big data analytics is applied to analyze energy consumption patterns across different sectors [6]. This information helps businesses and organizations identify opportunities for energy savings. Industry 5.0 principles promote the integration of humans and machines in monitoring and reducing carbon emissions. Big data analytics helps measure and analyze carbon footprints, identify emission sources and implement strategies for reducing greenhouse gas emissions.

3.5.5 Education and Learning Analytics

The concept or idea of integrating big data and Industry 5.0 concepts or features of both in education and learning analytics is revolutionizing the way educational institutions operate, adapt to individual student needs and enhance overall learning outcomes. Big data analytics is used to analyze student performance data and learning styles. This information is leveraged to create personalized learning paths, tailored to individual student needs [7].

Industry 5.0 principles motivates the integration of humans and machines in adaptive learning platforms and these platforms are used to dynamically adjust content, difficulty levels and learning experiences based on real-time student results.

3.5.6 Case Studies 1: Successful Implementations of Big Data and Analytics in Industry 5.0

3.5.6.1 Case Study

The following is an illustration for IOT-based intelligent shopping trolley, and Figure 3.5 illustrates proposed architecture for a smart shopping trolly system.

The trolley will consist of an Arduino Uno, RFID reader and a 16x2 LCD screen. All the medicines will have RFID tags attached to them. When an item drops into the cart, it is scanned by the RFID scanner and then added to the list of items purchased by customer. If customer puts two items at a time in the trolley, it gets scanned because of the RFID scanner; usually, the range of the RFID scanner is 13.56MHz. Such high frequency RFID tags are used here. Maximum reading distance is of 1.5 meters (4 foot 11 inches)—generally less than 1 meter (3 feet). To extend the read range we can use a custom antenna or a wider RFID reader zone.

Every customer has a connection to the shop Wi-Fi. Connecting to the Wi-Fi through an ESP 8266 wireless module on the Arduino Uno, it is able to notify the module at 366 metres with the PCB antenna or at 479 metres with a huge rubber duck antenna soldered on.

Whenever a user enters into the shop, the user is assigned with a unique list number. This list number is auto-incremented [7]. By entering the list number, the customer can view the details such as cost, item number and quantity of items in the list, which is prepared by adding desired items to the trolley. An even-odd concept is used in the project, which means that, for every EVEN count, an item will be there in the trolley, and it will get scanned by the RFID scanner; and for every ODD count, an item will be not be scanned by the RFID scanner. For generating the bill, the user will just have to go to the website and simply enter the list number. Then user will be asked to scan the QR code which is present on the trolley [1].

3.5.6.2 Architecture

Every product in the store is attached with RFID tags. The IOT-Based Intelligent Shopping Trolley System contains a RFID reader circuit, which will scan the RFID tags attached to the items. The RFID reader consists of Arduino Uno (microcontroller), RFID reader, LCD screen. There is a centralized database which stores all the information of the items and bills. Payment of the money should be done by customer by using their smart phone. Once the payment is completed, the IOT-Based Intelligent Shopping Trolley System resets. Some of the important components are listed as follows to have a clear understanding of its working [1].

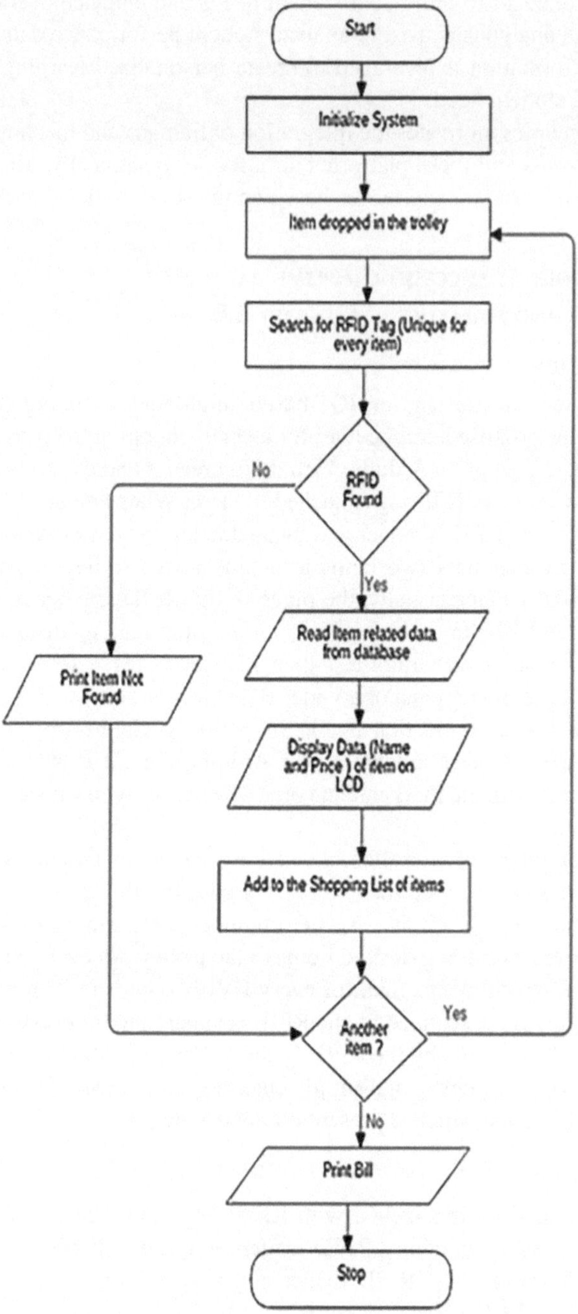

FIGURE 3.5 Proposed Architecture for IOT-Based Intelligent Shopping Trolley.

3.5.6.2.1 RFID Tags

RFID tags mainly consist of antenna and Integrated Circuit (IC). Antenna receives radio frequency signals and IC stores the tag's unique identification number and also modulates and demodulates the radio frequency signal.

3.5.6.2.1.1 RC 522 RFID Reader A RFID reader contains a RF module; it is both transmitter and receiver of radio frequency signals. The transmitter comprises of an oscillator to make the carrier frequency; a modulator to encroach information commands upon the transporting signal and an enhancer to support the sign enough to stir the tag. The receiver has a demodulator to separate the returned information and furthermore contains an amplifier to reinforce the sign for preparing.

3.5.6.2.1.2 Arduino UNO The Arduino UNO is an open-source microcontroller board based on the Microchip ATmega328P microcontroller and developed by Arduino. Arduino UNO structures the control unit, which utilizes an operating system and memory to channel and store the information. The information is presently fit to be shipped off the network.

3.5.6.2.1.3 16x2 LCD screen A good communication present between human world and machine world, display units play an important role. It shows the details such as ITEMTOTALCOST and also displays the information related to the item when the item is inserted or removed.

3.5.6.2.1.4 Centralized Database The database is stored at a particular location such as a mainframe computer. Then, it is accessed using an internet connection such as a LAN or WAN. It stores the data of item and stores the information related to bills in the database, as shown in Figure 3.6.

3.6 ETHICAL ISSUES AND PRIVACY IN BIG DATA ANALYTICS

The concern related to data is to decide privacy protection and ensure trustable data security practices as organizations and institutions collect, process and analyze large volumes of data,

3.6.1 Privacy Protection and Data Security

Implement strong encryption mechanisms for both data in transit and data at rest. Encryption helps to provide security for sensitive information from unauthorized access and ensures that, even if a breach occurs, the data remains unintelligible without the proper decryption keys. Implement strict access controls and authorization mechanisms. Limit access to sensitive data only to authorized personnel based on their roles and responsibilities. Regularly review and update access permissions to minimize the risk of unauthorized access. Integrate privacy considerations into the design and development of big data analytics systems from the outset. This "privacy

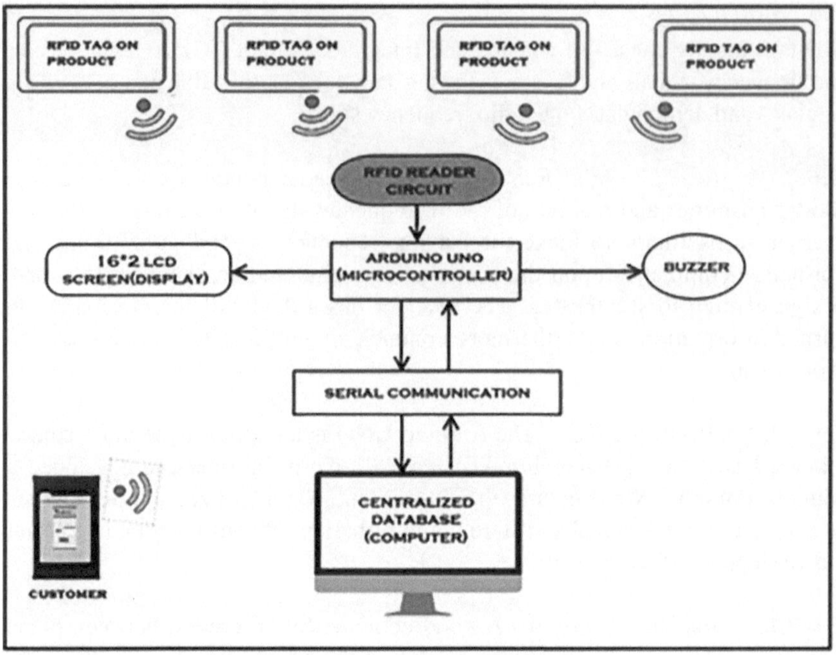

FIGURE 3.6 Architecture of IOT-Based Intelligent Trolley System.

by design" approach ensures that privacy measures are part of the system architecture and not added as an afterthought.

3.6.2 Transparency and Explainability in Analytics

Transparency and explainability are main features of analytics in contexts where decisions affect individuals, organizations or society.

Transparency refers to clarity in the way analytics processes data and takes decision. It involves making information, methods and decisions allowed to check only to authority stakeholders. Transparent analytics processes are a base of trust among stakeholders, including users, customers.

Explainability is the ability to articulate and communicate how analytics models arrive at specific decisions or predictions in a way that is understandable to non-experts. Explainable analytics helps users and stakeholders understand the rationale behind decisions made by models. Users are more likely to have confidence in analytics outcomes if they can comprehend the reasoning behind them.

3.6.3 Ethical Use of Big Data and Algorithmic Bias

The ethical use of big data and addressing algorithmic bias are crucial considerations in the development and deployment of data-driven technologies. Some key principles and practices for ensuring ethical use and mitigating bias in the context of big data and algorithms are listed as follows:

- Complete understanding of ethical considerations and algorithmic bias among data scientists, developers and decision-makers. So, provide ongoing education and training on ethical data practices, the implications of bias and the societal impact of algorithms.
- Conducting assessments to identify and address potential biases in algorithms. So, need to regularly assess the impact of algorithms on different demographic groups and take corrective actions if bias is detected.
- Establish ethical review boards to evaluate the potential ethical implications of data use and algorithmic decision-making and form interdisciplinary teams to review and assess the ethical dimensions of big data projects, especially those with potential societal impacts

3.6.4 Regulatory and Legal Frameworks

The following are some key aspects and considerations related to regulatory and legal frameworks for big data:

Data Protection Laws

Enforced in the European Union (EU), GDPR is a comprehensive regulation governing the handling of personal data. It allows individuals control over their personal data and applies strict requirements on organizations managing such type data.

Intellectual Property Laws

Big data often involves the analysis of vast datasets, and intellectual property laws may come into play. It's crucial to respect copyrights, trademarks.

Cybersecurity Laws

Various jurisdictions have enacted laws that require organizations to implement cybersecurity measures and to notify individuals in the event of a data breach. Compliance with these laws is critical for data security.

Sector Specific Regulations

Certain industries may have specific regulations governing the use of data. For example, in healthcare, the Health Insurance Portability and Accountability Act (HIPAA) in the U.S. regulates the use and protection of health data.

Government Surveillance

Laws related to government surveillance and national security may impact big data practices, particularly when dealing with sensitive information. Organizations must navigate the balance between privacy and security.

3.7 CHALLENGES AND OPPORTUNITIES IN BIG DATA ANALYTICS

3.7.1 Data Volume, Velocity, and Variety

3.7.1.1 Data Volume

Dealing with the tremendous volume of data can be overwhelming. Traditional databases and processing systems may find difficulty to handle the such huge data generated by modern applications and advanced devices.

The huge tons of data provide an opportunity to make more informed decisions. Organizations can uncover valuable information and trends by analyzing large datasets that were previously difficult to manage.

3.7.1.2 Data Velocity

The requirement for real-time data poses challenges in terms of designing systems that can ingest and analyze data at high speeds. Traditional batch processing may not be sufficient. Organizations can gain a competitive edge by extracting insights in real time. This is particularly valuable in scenarios such as fraud detection and IoT.

3.7.1.3 Data Variety

Big data comes in various formats and structures including categories of data such as structured, semi-structured and unstructured data. Integrating and analyzing diverse data sources can be complex. Embracing data variety allows organizations to gain a 360-degree view of their operations. Insights from structured databases, social media, sensor data and more contribute to a comprehensive understanding

3.7.2 Data Integration and Interoperability

Data integration and interoperability are related concepts in the field of IT in context of managing and leveraging data across diverse systems.

Data integration involves combining and unifying data from different sources into a cohesive and meaningful view.

The key components are:

Data extraction:

These can be databases, flat files, APIs or any other systems that store and manage data.

ETL (Extraction, Transformation and Loading)

ETL processes which are vast and time-consuming processes for organization used to extract data from source systems, transform it into a stable format and then load it into a target system.

DWH (Data Warehousing)

Data warehouses serve as centralized repositories for integrated data that makes it easier for organizations to analyze and create report on information.

Data Virtualization

This technique allows users to access and query data without the need for physical consolidation. Virtualization can provide a unified view of data distributed across various sources.

Interoperability

Interoperability refers to the ability of different systems or components to work together, exchange information and use the exchanged information seamlessly. In the context of data, interoperability ensures that data can be shared and utilized across various systems and applications.

Types

Technical Interoperability

Involves the ability of systems to exchange data and operate collaboratively. This often requires adherence to common standards and protocols.

Organizational Interoperability

Addresses the alignment of business processes and practices to facilitate effective collaboration between different organizations.

Some of the challenges when dealing with integration and interoperability are listed as follows:

- Different systems may use different data formats and standards, requiring careful consideration for compatibility.
- Sharing data between systems raises concerns about data security and privacy. Ensuring secure data exchange is crucial.

3.7.3 Research Gap in Skills and Talent

Big data is one of the adopting fields, and many researchers are trying to identify new gaps and areas of interest to gain more knowledge about many trends in data. There are some potential research gaps in the context of skills and talent:

- Use of the long-term impact of learning initiatives on skill development and career progression. Use of big data analytics to trace the impact of training programmes with time and identify factors that contribute to sustained skill development.
- Understanding the role of big data analytics in enhancing the overall employee experience, including factors such as job satisfaction, engagement and the relationship between positive employee experiences and skill retention.
- Research the ways to integrate qualitative insights with big data analytics in talent management. This could involve combining traditional methods like interviews and surveys with advanced analytics to gain a more comprehensive understanding of skills and talent dynamics

3.7.4 Trust and Data Sharing

Trust and data sharing are sensitive and critical points, and the need to deal with the proper care and understanding of maximum security should be given so that customers can trust easily. We can embed two areas, such as artificial intelligence and machine learning, to shared data to make sure ethical considerations are integrated into algorithms to prevent loss or make sure any other unfair activities can be avoided. By implementing valid data governance practices to make sure that data is organized and managed ethically across the organization, this originates creating roles and defining responsibilities and processes. Embedding security, governance, agreements, defining policies and following ethics, we can make sure about responsible and trustworthy data sharing process and techniques.

3.7.5 Data-Driven Decision Making and Governance

Data-driven decision-making is the process of using data analytics to understand and take business decisions. Effective governance ensures data-driven decision-making is executed with responsibility, ethically and followed with the regulations. It writes or explains data ownership responsibilities and create roles and assign responsibility for the accuracy, security and required use of data.

3.8 FUTURE TRENDS AND EMERGING TECHNOLOGIES

Big data analytics is using ML for more advanced information in support with AI. Predictive analytics, anomaly detection and automated decision-making are areas where AI is giving valuable contributions. The need for real-time data analytics is increasing, especially in industries such as finance, healthcare and the stock market; for technologies like Apache, Kafka and Spark are mostly used to process and analyze data in real time. Social networking sites motivate the use of graph databases, which are popular for handling complex relationships in data and find useful information as well used in fraud detection and recommendation systems which are important need of daily life.

3.8.1 Edge Analytics and Real-time Insights

Edge analytics is used to process and analyze data near the source of data generation, typically at the edge of the network or on IoT devices instead of transmitting all the data to a centralized data storage.

3.8.2 Internet of Things (IoT) and Big Data Integration

Just consider the scenario of searching the items in a mall when we go for physical shopping, and the main purpose of physical shopping is to have fun going out and choosing the required product in real time; otherwise, many applications are available to do online shopping. So, to make it more comfortable and easier, we can have the desk at some distance that will give information about which products are placed on which shelf and what is cost of the product as well as which are the alternative options available if that product is not available. Sometimes, it happens that the customer needs quick service and easy access. If the customer needs the product that he or she can't find, then there is need of a helper from the respective compartment, and the customer has to wait for the acknowledgement.

3.8.3 Federated Learning and Privacy-Preserving Analytics

Federated learning and privacy-preserving analytics can provide a comprehensive approach to protecting sensitive data during model training and analytics. By combining decentralized training with techniques like differential privacy, homomorphic encryption and secure multi-party computation, organizations can build robust systems that enable data analysis while safeguarding individual privacy. This is particularly crucial in industries where privacy regulations are stringent, such as healthcare and finance.

3.9 CONCLUSION

Essentially, the addition of big data to the Industry 5.0 landscape represents a revolutionary development in business, characterized by a seamless fusion of human ingenuity and technological prowess. This paradigm shift not only increases efficiency and competitiveness in various sectors such as manufacturing, supply chain management and healthcare but also highlights the potential of data-based decision-making in Industry 5.0. Valuable insights gained through big data analysis enable organizations to navigate consumer behaviour, anticipate market trends and optimize operations, enabling a proactive and flexible response to dynamic market changes. In manufacturing, the use of real-time data from sensors and IoT devices is a catalyst for process optimization. It facilitates preventive maintenance, improved quality control and resource optimization, which not only increases operational efficiency but also encourages commitment to sustainable practices. Similarly, big data acts as a catalyst for better visibility and traceability in supply chain management. The ability to track goods in real time, proactively predict disruptions and optimize logistics not only increases the efficiency of the entire supply chain but also strengthens relationships with stakeholders. In healthcare, big data plays a key role in advancing personalized medicine, accelerating drug development and improving patient care. Complex analysis of large datasets containing genomic information and patient data creates the basis for tailored treatments and targeted interventions that significantly increase the efficiency of health practices. Ultimately, the integration of big data in Industry 5.0 will create a better-connected, smarter and more responsive industrial landscape. This synergy between human intelligence and technological insight is transforming industries towards resilience, efficiency and sustainability. Exploiting the full potential of big data is the key to successfully navigating the complex challenges of Industry 5.0 and encouraging industries to progress.

REFERENCES

[1] Gaikwad S, Nagpure S, Nair B, Sawant P, Mhaske M. Smart Shopping Trolly and Billing System Using NODMCU and FID Technology. Publication Date 2020/9/25 Patent Office Indian IN. Application Number 201921010852A

[2] Akundi A, Euresti D, Luna S, Ankobiah W, Lopes A, Edinbarough I. State of Industry 5.0—Analysis and Identification of Current Research Trends. Applied System Innovation. 2022; 5(1):27. https://doi.org/10.3390/asi5010027

[3] Shetty V, Singh M, Salunkhe S, Rathod N. Comparative Analysis of Different Classification Techniques. SN Computer Science. 2022; 3(1):50.

[4] Verma S. Mapping the Intellectual Structure of the Big Data Research in the IS Discipline: A Citation/Co-Citation Analysis. In I. Management Association (ed.), Research Anthology on Big Data Analytics, Architectures, and Applications (pp. 1923–1957). IGI Global, 2022. https://doi.org/10.4018/978-1-6684-3662-2.ch094

[5] Patil P, Gaikwad S. Review on Integration of Big Data in IOT Agriculture System. International Journal Trendy Research in Engineering and Technology (IJTRET). 2022; 6. https://doi.org/10.54473/IJTRET.2022.6101

[6] Gaikwad S, Patil J. Big Data Platforms. In Advanced Analytics and Deep Learning Models, Publisher John Wiley & Sons, Inc. (pp. 283–310). 2022.

[7] Gaikwad S, Patil J. Malware Detection in Deep Learning. In Convergence of Deep Learning in Cyber-IoT Systems and Security. Journal Publisher John Wiley & Sons, Inc. (pp. 269–284). 2022.

4 Machine Learning–Enabled Predictive Analytics for Quality Assurance in Industry 4.0 and Smart Manufacturing

A Case Study on Red and White Wine Quality Classification

J. Olalekan Awujoola, T. Aniemeka Enem, F. N. Ogwueleka, O. Abioye, and R. Olayinka Adelegan

4.1 INTRODUCTION: NAVIGATING INDUSTRY 4.0 FOR WINE QUALITY ASSURANCE

Amid the relentless forces of globalization and escalating market demands in today's economic landscape, industries find themselves compelled to enhance the effectiveness and productivity of their manufacturing processes. This drive is essential to bolster competitiveness and meet the heightened expectations of customers. The convergence of connectivity, data utilization, innovative devices, inventory streamlining, customization, and controlled production has ushered in the era known as Industry 4.0, an unstoppable transformation (Achouch et al., 2022). Industry 4.0 embodies a hopeful vision for progressing the manufacturing sector by leveraging recent advancements in Information and Communication Technologies (ICT). These advancements facilitate the gathering, retention, and analysis of precise and comprehensive data concerning industrial processes (Bhardwaj et al., 2022). This data enables manufacturers to make decisions based on data analysis, resulting in substantial enhancements to their operations and profitability. Among the entities undergoing this transformation, large manufacturing enterprises are poised to gain the greatest advantage, given their capacity to accumulate more data for improving decision-making processes. Consequently, there is a pressing requirement to employ automation techniques for the seamless integration of emerging technologies, ultimately contributing to heightened productivity (Jasiulewicz-Kaczmarek and Gola, 2019).

DOI: 10.1201/9781003473886-4

In the pursuit of quality excellence, companies recognize the pivotal role that the overall quality of manufactured goods plays in achieving success within the market. Ensuring that each product meets both customer expectations and complies with legal requirements becomes imperative for sustained competitiveness. Concurrently, contemporary industries encounter a spectrum of challenges, including rising costs, operational safety, performance optimization, data security, and workforce turnover. In response to these complex challenges, industries are rapidly embracing smart manufacturing as a transformative solution. This paradigm shift integrates advanced technologies like the Industrial Internet of Things (IIoT), analytics of large datasets, cloud computing, and cyber-physical systems. The overarching objective is to instigate agile and efficient operations within industries, as outlined by Lee et al. in 2015.

Within this dynamic framework of Industry 4.0, the integration of wine quality-assurance classification-analysis stands as a pivotal advancement. By marrying the principles of smart manufacturing with the intricacies of winemaking, this research explores how cyber-physical systems, data analytics, and autonomous technologies contribute to the optimization of wine quality assurance. As we delve into the convergence of technology and oenology, the goal is not solely to attain decentralized production through shared facilities but also to create a unified global industrial system customized for on-demand manufacturing. This holistic approach not only upholds the principles of customization and optimal utilization of resources intrinsic to Industry 4.0 but also underscores the role of wine quality assurance as a cornerstone in this era of innovation, driven by data analytics and smart, autonomous systems.

The significance of wine quality in contemporary society, marked by its widespread consumption, is crucial for both consumers and producers. In today's fiercely competitive market, wine quality directly influences revenue generation (Dahal et al., 2021). Historically, assessing wine quality occurred post-production, demanding substantial time and financial investments (Dahal et al., 2021). Yet, deploying such a meticulous and intricate wine-tasting process on a broad scale poses challenges and proves resource-intensive. Each wine's unique combination of ingredients, coupled with consistent external conditions, suggests that wines with similar compositions should exhibit comparable quality and taste profiles. Emphasizing the significance of wine quality in both production and consumption, it is closely tied to consumer health considerations. The prediction of wine quality contributes significantly to uplifting wine production, instilling consumer confidence in the quality of wines available today (Bhardwaj et al., 2022).

To enhance product quality, testing is pivotal for ensuring consistency and excellence. Modern companies are increasingly adopting new technologies to verify and assess product quality. However, relying solely on human expertise for quality evaluation proves to be costly and time-consuming (Korade, 2022). Delving into the intricacies of wine tasting, the Wine & Spirit Education Trust (WSET) offers a tasting grid evaluating various aspects of wines. This includes assessing the appearance, aroma, and taste, with specific criteria for sweetness, acidity, tannins, alcohol content, body, and intensity. Furthermore, the taste of wine is compared to discerning specific attributes by exploring the tastes of fruits, spices, or herbs (Jiang et al., 2023). Subsequently, a numerical value is assigned as the final quality level. This has led to a substantial increase in wine consumption in both domestic and international markets, fueled by the availability of high-quality wines and heightened competition. Within the framework of Industry 4.0 and intelligent manufacturing, predicting wine quality based on chemical properties becomes pivotal for quality assurance in the wine industry. Understanding the relationship between various chemical concentrations and resulting wine quality can optimize production processes and reduce costs (Gupta et al., 2020). Traditionally, this assessment relied on taste experts, a challenging and time-consuming task incurring significant expenses (Gupta et al., 2020).

This research centers on enhancing the accuracy of predicting wine quality through the application of machine learning techniques, aligning with the industry's goal of efficiently producing high-quality wine. Machine learning emerges as a valuable tool in quality assurance, enhancing efficiency, accuracy, and reducing reliance on expensive manpower (Aich et al., 2018). The growing demand for wine and the industry's objective to produce necessity to comprehend exact chemical concentrations in various types of wine is emphasized by the desire to produce excellent quality wine at a reduced expense (Kumar et al., 2020). Leveraging machine learning enables the analysis and prediction of fine chemical concentrations, empowering wine producers to optimize processes and maintain consistent quality. Recent advancements in machine learning techniques and information technologies have transformed the handling of vast and intricate datasets. This progress facilitates the classification of wines and identifies the importance of every chemical variables in determining standard (Gupta, 2018). Machine learning algorithms analyze extensive datasets, extracting valuable insights that aid in understanding factors contributing to wine quality. Researchers and industry professionals can leverage machine learning to make informed decisions, optimizing wine production processes and enhancing overall quality. The study specifically investigates different machine learning algorithms, including Decision Tree, Random Forest, Support Vector Machine (SVM), and Logistic Regression, artificial neural network, Xgboost, and KNN neighborhood. These approaches automate the quality assurance process, reducing human intervention and leveraging available product characteristics.

4.2 INDUSTRY EVOLUTIONS

The evolution of industry from 1.0 to 4.0, as presented by Markatos and Mousavi (2023), is a dynamic journey marked by transformative shifts in technological, operational, and organizational paradigms. From the inception of mechanization to the age of cyber-physical systems (CPS), this evolution, spanning Industry 1.0 to Industry 4.0, mirrors the continuous quest for efficiency, innovation, and adaptability within the industrial landscape. Each phase brings forth groundbreaking changes, shaping the way products are manufactured, services are delivered, and businesses operate. In this brief exploration, we delve into the key milestones and advancements that characterize the fascinating story of industry evolution.

4.2.1 Industry 1.0

The inaugural phase of the Industrial Revolution, often referred to as Industry 1.0, which occurred nearly between 1760 and 1840, is notable for the widespread adoption of steam energy and the industrialization of manufacturing (Vinitha et al., 2020). This transformative period witnessed a pivotal shift from manual labor to mechanized production processes, with steam power and water power playing instrumental roles in driving this industrial metamorphosis.

Various sectors underwent substantial transformations during this epoch, encompassing innovations in machine tools, textile manufacturing, the iron industry, mining, canals, improved waterways, railways, roads, steam power utilization, chemical advancements, cement production, gaslighting, glassmaking, agriculture, paper machines, transportation., and the establishment of railroads (Markatos and Mousavi, 2023). These pioneering developments left an indelible mark on the initial sectors, propelling them towards a new era characterized by groundbreaking industrial practices and technological advancements (Vinitha et al., 2020).

4.2.2 Industry 2.0

The second phase of the Industrial Revolution, often referred to as Industry 2.0., distinguished by its revolutionary incorporation of technologies, inventive management strategies, and the extensive embrace of electricity, collectively facilitated mass production in the early twentieth century (Zonnenshain and Kenett, 2020). This era saw a significant shift in manufacturing practices, fueled by advancements that harnessed the power of electricity to drive production processes.

A pivotal contributor to this revolution was Walter Shewhart, a distinguished engineer, statistician, and physicist. Shewhart brought in control charts, a statistical tool that transformed the supervision of production processes (Shewhart, 1926). The incorporation of control charts into the manufacturing process had profound implications—it minimized the need for extensive inspection, resulting in notable time and cost savings while concurrently enhancing the overall quality of produced goods.

One of the key transformations during this period was the shift in emphasis within data analysis. It evolved from a focus solely on inspection to a comprehensive understanding of process performance and variation. Statistical models and probability emerged as pivotal elements in quality assessment tools, enabling industries to adopt a more sophisticated and data-driven approach to ensuring the consistency and excellence of their products (Zonnenshain and Kenett, 2020). This integration of statistical methodologies marked a fundamental change in the way industries approached quality control and paved the way for more efficient and effective manufacturing processes during Industry 2.0 (Markatos and Mousavi, 2023).

4.2.3 Industry 3.0

The commencement of the Third Industrial Revolution (Industry 3.0) took place in the 1970s and was fueled by the incorporation of computers, leading to the partial automation of industries (Rifkin, 2011). This era marked a pivotal shift as computers became instrumental in transforming industrial processes, paving the way for what would later be termed as 'mass customization' (Davis, 1997). The integration of computers brought about a revolutionary blend—the expansiveness found in extensive, uninterrupted production systems with the flexibility of job shops.

The introduction of computers during this period had far-reaching implications, allowing industries to explore the concept of 'mass customization'—an innovative approach that merged the efficiency of large-scale production with the flexibility of smaller, more adaptable manufacturing setups. This transformation laid the foundation for a notable expansion in engineering within the production sector. Industries witnessed a continuous and comprehensive automation of the entire production process, effectively eliminating the need for extensive human intervention (Markatos and Mousavi, 2023).

Automation, at this stage, heavily relied on the integration of electronics and computer-controlled hardware. This infusion of technology played a pivotal role in improving the dependability and efficiency of industrial systems. Integration of computers, electronics, and automated processes during Industry 3.0 not only streamlined manufacturing operations but also set the stage for further advancements in industrial technology, leading the way into an era where processes became increasingly self-reliant and technology-driven.

4.2.4 Industry 4.0

The concept of 'Industry 4.0' originated in Germany in 2011, introduced by the German National Academy of Science and Engineering. It was later translated into English as 'Industry 4.0' (Kagermann et al., 2013). This framework signifies the onset of the Fourth Industrial Revolution, driven by the seamless integration of cyber-physical systems (CPS) and the Internet of Things and Services. With nearly two decades of experience in CPS, Germany stands as a pioneering force at the forefront of this transformative revolution.

At the heart of Industry 4.0 is the fusion of physical and digital technologies, creating interconnected systems that communicate and collaborate intelligently. Cyber-physical systems, a hallmark of Industry 4.0, seamlessly integrate the physical and digital realms, enabling real-time data exchange and decision-making. This integration extends to the Internet of Things (IoT) and other services, where interconnected devices and systems communicate to enhance operational efficiency and provide valuable insights.

Germany's leadership in Industry 4.0 is not only a testament to its technological prowess but also highlights its forward-thinking approach to industrial innovation. The integration of cyber technologies connecting products to the internet paves the way for numerous cutting-edge services. These services include cost-effective, Internet-based diagnostics, maintenance, and operation, ushering in a new era of intelligent, data-driven industrial practices.

Beyond just technological progress, Industry 4.0 stimulates the creation of new business models, operational concepts, smart controls, and user-centric solutions (Jazdi, 2014). This shift transcends traditional manufacturing approaches, emphasizing adaptability, customization, and efficiency through interconnected systems. Figure 4.1 illustrates the transformative journey from the industrialization of the first industrial era to the integration of cyber-physical systems in Industry 4.0, showcasing the evolution and progress achieved over the course of industrial history. Industry 4.0 represents not just a technological leap but a paradigm shift, shaping the future of manufacturing and industrial processes on a global scale (Markatos and Mousavi, 2023).

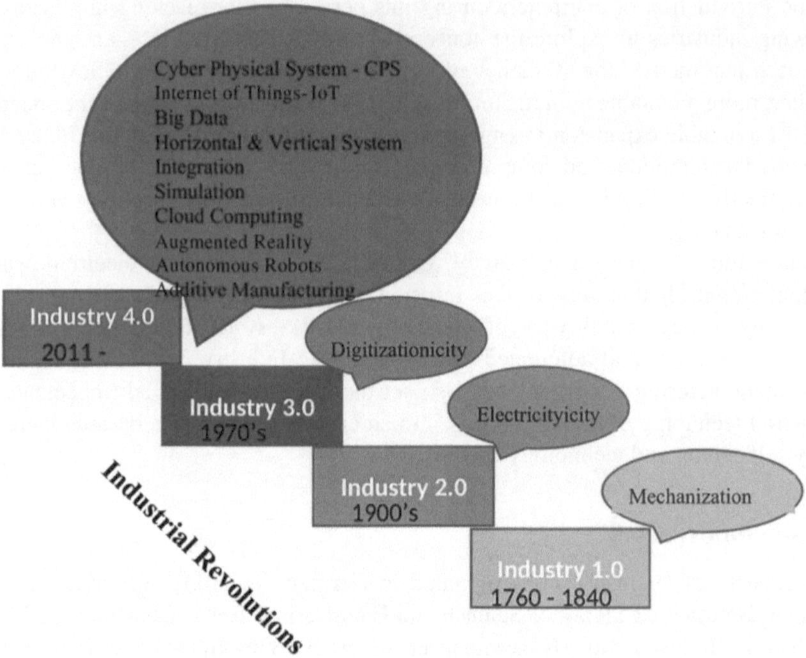

FIGURE 4.1 Industrial History Adapted from Markatos and Mousavi (2023).

4.3 INDUSTRY 4.0 AND SMART MANUFACTURING

Industry 4.0 is reshaping how companies manufacture, improve, and bring their products to market. Manufacturers are integrating state-of-the-art technologies, including the Internet of Things (IoT), cloud computing, analytics, and artificial intelligence (AI) and machine learning, into their production facilities and operational processes (Ghobakhloo, 2020). The integration of these technologies into the manufacturing process results in smart factories that exhibit greater efficiency, flexibility, and adaptability. Considered a significant driver for the future of manufacturing, Industry 4.0 involves the infusion of advanced technologies like AI, IoT, and robotics, promising revolutionary changes in how manufacturing is conducted and increasing global competitiveness. However, the effective execution of Industry 4.0 necessitates substantial investments in technology, infrastructure, and the adoption of new processes, posing a challenge for some companies (Frank et al., 2019).

Smart factories incorporate sophisticated sensors, embedded software, and robotics to collect and analyze data, enabling well-informed decision-making. The real benefits arise when data from production operations are integrated with information from ERP, supply chain, customer service, and other enterprise systems, providing a holistic view and valuable insights (Soori et al., 2023).

Advanced digital technologies propel heightened automation, predictive maintenance, self-optimization of process enhancements, and, notably, elevated levels of efficiency and responsiveness to customer needs. Smart factories, a consequence of integrating Industry 4.0, present a distinct opportunity for the manufacturing sector to engage in the Fourth Industrial Revolution. The analysis of extensive volumes of big data from sensors on the factory floor guarantees real-time visibility of manufacturing assets, enabling predictive maintenance strategies to minimize equipment downtime (Annanth et al., 2021).

Implementing advanced IoT devices in smart factories enhances productivity and elevates product quality. The adoption of AI-powered visual insights in lieu of manual inspection models minimizes manufacturing errors, resulting in time and cost savings. Quality control personnel can use smartphones connected to the cloud to monitor manufacturing processes remotely with a modest investment. Through the application of machine learning algorithms, manufacturers can promptly identify errors, averting the need for more costly repairs at later stages (Markatos and Mousavi, 2023). The principles and technologies of Industry 4.0 can be implemented in various industrial fields, spanning from discrete and process manufacturing to sectors such as mining, oil and gas, and others.

Smart manufacturing, commonly linked with Industry 4.0, specifically highlights the use of cutting-edge technologies like sensors, data analytics, and machine learning to enhance and streamline manufacturing operations. While Industry 4.0 covers the broader scope of the Fourth Industrial Revolution, smart manufacturing places a more targeted emphasis on the technologies and methods employed to boost efficiency and maximize data utilization in the manufacturing process (Zhong et al., 2017; Annanth et al., 2021). Essentially, Industry 4.0 represents the overarching revolution, whereas smart manufacturing focuses on the precise technologies and methods designed to fulfill the goals set by Industry 4.0. These two concepts, closely connected, symbolize the incorporation of advanced technologies into the manufacturing process, leading to smart factories that are not only more efficient but also more flexible, adaptive, and adaptable to market changes and customer demands.

4.4 QUALITY ASSURANCE WITHIN THE INDUSTRY 4.0 FRAMEWORK

Quality Assurance (QA) in the age of Industry 4.0, equivalent to smart manufacturing, is undergoing a profound revolution driven by the assimilation of cutting-edge technologies. This transformative process involves the adoption of disruptive digital technologies, coupled with the application of data analytics and machine learning to enhance and streamline manufacturing operations and improve the standard of products. The integration of these advanced technologies is reshaping the landscape of quality control within Industry 4.0, offering numerous advantages, especially with the widespread implementation of sensor technologies on shop floors. This facilitates comprehensive data collection, enabling the identification of defects that might otherwise remain undetected. This paradigm shift opens the possibility of documenting 100% of measured data, potentially eliminating the necessity for specific statistical tools in real-time quality control (Foidl and Felderer, 2015).

Beyond the refinement of individual piece measurements, Industry 4.0 makes a substantial contribution to overall quality across various stages of the production process. This effort involves improving the standard of information essential for planning and execution, enhancing forecasting accuracy, simulations, prototyping, and predicting heightened worker participation and engagement In essence, Industry 4.0 not only transforms manufacturing processes but also holds extensive implications for the entire quality management landscape.

In Industry 4.0, quality control is embodied by the integration of cutting-edge technologies. This integration involves incorporating advanced elements such as the IoT, cloud computing, analytics, AI, and implementing machine learning in both manufacturing facilities and business operations (Shahin et al., 2020). It transcends traditional quality control by providing enhanced safety, reliability, and confidence in production through process automation (Godina and Matias, 2019). The term 'Quality 4.0' envisions the future of high standards and organizational achievement within the Industry 4.0 context, aligning management practices with Industry 4.0 technologies to enhance organizational outcomes (Godina and Matias, 2019).

Furthermore, Industry 4.0 introduces predictive maintenance and data-driven decision-making, enabling self-optimization of process improvements and achieving new levels of efficiency and responsiveness to customer needs (Bousdekis et al., 2021). Despite the significant potential for quality enhancement, challenges such as the digital skills talent gap and data-related barriers are recognized as impediments to the seamless integration of Industry 4.0 in the domain of quality assurance (Bousdekis et al., 2021).

In the context of wine production, the fundamental concept of Industry 4.0 can be specifically utilized for red and white wine manufacturing, ensuring the highest standards of quality and efficiency throughout the entire production process.

4.5 MACHINE LEARNING IN INDUSTRY 4.0 AND SMART MANUFACTURING

Machine Learning (ML) assumes a pivotal role in the dynamic landscape of Industry 4.0 and smart manufacturing, contributing to diverse applications and yielding substantial benefits. The integration of machine learning in this context encompasses various factors that underpin its significance and impact.

4.5.1 Smart Factories and Data Collection

Industry 4.0 advocates for the adoption of advanced sensors, tools, and machinery, resulting in the emergence of intelligent factories that consistently accumulate data (Rai et al., 2021). This sophisticated data collection lays the groundwork for information-based decision making, forming the backbone of an intelligent and responsive manufacturing ecosystem. In the context of wine production, ML can be harnessed to optimize grape harvesting processes and monitor fermentation conditions.

4.5.2 Quality Control and Overall Equipment Effectiveness (OEE)

ML assumes a pivotal role in improving Overall Equipment Effectiveness (OEE) by assessing the efficiency, availability, and integrity of manufacturing equipment.

Swift identification of machine weaknesses enables proactive measures to minimize inefficiencies and boost productivity (Bajic et al., 2018). In the wine industry, ML applications can ensure quality control throughout the production stages, from grape sorting to bottling.

4.5.3 Predictive Maintenance and Process Optimization

ML proves indispensable for predictive maintenance, enabling self-adjustment of process enhancements and ushering in heightened efficiency and customer responsiveness. Rapid error detection prevents costly repair work in later stages, aligning with the overarching goals of Industry 4.0 (Bajic et al., 2018). In wine production, ML can be applied to optimize fermentation conditions and predict equipment maintenance needs.

4.5.4 Challenges and Opportunities

The application of ML in manufacturing, while advantageous, presents challenges and opportunities, encompassing considerations related to data security, model interpretability, and the need for skilled personnel (Chen et al., 2023; Markatos and Mousavi, 2023). Addressing these factors is critical to unlocking ML's full potential in wine production and ensuring the excellence and safety of the final outcome.

4.5.5 Increased Productivity and Quality Control

The integration of ML algorithms has yielded tangible impacts on manufacturing processes, including increased productivity, reduced waste, improved quality control, and optimized asset management (Markatos and Mousavi, 2023). In the wine industry, ML applications can enhance the precision of grape quality assessment, leading to improved wine quality.

Therefore, the transformative role of machine learning in the context of Industry 4.0 and smart manufacturing extends to diverse sectors, including the nuanced landscape of wine production. Its contributions encompass predictive maintenance, quality control enhancement, and overall equipment optimization. While offering substantial benefits, the application of ML in wine manufacturing necessitates careful consideration of challenges and opportunities to ensure successful implementation and sustained advancements in the field.

4.6 MACHINE LEARNING ALGORITHMS

In the field of machine learning, there's a range of algorithms tailored for the learning process, including linear regression, logistic regression, support vector machines, kernel methods, neural networks, and more (Dahal et al., 2021). Each algorithm has its unique strengths and weaknesses. In this study, we employ a set of supervised learning algorithms to predict wine quality, taking advantage of their specific capabilities and features.

4.6.1 The Decision Tree

This functions as a strong, supervised learning technique adept at addressing both classification and regression problems, particularly effective in handling challenges related to categorization. It adopts a tree-like structure, where the leaf nodes represent outcomes, interior nodes indicate dataset attributes, and branches depict decision rules. Decision trees offer an advantage by requiring less effort for data preparation during the pre-processing stage compared to alternative methods. Unlike certain approaches, they don't require data normalization or scaling, and the impact of missing values on the tree-building process is minimal. However, it's crucial to acknowledge that even a slight change in the data can lead to a significant alteration in the optimal decision tree structure. The complexities can escalate, especially when dealing with uncertain values or multiple interconnected outcomes. Effectively implementing a decision tree poses a challenging task that demands careful consideration (Saini and Sharma, 2019).

4.6.2 Support Vector Machines (SVM)

The SVM classifier functions as a supervised classification algorithm based on the principle of structural risk minimization (SRM). Originally introduced by Vladimir Vapnik (Cortes and Vapnik, 1995), the SVM classifier was initially designed for binary classification problems and has been further adapted for multiclassification by various researchers. When dealing with feature vectors that do not exhibit linear separation, nonparametric functions are employed. The primary goal is to optimize the distance of the separating boundary between violent and nonviolent events by maximizing the distance of the separating plane from each feature vector (Bhattacharyya et al., 2021).

4.6.3 Random Forest

Random Forest stands out as a sophisticated ensemble learning technique applicable to both classification and regression tasks. The methodology underlying Random Forest involves the generation of multiple decision trees in the training phase. Each tree undergoes training on a randomly selected subset of the dataset and a random subset of features. In the classification phase, the outputs of all individual trees are amalgamated to formulate a conclusive decision. The strength of Random Forest lies in its capacity to effectively manage intricate datasets, mitigate overfitting, and offer insights into feature importance rankings. By leveraging the diverse set of decision trees, Random Forest enhances the robustness and accuracy of predictions. This ensemble approach contributes to the model's adaptability, making it particularly beneficial in scenarios where complex relationships within the data need to be unraveled. Additionally, the incorporation of feature importance rankings aids in identifying the most influential factors contributing to the predictive outcomes, providing valuable insights for data interpretation and decision-making (Gao et al., 2021).

4.6.4 Linear Regression

Linear regression is a fundamental statistical method widely used to model the relationship between a dependent variable and one or more independent variables. It plays a crucial role in predictive modeling and regression analysis, proving to be a versatile tool in various fields such as economics, finance, biology, and the social sciences. The core idea of linear regression involves fitting a linear equation to observed data, with the goal of quantifying and understanding the relationship between variables. In a straightforward linear regression context, there exists a lone independent variable, whereas multiple independent variables come into play in the realm of multiple linear regression.

The linear equation is represented by $y = mx + b$, where 'y' denotes the dependent variable, 'x' signifies the independent variable, 'm' represents the slope, and 'b' corresponds to the y-intercept. The primary objective is to ascertain the optimal values for the slope and y-intercept that minimize the disparity between predicted and actual values of the dependent variable. This methodological approach, outlined by James et al. (2023), forms the basis for understanding and quantifying associations between variables, making linear regression an indispensable tool for researchers, statisticians, and data analysts alike. In simple terms, linear regression provides useful information about how variables are connected, helping us predict outcomes and test hypotheses. It serves as a fundamental method that lays the groundwork for more advanced modeling techniques, making it a crucial tool for data analysts, statisticians, and researchers.

4.6.5 k-Nearest Neighbors

The k-Nearest Neighbors (KNN) algorithm is a non-parametric classification technique, implying that it refrains from making assumptions about the underlying dataset structure. Renowned for its simplicity and effectiveness, KNN falls under the category of supervised learning algorithms. In this approach, a labeled training dataset is furnished, wherein data points are already assigned to specific classes. This pre-existing categorization allows the algorithm to predict the class of unlabeled data based on its proximity to the labeled instances in the feature space. The fundamental principle behind KNN is to classify a data point by considering the classes of its nearest neighbors in the training dataset. The flexibility and intuitive nature of KNN make it a popular choice in various applications, particularly when dealing with datasets with complex structures or when assumptions about data distribution are challenging (Taunk et al., 2019).

4.7 LITERATURE REVIEW: UNRAVELING INSIGHTS AND CONTEXT

In a study conducted by Gupta (2018), the focus was on predicting wine quality using two machine learning techniques: neural networks and support vector machines (SVM). The evaluation encompassed considering all features of the dataset and employing selected relevant features. The neural network algorithm surpassed SVM, achieving an accuracy of 83.05% when utilizing all features, while SVM achieved

75.66%. Similarly, with selected features, the neural network algorithm achieved 85.97% accuracy, compared to SVM's 81.03%.

Kumar et al. (2020) directed their efforts and focused on predicting the quality of red wine, employing random forest, support vector machine, and naive Bayes techniques. The assessment criteria encompassed precision, recall, F1-score, accuracy, specificity, and misclassification error. Among the three methods, support vector machine demonstrated superior performance, achieving the highest accuracy at 67.25%. This research specifically emphasized the prediction of red wine quality and the comparison of different machine learning models. In contrast, in 2020, Ye and collaborators introduced a novel wine prediction framework utilizing the XGBoost and LightGBM machine learning classification algorithms, with both achieving an identical accuracy of 91.04%.

Dasgupta et al. (2020) introduced hybrid machine learning models, combining Support Vector Machines and Genetic Algorithms for wine classification, aiming to enhance accuracy and overcome algorithmic limitations. Yan et al. (2020) presented an ensemble learning method based on Random Forest to improve predictive performance through the diversity of multiple classifiers. Ray and Sharif Ullah (2021) conducted a comparative study evaluating Random Forest, Decision Trees, and k-Nearest Neighbors for wine classification to identify the most suitable algorithm.

Pathak and Sharma (2020) explored the use of deep learning neural networks, including Multilayer Perceptrons and Convolutional Neural Networks, to enhance classification accuracy. Bianchi et al. (2020) investigated wine classification using Convolutional Neural Networks and feature extraction techniques for improved model performance. Vani and Sathya (2020) proposed an ensemble learning approach with feature selection techniques for enhanced wine classification. Hassanzadeh and Pourahmad (2020) introduced an ensemble-based feature selection method to enhance wine classification model accuracy. Masoudnia et al. (2020) explored feature selection techniques combined with machine learning algorithms to optimize wine classification performance. Abdi et al. (2020) explored traditional machine learning algorithms and deep learning techniques for wine classification, comparing Random Forest, Support Vector Machines, and deep neural networks.

On a different note, Dahal and team (2021) delved into predicting wine quality using multiple parameters, employing diverse machine learning models like ridge regression, support vector machine, gradient boosting regressor, and a multi-layer artificial neural network. The gradient boosting regressor demonstrated superior performance, recording MSE, R, and MAPE values of 0.3741, 0.6057, and 0.0873, respectively. In a separate study, Pavlou et al. (2021) conducted a comparative analysis to assess the suitability of various machine learning algorithms for wine classification within Industry 4.0 applications. Chakraborty and Sengupta (2021) provided a review article discussing different machine learning algorithms used for wine classification and their comparative strengths and weaknesses.

Furthermore, in their work, Jain et al. (2023) tackled the crucial task of certifying wine quality within the wine industry. The main objective was to develop a machine learning model to predict wine quality, using a dataset made up of samples from the red wine dataset (RWD) that included eleven physiochemical properties. The research assessed five machine learning models, with Random Forest (RF) and Extreme Gradient Boosting (XGBoost) identified as the most accurate algorithms. The exploration of feature importance led to the identification of the top three features, which were subsequently analyzed in-depth. The study emphasized the significance of essential characteristics, identified through feature selection, in predicting wine quality. Notably, the XGBoost classifier demonstrated excellent accuracy when trained and tested with these essential features, surpassing other models. The Random Forest classifier, especially with essential variables, exhibited superior performance. To refine prediction precision, the authors further refined, validated, and optimized the RF classifier by adjusting hyperparameters. Additionally, cluster analysis was applied to address collinearity and reduce predictors while maintaining model accuracy. Overall, the study provides valuable insights into the successful application of machine learning in predicting wine quality, with a focus on essential physiochemical properties. Also, the investigation conducted by Török (2023) addresses the drawbacks of traditional methods for evaluating wine, which heavily rely on subjective human assessments, resulting in inconsistencies and a lack of standardization. To overcome these issues, the study endeavors to create a machine learning model for the objective and efficient prediction of wine quality based on physicochemical properties. Eight different machine learning models were employed, yielding impressive accuracy scores. The analysis underscored the positive impact of alcohol content on quality, while volatile acidity exhibited a negative correlation. Key factors influencing wine quality, as identified by SHAP values, included sulphates, alcohol, and volatile acidity. Following hyperparameter optimization, the optimal model achieved an accuracy of 91.15%, highlighting the importance of sulphates, alcohol, and volatile acidity. This approach provides an objective and scalable means of evaluating wine, benefiting distributors, consumers, and producers seeking product improvement. Despite minor adjustments post-optimization, the model maintained high accuracy, reinforcing the significance of the identified key factors in determining wine quality.

Chhikara et al. (2023), in their study, addressed the significance of evaluating wine quality, considering its association with human health and the time-consuming process of wine production. The paper emphasizes the need for a cost-friendly and accessible procedure for assessing wine quality, benefiting both consumers and wine makers. The study explores various machine learning algorithms and develops a model, using the Random Forest Classifier to predict wine quality as either bad or good. The findings suggest the applicability of such predictions to other products, showcasing the potential for streamlining human work in quality analysis.

4.8 METHODOLOGY

For this research, we utilize the openly accessible wine quality dataset sourced from the UCL Machine Learning Repository. This repository houses a diverse range of datasets widely employed by the machine learning community, enhancing the transparency and reproducibility of our study. The datasets underwent a preprocessing stage, including data cleaning, handling missing values, and feature transformation, to ensure their quality and suitability for the prediction task. The pre-processed dataset was then divided into a train set and a test set. Various machine learning algorithms, such as Decision Tree, Random Forest, Support Vector Machine (SVM), Naïve Bayes, KNN neighborhood, Logistic Regression, and artificial neural network, were trained and fitted on the train set. This approach allowed the researchers to evaluate the performance of different algorithms in predicting wine quality based on the selected features.

The performance of these models was evaluated on the independent test set to assess their generalization ability and effectiveness in predicting wine quality. The use of an open-sourced dataset, comprehensive preprocessing, exploration of various machine learning algorithms, and the choice of Python as the programming language contributed to the transparency and reproducibility of this research work. The findings of this study provide insights into wine quality prediction and help identify the most suitable algorithm for this specific task.

The objective of this research work was to make a valuable contribution to the field of wine prediction. To achieve this, the study utilized both red and white wines, merging their respective datasets. Rigorous preprocessing techniques were applied to ensure data quality and suitability for analysis. The datasets were then divided into separate sets for training and evaluation purposes. By adopting an open-sourced dataset, the research work emphasized transparency and reproducibility, while the merging of red and white wine data enabled a comprehensive assessment of the models' performance across various wine varieties. This approach facilitated a deeper understanding of wine prediction and enhanced the reliability of the study's findings.

4.8.1 Dataset Source

This study employs datasets for red and white wines sourced from the UCI Machine Learning Repository, a comprehensive resource for the machine learning community. The dataset comprises two Excel files, each representing the red and white variants of Portuguese 'Vinho Verde' wine (Cortez and Silva, 2009). The red wine dataset contains 1599 instances, while the white wine dataset comprises 4898 instances. Each dataset features 11 input variables derived from physicochemical tests, including fixed acidity, volatile acidity, citric acid, residual sugar, chlorides, free sulfur dioxide, total sulfur dioxide, density, pH, sulfates, and alcohol. Additionally, there is one output variable based on sensory data, indicating the quality of the wine. The sensory data is categorized into 11 quality classes, ranging from 0 to 10, denoting very bad to very good quality. Specifically, this research centers on analyzing the red and white wine variations of the Vinho Verde varietal from Portugal using the 'Wine Quality' dataset accessible in the UCI repository.

4.9 EXPERIMENTAL RESULT AND ANALYSIS

In this study, the red and white wine datasets were amalgamated to create the red-whitewine dataset. The prediction of wine quality for the experiments was conducted utilizing the scikit-learn library in Python.

4.9.1 Red and White Wine Quality Prediction Results

The section offers a comprehensive analysis and evaluation of the outcomes obtained through predictive modeling to assess the quality of both red and white wines. This segment unveils the findings and insights derived from the application of quality prediction methodologies, shedding light on the effectiveness of the models employed and their implications for the wine production domain. The results obtained are presented in Tables 4.1, 4.2, 4.3, 4.4, 4.5, 4.6, and 4.7, while the confusion matrix for each classifier's results is visualized in Figures 4.2, 4.3, 4.4, 4.5, 4.6, 4.7, and 4.8.

The classification report for the wine quality prediction using linear regression provides a detailed analysis of the model's performance across different quality categories, as shown in Table 4.1. The model aimed to categorize wines as either 'Good' or 'Bad' based on a predefined threshold. The report includes precision, recall, F1-score, and support values for each class, shedding light on the model's ability to make accurate predictions. The precision metric measures the accuracy of the model's positive predictions. For wines categorized as 'Bad,' the precision was an impressive 94%, indicating that when the model predicted a wine as 'Bad,' it was correct approximately 94% of the time. Similarly, for 'Good' quality wines, the precision was 99%, showcasing a high level of accuracy in identifying 'Good' quality wines.

The recall metric assesses the model's ability to correctly identify positive instances from the entire dataset. The recall for 'Bad' quality wines was 98%, signifying that the model successfully identified 98% of the actual 'Bad' quality wines. In the case of 'Good' quality wines, the recall was 98%, indicating a high level of accuracy in identifying 'Good' quality wines.

The F1-score, a balanced measure of precision and recall, provides an overall assessment of the model's accuracy. For 'Bad' quality wines, the F1-score was 96%, reflecting a harmonious balance between precision and recall. Similarly, the F1-score for 'Good' quality wines was 98%, indicating a robust performance in this category. Support values in the report indicate the actual occurrences of each class in the dataset. The 'Bad' quality category had a support of 480, while the 'Good' quality category had a support of 1,470.

In summary, the classification report for linear regression demonstrates a high level of accuracy in predicting both 'Good' and 'Bad' quality wines. The model exhibited excellent precision and recall for both categories, leading to high F1-scores. This

TABLE 4.1
Logistic Regression Classification Report

Quality	Precision	recall	f1-score	support
Bad (0)	0.94	0.73	0.82	1047
Good (1)	0.41	0.80	0.54	253
Accuracy			0.74	1300
Macro Avg	0.68	0.76	0.68	1300
Weighted Avg	0.84	0.74	0.76	1300

suggests that linear regression is a reliable method for wine quality prediction, providing valuable insights for quality assessment in the wine industry.

In the context of wine quality prediction using the Linear Regression model, a detailed examination of the confusion matrix offers valuable insights into the model's performance. The confusion matrix, as presented in Figure 4.2, categorizes instances into four distinct outcomes: True Positive (Predicted Good and Actually Good), True Negative (Predicted Bad and Actually Bad), False Positive (Predicted Good but Actually Bad), and False Negative (Predicted Bad but Actually Good).

The confusion matrix reveals that the Linear Regression model successfully identified 202 instances of 'Good' quality wines (True Positives). Additionally, 760 instances were accurately classified as 'Bad' quality wines (True Negatives). These instances represent successful predictions where the model aligned with the actual quality labels. On the contrary, the model encountered challenges in certain predictions. There were 287 instances falsely predicted as 'Good' quality wines, though they were, in fact, of 'Bad' quality (False Positives). Moreover, 51 instances were incorrectly labeled as 'Bad' quality wines when they were 'Good' quality (False Negatives). These inaccuracies represent instances where the model did not align with the true quality labels.

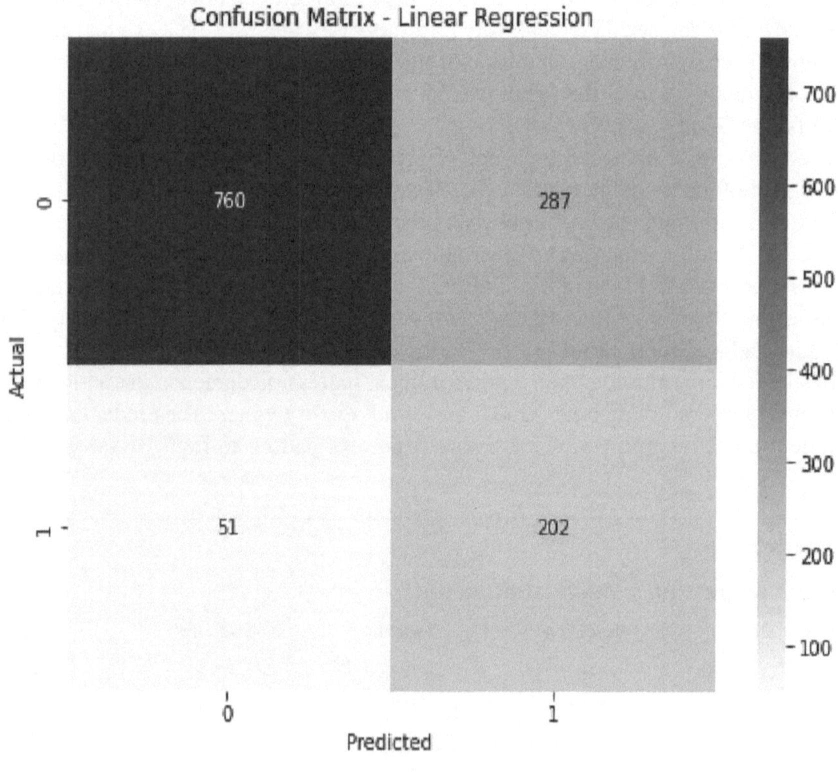

FIGURE 4.2 Confusion Matrix for Linear Regression.

The presence of False Positives and False Negatives highlights specific areas for improvement in the model. The 287 instances falsely identified as 'Good' quality wines suggest that the model may be overly optimistic in predicting positive outcomes. On the other hand, the 51 instances misclassified as 'Bad' quality wines indicate a tendency to be overly conservative in certain predictions. In conclusion, a nuanced analysis of the confusion matrix provides a deeper understanding of the model's strengths and weaknesses. It serves as a foundation for targeted improvements to enhance the overall accuracy and reliability of the wine quality prediction system

Table 4.2 provides a detailed breakdown of precision, recall, F1-score, and support metrics for each class predicted by the Random Forest model. The Random Forest model demonstrates strong predictive capabilities for identifying 'Bad' quality wines, achieving a precision of 0.93. This indicates that when the model predicts a wine as 'Bad,' it is correct approximately 93% of the time. The recall value of 0.90 signifies that the model successfully captures 90% of all actual 'Bad' quality instances. In predicting 'Good' quality wines, the model exhibits a precision of 0.62, indicating that when it predicts a wine as 'Good,' it is correct around 62% of the time. The recall value of 0.70 signifies that the model captures 70% of all actual 'Good' quality instances.

The model achieves an overall accuracy of 0.86, implying that it correctly classifies wine quality in 86% of instances. The macro-average F1-score of 0.79 provides a balanced measure of precision and recall across both classes, while the weighted average F1-score of 0.86 considers the class imbalance in the dataset. Therefore, the Random Forest classifier showcases a commendable performance in distinguishing between 'Bad' and 'Good' quality wines. Its ability to maintain a high precision for 'Bad' quality predictions contributes to the overall effectiveness of the model. As with any classifier, there is a trade-off between precision and recall, and tuning the model parameters could potentially enhance its performance further.

In the evaluation of the Random Forest classifier's performance in predicting wine quality, attention is directed to Figure 4.3, encapsulating the confusion matrix. This matrix serves as a comprehensive depiction of the model's ability to classify instances into 'Bad' (Class 0) and 'Good' (Class 1) quality wines. Within this matrix, instances are meticulously categorized based on the accuracy of predictions.

Notably, the Random Forest classifier excels in correctly classifying instances, demonstrating a nuanced understanding of both 'Bad' and 'Good' quality wines.

TABLE 4.2
Random Forest Classification Report

Quality	Precision	Recall	F1-Score	Support
Bad (0)	0.93	0.90	0.91	1047
Good (1)	0.62	0.70	0.66	253
Accuracy			0.86	1300
Macro Avg	0.78	0.80	0.79	1300
Weighted Avg	0.87	0.86	0.86	1300

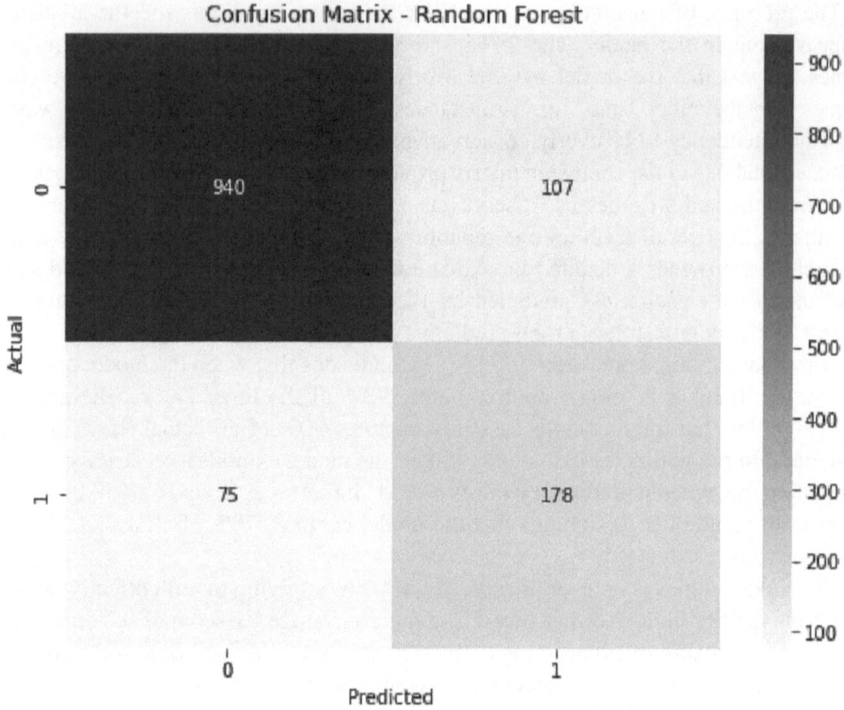

FIGURE 4.3 Random Forest Confusion Matrix.

Examining instances correctly classified, the model accurately identifies 940 instances of 'Bad' quality wines (Class 0) and 178 instances of 'Good' quality wines (Class 1). These instances represent successful predictions, showcasing the classifier's proficiency in recognizing distinct wine quality categories.

However, the model is not immune to misclassifications. In particular, 107 instances are falsely predicted as 'Good' quality wines when they actually belong to the 'Bad' quality category. Additionally, there are 75 instances where the model erroneously categorizes 'Good' quality wines as 'Bad.' This detailed breakdown of correct and incorrect predictions unveils the model's strengths and areas for improvement. While the overall accuracy is commendable, understanding the nature of misclassifications becomes crucial for refining the model's predictive capabilities.

In essence, the confusion matrix acts as a powerful tool for dissecting the Random Forest classifier's performance, guiding the path toward further optimization and precision in wine quality predictions.

In evaluating the performance of the Support Vector Machine (SVM) classifier for predicting wine quality, the results are summarized in Table 4.3. This table presents a comprehensive classification report, including precision, recall, F1-score, and support metrics for both 'Bad' (Class 0) and 'Good' (Class 1) quality wines.

The SVM classifier demonstrates notable precision in classifying 'Bad' quality wines, achieving a value of 95%. This indicates a high level of accuracy in correctly

TABLE 4.3
Support Vector Machine Classification Report

Quality	Precision	Recall	f1-Score	Support
Bad (0)	0.95	0.71	0.81	1047
Good (1)	0.41	0.84	0.55	253
Accuracy			0.73	1300
Macro Avg	0.68	0.77	0.68	1300
Weighted Avg	0.84	0.73	0.76	1300

identifying instances belonging to the 'Bad' quality category. However, the precision for predicting 'Good' quality wines is comparatively lower, standing at 41%. This implies that while the SVM model excels at discerning 'Bad' quality instances, it tends to produce more false positives for 'Good' quality wines.

The recall metric, which measures the classifier's ability to capture instances of each class, shows that the SVM classifier successfully identifies 71% of 'Bad' quality instances. On the other hand, the recall for 'Good' quality instances is notably higher at 84%. This suggests that the model effectively captures a significant portion of instances belonging to the 'Good' quality category.

The F1-score, a harmonic mean of precision and recall, serves as a balanced measure of the classifier's performance. For 'Bad' quality wines, the SVM classifier attains an F1-score of 81%, reflecting a harmonious balance between precision and recall. However, for 'Good' quality wines, the F1-score is 55%, indicating a trade-off between precision and recall in this category.

The overall accuracy of the SVM classifier is reported as 73%, representing its ability to correctly classify instances into 'Bad' and 'Good' quality wines. The macro-average, considering equal weight for both classes, and the weighted average, accounting for the imbalance in class distribution, provide a comprehensive evaluation of the classifier's overall performance.

In the evaluation of the Support Vector Machine (SVM) classifier's predictive performance for wine quality, attention is directed to Figure 4.4, which visualizes the confusion matrix. The confusion matrix serves as a detailed representation of the SVM model's ability to categorize instances into 'Bad' (Class 0) and 'Good' (Class 1) quality wines.

Examining instances correctly classified, the SVM model accurately identifies 739 instances of 'Bad' quality wines (Class 0) and 212 instances of 'Good' quality wines (Class 1). These instances represent successful predictions, underscoring the classifier's efficacy in recognizing distinct wine quality categories.

However, the model is not without misclassifications. Specifically, 308 instances are falsely predicted as 'Good' quality wines when they actually belong to the 'Bad' quality category. Conversely, there are 41 instances where the model erroneously categorizes 'Good' quality wines as 'Bad.' This detailed breakdown of correct and incorrect predictions provides insights into the strengths and areas for improvement in the SVM classifier's performance.

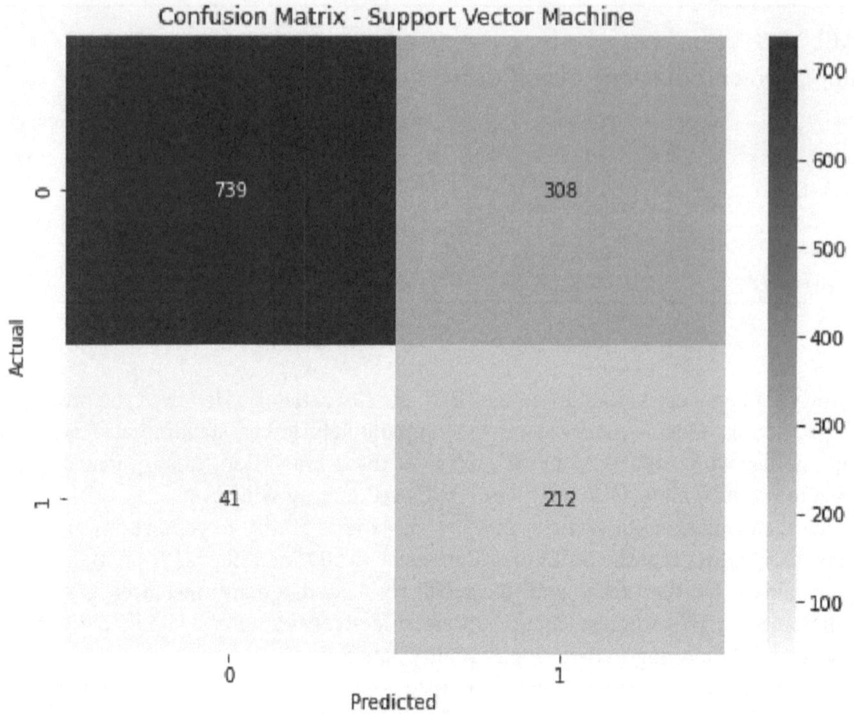

FIGURE 4.4 Support Vector Machine Confusion Matrix.

In essence, the confusion matrix acts as a powerful tool for dissecting the SVM classifier's performance, offering valuable insights into the nature of predictions and guiding further refinements to enhance the model's precision in wine quality predictions.

In assessing the performance of the K-Nearest Neighbors (KNN) classifier for predicting wine quality, the focus turns to the classification report, as depicted in the Table 4.4. This report provides a detailed overview of precision, recall, F1-score, and support metrics for both 'Bad' (Class 0) and 'Good' (Class 1) quality wines.

The KNN classifier demonstrates commendable precision in classifying 'Bad' quality wines, achieving an accuracy of 93%. However, it exhibits lower precision in predicting 'Good' quality wines, with a value of 44%. This indicates that while the KNN model is adept at identifying instances of 'Bad' quality, it may generate more false positives for 'Good' quality wines. The recall metric reflects the classifier's ability to capture instances of each class. The KNN classifier achieves a recall of 77% for 'Bad' quality wines, indicating that it successfully identifies a significant portion of instances belonging to this category. On the other hand, the recall for 'Good' quality wines is notably higher at 75%, suggesting that the model effectively captures instances of this class.

The F1-score, a harmonic mean of precision and recall, provides a balanced measure of a classifier's performance. For 'Bad' quality wines, the KNN classifier attains

TABLE 4.4
KNN Classification Report

Quality	Precision	Recall	f1-Score	Support
Bad (0)	0.93	0.77	0.84	1047
Good (1)	0.44	0.75	0.56	253
Accuracy			0.77	1300
Macro Avg	0.68	0.76	0.70	1300
Weighted Avg	0.83	0.77	0.79	1300

an F1-score of 84%, while, for 'Good' quality wines, the F1-score is 56%. These values underscore the trade-off between precision and recall and highlight the model's overall effectiveness in classifying different quality categories.

The weighted average of precision, recall, and F1-score is crucial for assessing the overall performance of the KNN classifier. With an accuracy of 77%, the model demonstrates a reasonable ability to classify instances into 'Bad' and 'Good' quality wines. The macro-average, considering equal weight for both classes, and the weighted average, considering the imbalance in class distribution, provide a comprehensive evaluation of the classifier's performance.

In conclusion, the KNN classifier exhibits strengths in precision and recall, showcasing its proficiency in discerning 'Bad' and 'Good' quality wines. However, the trade-off between precision and recall, especially for 'Good' quality wines, emphasizes the need for further refinement to achieve a more balanced and precise predictive model.

In scrutinizing the confusion matrix presented in Figure 4.5, which encapsulates the outcomes of the K-Nearest Neighbors (KNN) classifier in predicting wine quality, we delve into a detailed analysis of instances that were correctly and wrongly classified. This matrix, an essential tool in evaluating the model's performance, delineates instances into the categories of 'Bad' (Class 0) and 'Good' (Class 1) quality wines.

Examining instances that were correctly classified, we find that the KNN classifier excels in identifying instances labeled as 'Bad' quality wines, accurately predicting 804 instances within this category. Likewise, instances labeled as 'Good' quality wines see success, with 191 instances correctly classified as such. However, the model encounters challenges in certain scenarios. Instances labeled as 'Bad' quality wines, amounting to 243 instances, are erroneously predicted as 'Good.' Conversely, instances labeled as 'Good' quality wines experience misclassification, with 62 instances mistakenly identified as 'Bad.'

This comprehensive breakdown offers valuable insights into the nuances of the KNN classifier's performance. While the overall accuracy stands at a respectable 77%, the detailed examination of the confusion matrix allows us to pinpoint specific areas that may benefit from refinement. Understanding both the instances correctly identified and those subject to misclassification is crucial for enhancing the model's predictive prowess.

In essence, the confusion matrix serves as a diagnostic tool, guiding efforts to optimize the KNN classifier's precision in distinguishing between different wine

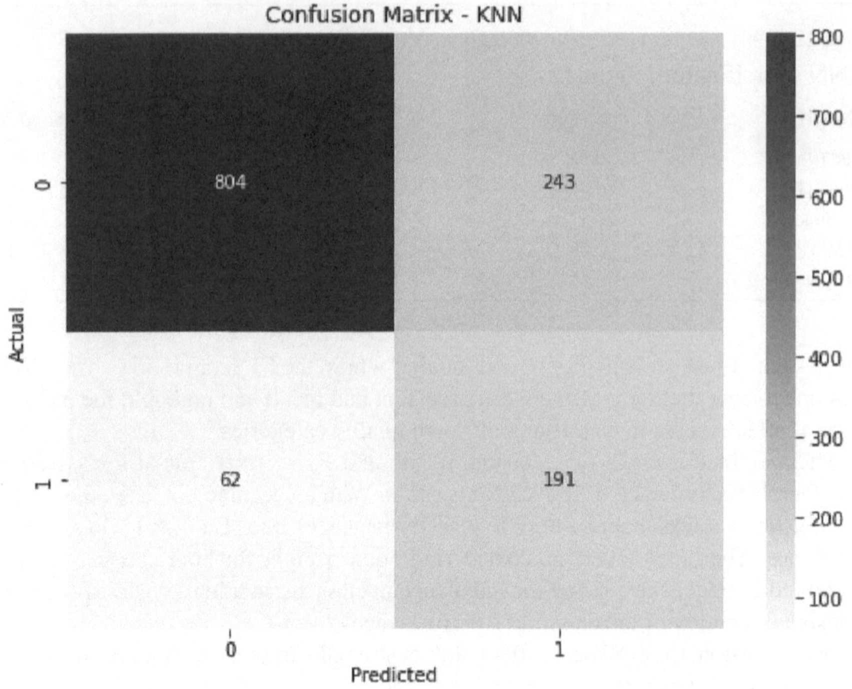

FIGURE 4.5 KNN Confusion Matrix.

quality categories. This nuanced evaluation goes beyond accuracy metrics, providing a deeper understanding of the model's strengths and areas for improvement.

Examining the Decision Tree classifier's performance, as depicted in Table 4.5, reveals insights into its predictive capabilities regarding wine quality. The classification report provides a thorough examination of precision, recall, F1-score, and support metrics, elucidating the model's effectiveness in categorizing instances.

With a precision of 90%, the classifier demonstrates accuracy in identifying instances of 'Bad' (Class 0) quality wines, instilling confidence in its ability to make correct predictions within this category. Conversely, the precision drops to 63% for instances of 'Good' (Class 1) quality wines, suggesting a more moderate level of accuracy in this classification.

The recall metric, indicating the model's capacity to capture instances of each class, is noteworthy. Achieving a recall of 92% for 'Bad' quality wines signifies the model's success in identifying a substantial portion of instances within this category. However, the recall for 'Good' quality wines is slightly lower at 55%, suggesting potential oversight of some instances within this class.

The F1-score, a balanced measure of precision and recall, reflects the classifier's overall performance. For 'Bad' quality wines, the Decision Tree classifier attains an F1-score of 91%, indicating a harmonious balance between precision and recall in this category. Conversely, the F1-score drops to 59% for 'Good' quality wines, highlighting a trade-off between precision and recall.

TABLE 4.5
Decision Tree Classification Report

Quality	Precision	Recall	f1-Score	Support
Bad (0)	0.90	0.92	0.91	1047
Good (1)	0.63	0.55	0.59	253
Accuracy			0.85	1300
Macro Avg	0.76	0.74	0.75	1300
Weighted Avg	0.84	0.85	0.85	1300

The weighted average of precision, recall, and F1-score provides a holistic evaluation of the Decision Tree classifier's overall performance, achieving an accuracy of 85%. This reflects the model's commendable ability to classify instances into 'Bad' and 'Good' quality wines. The macro-average and weighted average offer comprehensive insights, considering equal weight for both classes and accounting for the class distribution imbalance.

In conclusion, the Decision Tree classifier excels in precision, particularly for 'Bad' quality wines but reveals a potential trade-off in precision-recall balance for 'Good' quality wines. This nuanced evaluation aids in understanding the overall effectiveness of the classifier, guiding further refinement for enhanced predictive capabilities.

The evaluation of the Decision Tree (DT) classifier's performance in predicting wine quality unveils valuable insights through the analysis of the confusion matrix, visually represented in Figure 4.6. This matrix serves as a comprehensive tool, dissecting instances into 'Bad' (Class 0) and 'Good' (Class 1) quality wines and providing a nuanced perspective on the model's capabilities.

Upon scrutinizing instances correctly classified, the DT classifier demonstrates notable precision in identifying instances of 'Bad' quality wines, achieving a commendable count of 964. Moreover, the model successfully identifies 140 instances of 'Good' quality wines, showcasing its accuracy in categorizing instances within this class.

However, instances wrongly classified reveal certain challenges faced by the model. Specifically, 83 instances categorized as 'Bad' quality wines are mispredicted as 'Good,' highlighting a potential area for improvement in precision. Additionally, 113 instances of 'Good' quality wines are inaccurately classified as 'Bad,' indicating a need for enhanced precision within this category.

This detailed breakdown of correct and incorrect predictions within the confusion matrix provides valuable insights into the Decision Tree classifier's performance. While the model excels in certain aspects, the misclassifications pinpoint areas for refinement, directing attention toward improving predictive accuracy and precision for both 'Bad' and 'Good' quality wines. The intricate analysis of the confusion matrix serves as a guide for refining the model and enhancing its effectiveness in making accurate wine quality predictions.

In the examination of the Naive Bayes classifier's performance for wine quality prediction, as outlined in Table 4.6, the classification report provides a comprehensive

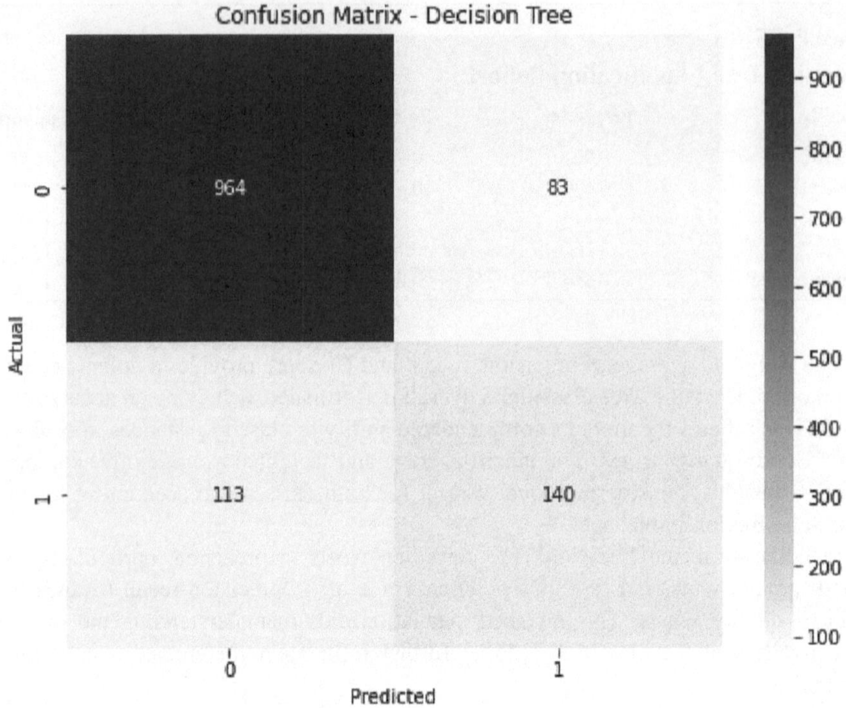

FIGURE 4.6 Decision Tree Confusion Matrix.

TABLE 4.6 NAIVE BAYES CLASSIFICATION REPORT

Quality	Precision	Recall	f1-Score	Support
Bad (0)	0.93	0.62	0.74	1047
Good (1)	0.34	0.82	0.48	253
Accuracy			0.66	1300
Macro Avg	0.64	0.72	0.61	1300
Weighted Avg	0.82	0.66	0.69	1300

overview of precision, recall, F1-score, and support metrics for both 'Bad' (Class 0) and 'Good' (Class 1) quality wines.

The Naive Bayes classifier demonstrates a precision of 93% in classifying 'Bad' quality wines, indicating a substantial ability to accurately identify instances within this category. However, its precision for 'Good' quality wines is relatively lower, at 34%, suggesting a higher likelihood of false positives within this class. This trade-off between precision and recall is further highlighted in the recall metrics, where the classifier achieves a recall of 62% for 'Bad' quality wines, but notably higher, at 82%, for 'Good' quality wines. This indicates that the model effectively captures a significant portion of instances belonging to the 'Good' quality category.

The F1-score, representing the harmonic mean of precision and recall, provides a balanced evaluation of the classifier's performance. For 'Bad' quality wines, the Naive Bayes classifier attains an F1-score of 74%, while for 'Good' quality wines, the F1-score is 48%. These values underscore the model's overall effectiveness in classifying different quality categories, acknowledging the inherent trade-offs involved.

The weighted average, considering the imbalance in class distribution, provides an overall assessment of the Naive Bayes classifier's performance. With an accuracy of 66%, the model exhibits a reasonable ability to classify instances into 'Bad' and 'Good' quality wines. The macro-average, considering equal weight for both classes, further contributes to a comprehensive understanding of the classifier's performance across different quality categories.

In summary, the Naive Bayes classifier showcases strengths in precision for 'Bad' quality wines and recall for 'Good' quality wines. However, the observed trade-off necessitates careful consideration in refining the model for a more balanced predictive performance across both quality categories.

The analysis of the Naive Bayes classifier's performance extends to the examination of the confusion matrix, as visualized in Figure 4.7. This matrix offers a detailed breakdown of correctly classified and misclassified instances for 'Bad' (Class 0) and 'Good' (Class 1) quality wines.

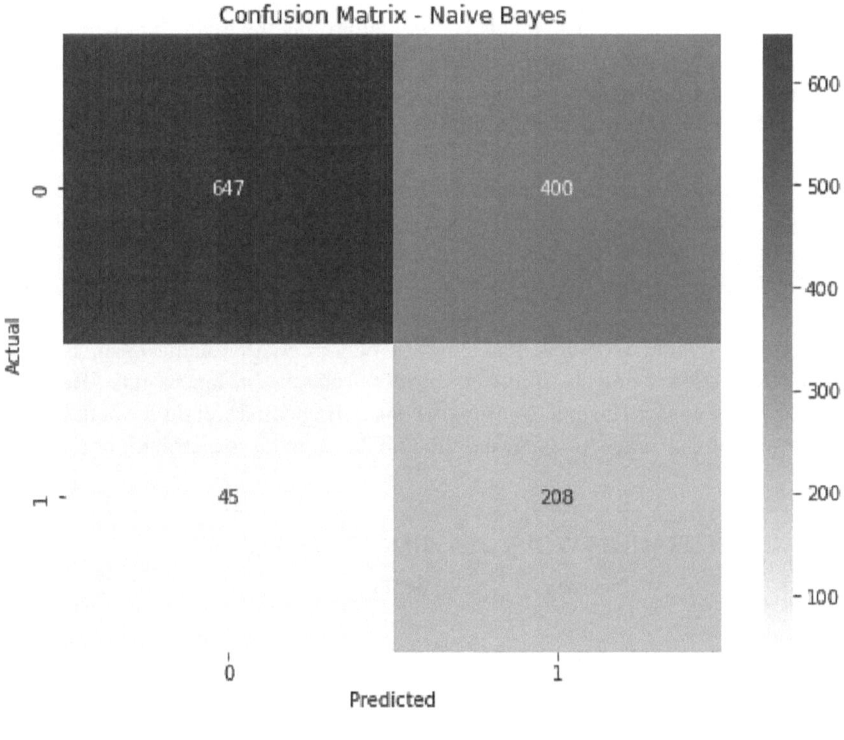

FIGURE 4.7 Naïve Bayes Confusion Matrix.

Instances correctly classified reveal that the Naive Bayes classifier accurately identifies 647 instances of 'Bad' quality wines (Class 0) and 208 instances of 'Good' quality wines (Class 1). These instances represent successful predictions, highlighting the classifier's proficiency in recognizing distinct wine quality categories.

However, the model is not immune to misclassifications. Notably, 400 instances are falsely predicted as 'Good' quality wines when they actually belong to the 'Bad' quality category. Additionally, there are 45 instances where the model erroneously categorizes 'Good' quality wines as 'Bad.' This detailed breakdown of correct and incorrect predictions unveils the model's strengths and areas for improvement. While the overall accuracy is noteworthy, understanding the nature of misclassifications becomes crucial for refining the model's predictive capabilities.

In essence, the confusion matrix serves as a powerful tool for dissecting the Naive Bayes classifier's performance, providing insights into the classifier's ability to distinguish between 'Bad' and 'Good' quality wines. This nuanced examination guides the path toward further optimization and precision in wine quality predictions.

In scrutinizing the classification report derived from the Neural Network classifier, as depicted in Table 4.7, a comprehensive evaluation of precision, recall, F1-score, and support metrics for both 'Bad' (Class 0) and 'Good' (Class 1) quality wines emerges.

For 'Bad' quality wines, the Neural Network classifier achieves a commendable precision of 93%, indicating a high proportion of accurate predictions among instances predicted as 'Bad.' The recall metric, denoting the classifier's ability to capture instances of this class, stands at 79%, signifying the model's efficacy in correctly identifying a substantial portion of 'Bad' quality instances. The F1-score, a harmonized measure of precision and recall, is calculated at 85%, reflecting the classifier's balanced performance in classifying 'Bad' quality wines.

Conversely, the classifier exhibits a lower precision of 46% for predicting 'Good' quality wines, indicating a higher rate of false positives among instances predicted as 'Good.' However, the recall for 'Good' quality wines is notably higher at 75%, suggesting that the model effectively captures instances of this class. The F1-score for 'Good' quality wines stands at 57%, indicating a balanced trade-off between precision and recall.

In terms of overall accuracy, the Neural Network classifier achieves an accuracy rate of 78%, showcasing its ability to correctly classify instances into 'Bad' and 'Good' quality wines. The macro-average, considering equal weight for both classes, and the weighted average, considering the imbalance in class distribution, provide a

TABLE 4.7 NEURAL NETWORK CLASSIFICATION REPORT

Quality	Precision	Recall	f1-Score	Support
Bad (0)	0.93	0.79	0.85	1047
Good (1)	0.46	0.75	0.57	253
Accuracy			0.78	1300
Macro Avg	0.69	0.77	0.71	1300
Weighted Avg	0.84	0.78	0.80	1300

holistic evaluation of the classifier's performance. With a macro-average F1-score of 71%, the Neural Network model demonstrates a balanced performance across both quality categories.

In conclusion, the Neural Network classifier exhibits strengths in precision and recall, providing a nuanced understanding of both 'Bad' and 'Good' quality wines. The balanced evaluation across multiple metrics contributes to a comprehensive assessment of the model's efficacy in predicting wine quality.

Analyzing the confusion matrix from the Neural Network classifier, as depicted in Figure 4.8, sheds light on the model's performance in predicting 'Bad' (Class 0) and 'Good' (Class 1) quality wines. The matrix delineates instances that were correctly and incorrectly classified, offering a comprehensive understanding of the classifier's strengths and areas for improvement.

For 'Bad' quality wines (Class 0), the classifier accurately identifies 824 instances, showcasing its precision in recognizing this category. Similarly, instances of 'Good' quality wines (Class 1) correctly classified amount to 189, underscoring the model's proficiency in predicting instances belonging to this quality category.

Conversely, there are 223 instances of 'Bad' quality wines that the model incorrectly predicts as 'Good,' indicating instances of false positives in the 'Bad' quality category. Simultaneously, the classifier erroneously categorizes 64 instances

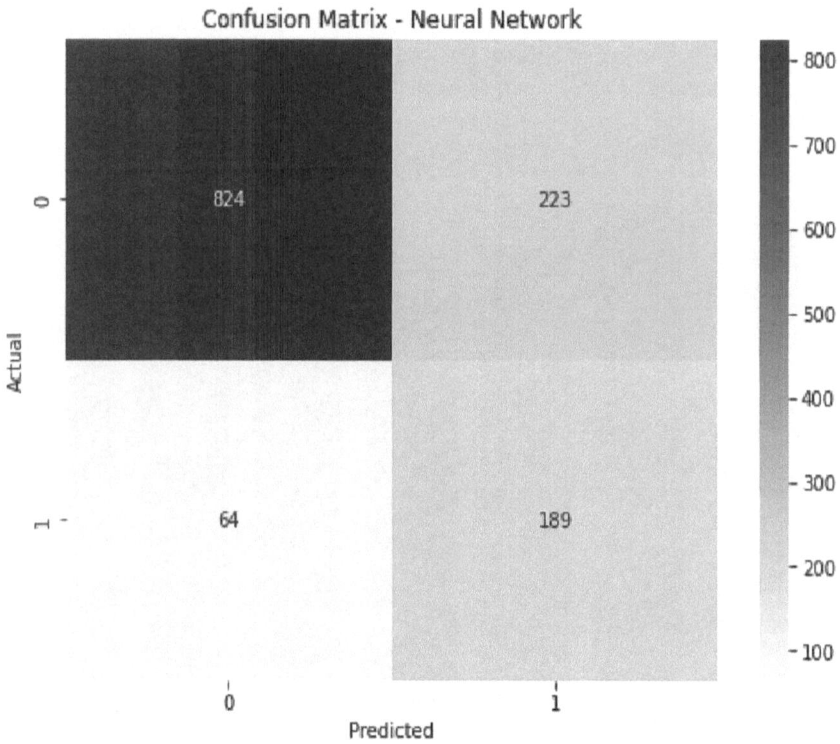

FIGURE 4.8 Neural Network Confusion Matrix.

of 'Good' quality wines as 'Bad,' representing instances of false negatives in the 'Good' quality category.

This detailed breakdown of correctly and wrongly classified instances provides nuanced insights into the Neural Network model's predictive abilities. While the model excels in accurately identifying a significant number of instances for both quality categories, addressing instances of misclassification becomes imperative for refining its predictive capabilities.

4.10 COMPARISON OF ACCURACY BY DIFFERENT ALGORITHM

The comparison of holdout accuracy across seven evaluated algorithms constitutes a critical exploration into the performance and efficacy of diverse machine learning models for predicting wine quality. This investigation entails evaluating the accuracy metrics derived from a holdout dataset, distinct from the training data, providing insights into how well each algorithm generalizes to new, unseen instances. The objective is to discern the algorithm that exhibits the highest holdout accuracy, signifying robustness and reliability in predicting the quality of red and white wines. Delving into these comparative analyses aims to identify the algorithm that excels in delivering accurate predictions and contributes to the advancement of wine quality prediction methodologies. Notably, Figure 4.9 visualizes the holdout set accuracy comparison, while Table 4.8 presents a comparison of Algorithm Performance.

In the comparison of algorithm performance, the table presents the Cross-Validated accuracy (CV Accuracy) and holdout set accuracy (Holdout Accuracy) for each evaluated model. Random Forest emerges as the top-performing model, boasting a CV

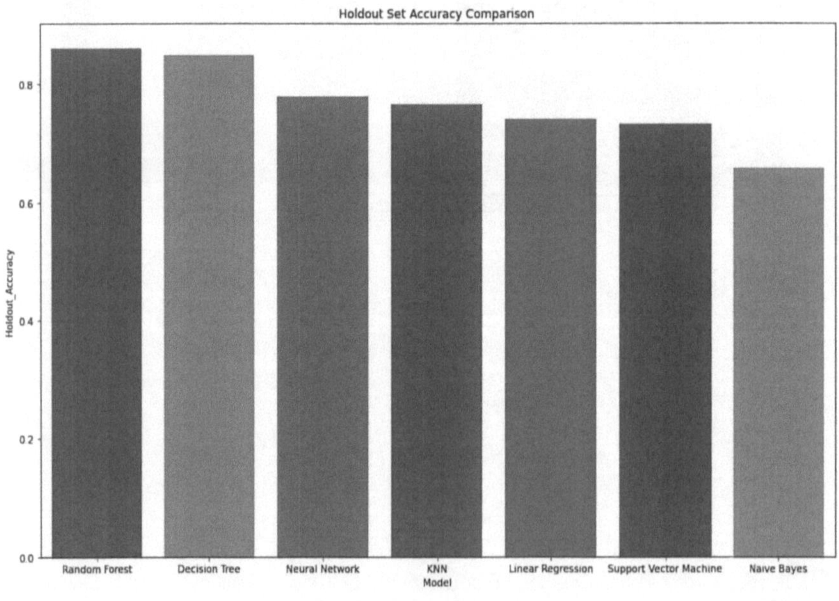

FIGURE 4.9 Holdout Set Accuracy Comparison.

TABLE 4.8
Comparison of Algorithm Performance

Models	CV Accuracy	Holdout Accuracy
Random Forest	0.929308	0.860000
Decision Tree	0.931465	0.849231
Neural Network	0.787203	0.779231
KNN	0.853342	0.765385
Linear Regression	0.743229	0.740000
Support Vector Machine	0.767792	0.731538
Naive Bayes	0.686317	0.657692

Accuracy of 92.93% and a Holdout Accuracy of 86.00%. Following closely is the Decision Tree model with a CV Accuracy of 93.15% and a Holdout Accuracy of 84.92%.

Neural Network, while achieving a solid CV Accuracy of 78.72%, maintains a comparable performance with a Holdout Accuracy of 77.92%. KNN, known for its proximity-based classification, exhibits a CV Accuracy of 85.33%, yet its holdout accuracy slightly decreases to 76.54%.

Linear Regression, a baseline model, demonstrates a CV Accuracy of 74.32% and a consistent Holdout Accuracy of 74.00%. Support Vector Machine and Naive Bayes, though delivering lower CV Accuracy values of 76.78% and 68.63% respectively, maintain respective Holdout Accuracy values of 73.15% and 65.77%.

This comprehensive comparison allows us to identify Random Forest as the most robust model, achieving high accuracy both in cross-validation and on the holdout set, making it a promising candidate for wine quality prediction.

4.11 SUMMARY CONCLUSION RECOMMENDATION FOR FURTHER WORK

In this study, the exploration into the implementation of machine learning-based predictive analytics within the Industry 4.0 paradigm focused specifically on quality assurance in the wine industry. Utilizing the Scikit-learn ML library, a comprehensive case study was conducted to evaluate the efficacy of various machine learning algorithms for classifying red and white wine quality. The primary objective was to enhance quality assurance practices by leveraging predictive analytics to accurately categorize attributes associated with different wines.

The findings underscore the transformative potential of machine learning-driven predictive analytics in revolutionizing quality assurance methodologies. The application of advanced algorithms yielded promising results, enabling accurate discernment and categorization of wine quality attributes. This not only contributes to optimizing production processes but also ensures the maintenance of high-quality standards within the wine industry, aligning seamlessly with the principles of Industry 4.0.

The observed variations in the algorithms' performance revealed that Random Forest and Decision Tree models exhibited high accuracy, precision, and recall, showcasing proficiency in wine quality prediction. Additionally, the Neural Network demonstrated competitive performance, achieving notable accuracy and maintaining a balanced trade-off between precision and recall.

The comparison of cross-validated accuracy and holdout set accuracy further emphasized the standout performance of Random Forest and Decision Tree models. These algorithms consistently proved to be top performers, demonstrating robustness and reliability, suggesting their suitability for predicting wine quality across diverse datasets. Moreover, the study extends beyond the wine industry, offering valuable insights into the broader applications of predictive analytics in diverse domains. The outcomes underscore the versatility and adaptability of machine learning algorithms, positioning them as powerful tools for driving advancements in quality assurance practices across various industries.

In conclusion, this work makes a significant contribution to the growing body of knowledge on integrating machine learning in quality assurance within the Industry 4.0 framework. The success in accurately predicting wine quality attributes highlights the potential for similar applications in different sectors. As industries increasingly embrace the digital transformation ushered in by Industry 4.0, the adoption of predictive analytics becomes pivotal for maintaining and enhancing product quality. Future work may involve exploring advanced modeling techniques and additional feature engineering to continually refine and improve predictive accuracy, addressing diverse challenges and opportunities in quality assurance across various industrial domains.

REFERENCES

Abdi, M., Adebiyi, A., Fasoli, A., Mannari, A., Labby, R., & Bozano, L. (2020, April 26–30). Model comparison of beer data classification using an electronic nose. In *Proceedings of the 8th international conference on learning representations (ICLR 2020)*, Addis Ababa.

Achouch, M., Dimitrova, M., Ziane, K., Sattarpanah Karganroudi, S., Dhouib, R., Ibrahim, H., & Adda, M. (2022). On predictive maintenance in Industry 4.0: Overview, models, and challenges. *Applied Sciences*, *12*(16), 8081.

Aich, S., Al-Absi, A. A., Hui, K. L., Lee, J. T., & Sain, M. (2018). A classification approach with different feature sets to predict the quality of different types of wine using machine learning techniques. In *2018 20th International conference on advanced communication technology (ICACT)* (pp. 139–143). IEEE.

Annanth, V. K., Abinash, M., & Rao, L. B. (2021, July). Intelligent manufacturing in the context of Industry 4.0: A case study of siemens industry. *Journal of Physics: Conference Series*, *1969*(1), 012019.

Bajic, B., Cosic, I., Lazarevic, M., Sremcev, N., & Rikalovic, A. (2018). Machine learning techniques for smart manufacturing: Applications and challenges in Industry 4.0. *Department of Industrial Engineering and Management Novi Sad, Serbia*, *29*, 18.

Bhardwaj, P., Tiwari, P., Olejar Jr, K., Parr, W., & Kulasiri, D. (2022). A machine learning application in wine classification. *Machine Learning with Applications*, *8*, 100261.

Bhattacharyya, S., Dutta, P., Samanta, D., Mukherjee, A., & Pan, I. (Eds.). (2021). *Recent trends in computational intelligence enabled research: Theoretical foundations and applications*. Academic Press.

Bianchi, A. C., Capobianco, F. L., & Papa, J. P. (2020). Wine classification using convolutional neural networks and feature extraction. In *2020 international joint conference on neural networks (IJCNN)* (pp. 1–7). Institute of Electrical and Electronics Engineers (IEEE). https://doi.org/10.1109/IJCNN48605.2020.9374175

Bousdekis, A., Lepenioti, K., Apostolou, D., & Mentzas, G. (2021). A review of data-driven decision-making methods for Industry 4.0 maintenance applications. *Electronics*, *10*(7), 828.

Chakraborty, I., & Sengupta, A. (2021). Wine classification using machine learning algorithms: A review and comparative analysis. *AI Communications*, *34*(1), 1–18.

Chen, T., Sampath, V., May, M. C., Shan, S., Jorg, O. J., Aguilar Martín, J. J., & Calaon, M. (2023). Machine learning in manufacturing towards Industry 4.0: From 'for now' to 'four-know'. *Applied Sciences*, *13*(3), 1903.

Chhikara, S., Bansal, P., & Malik, K. (2023, January). Wine quality prediction using machine learning techniques. In *International conference on smart trends in computing and communications* (pp. 137–148). Springer Nature.

Cortes, C., & Vapnik, V. (1995). Support-vector networks. *Machine Learning*, *20*, 273–297.

Cortez, P., & Silva, A. (2009). Modeling wine preferences by data mining from physicochemical properties. *Decision Support Systems*, *47*(4), 547–553.

Dahal, K. R., Dahal, J. N., Banjade, H., & Gaire, S. (2021). Classification of wine quality using machine learning algorithms. *Open Journal of Statistics*, *11*(2), 278–289.

Dasgupta, D., Chaudhuri, B. B., & Bandyopadhyay, S. (2020). Predicting wine quality using hybrid machine learning models. *Expert Systems with Applications*, *153*, 113418.

Davis, S. (1997). *Future perfect* (10th anniversary ed.). Perseus Books.

Foidl, H., & Felderer, M. (2015). Research challenges of Industry 4.0 for quality management. In *Innovations in enterprise information systems management and engineering* (pp. 121–137). Springer.

Frank, A. G., Dalenogare, L. S., & Ayala, N. F. (2019). Industry 4.0 technologies: Implementation patterns in manufacturing companies. *International Journal of Production Economics*, *210*, 15–26.

Gao, F., Zeng, G., Wang, B., Xiao, J., Zhang, L., Cheng, W., Wang, H., Li, H., & Shi, X. (2021). Discrimination of the geographic origins and varieties of wine grapes using high-throughput sequencing assisted by a random forest model. *Lebensmittel-Wissenschaft + Technologie/Food Science & Technology*, *145*, 111333. https://doi.org/10.1016/j.lwt.2021.111333

Ghobakhloo, M. (2020). Industry 4.0, digitization, and opportunities for sustainability. *Journal of Cleaner Production*, *252*, 119869.

Godina, R., & Matias, J. C. (2019, July 18–24). Quality control in the context of Industry 4.0. In *Industrial engineering and operations management II: XXIV IJCIEOM, Lisbon, Portugal* (pp. 177–187). Springer International Publishing.

Gupta, M. U., Patidar, Y., Agarwal, A., & Singh, K. P. (2020). Wine quality analysis using machine learning algorithms. In *Micro-electronics and telecommunication engineering* (pp. 11–18). Springer. https://doi.org/10.1007/978-981-15-2329-8_2

Gupta, Y. (2018). Selection of important features and predicting wine quality using machine learning techniques. *Procedia Computer Science*, *125*, 305–312.

Hassanzadeh, R., & Pourahmad, S. (2020). An ensemble-based feature selection for wine classification. *SN Applied Sciences*, *2*(12), 1–11.

Jain, K., Kaushik, K., Gupta, S. K., Mahajan, S., & Kadry, S. (2023). Machine learning-based predictive modelling for the enhancement of wine quality. *Scientific Reports*, *13*(1), 17042.

James, G., Witten, D., Hastie, T., Tibshirani, R., & Taylor, J. (2023). Linear regression. In *An introduction to statistical learning: With applications in Python* (pp. 69–134). Springer International Publishing.

Jasiulewicz-Kaczmarek, M., & Gola, A. (2019). Maintenance 4.0 technologies for sustainable manufacturing: An overview. *IFAC-PapersOnLine*, *52*(10), 91–96.

Jazdi, N. (2014). Cyber physical systems in the context of Industry 4.0. In *Proceedings of 2014 IEEE international conference on automation, quality and testing, robotics, AQTR*. https://doi.org/10.1109/AQTR.2014.6857843.

Jiang, X., Liu, X., Wu, Y., & Yang, D. (2023). White wine quality prediction and analysis with machine learning techniques. *Highlights in Science, Engineering and Technology, 39*, 321–326.

Kagermann, H., Helbig, J., Hellinger, A., & Wahlster, W. (2013). Recommendations for implementing the strategic initiative INDUSTRIE 4.0: Securing the future of German manufacturing industry. Final report of the Industrie 4.0 working group.

Korade, N. B. (2022). Analysis of a machine learning algorithm to predict wine quality. In *Machine vision for Industry 4.0* (pp. 245–262). CRC Press.

Kumar, S., Agrawal, K., & Mandan, N. (2020, January). Red wine classification using machine learning techniques. In *2020 International conference on computer communication and informatics (ICCCI)* (pp. 1–6). IEEE.

Lee, J., Bagheri, B., & Kao, H. A. (2015). A cyber-physical systems architecture for Industry 4.0-based manufacturing systems. *Manufacturing Letters*, *3*, 18–23.

Markatos, N. G., & Mousavi, A. (2023). Manufacturing quality assessment in the Industry 4.0 era: A review. *Total Quality Management & Business Excellence*, 1–27.

Masoudnia, S., Barani, G., & Shirazi, M. S. (2020). Wine classification using feature selection and machine learning algorithms. *Neural Computing and Applications*, *32*(21), 16089–16101.

Pathak, N., & Sharma, D. (2020). Classification of wine quality using deep learning neural networks. In *2020 international conference on smart electronics and communication (ICOSEC), Trichy* (pp. 651–655). IEEE. https://doi.org/10.1109/ICOSEC49089.2020.9215233

Pavlou, P., Papaevaggelou, G., & Dounias, G. (2021). A comparative study of machine learning algorithms for wine classification in the era of Industry 4.0. *Sensors*, *21*(2), 399.

Rai, R., Tiwari, M. K., Ivanov, D., & Dolgui, A. (2021). Machine learning in manufacturing and Industry 4.0 applications. *International Journal of Production Research*, *59*(16), 4773–4778.

Ray, S., & Sharif Ullah, A. M. (2021). A comparative study of machine learning algorithms for wine classification. *International Journal of Advances in Engineering Sciences and Applied Mathematics*, *13*(1), 85–92.

Rifkin, J. (2011). *The third industrial revolution: How lateral power is transforming energy, the economy, and the world*. http://ci.nii.ac.jp/ncid/BB08217222

Saini, A., & Sharma, A. (2019). Predicting the unpredictable: An application of machine learning algorithms in Indian stock market. *Annals of Data Science*. https://doi.org/10.1007/s40745-019-00230-7

Shahin, M., Chen, F. F., Bouzary, H., & Krishnaiyer, K. (2020). Integration of Lean practices and Industry 4.0 technologies: Smart manufacturing for next-generation enterprises. *The International Journal of Advanced Manufacturing Technology*, *107*, 2927–2936.

Shewhart, W. A. (1926). Correction of data for errors of measurement. *The Bell System Technical Journal*, 5(1), 11–26. https://doi.org/10.1002/j.1538-7305.1926.tb00106.x

Soori, M., Arezoo, B., & Dastres, R. (2023). Internet of things for smart factories in industry 4.0, a review. *Internet of Things and Cyber-Physical Systems, 3*, 192–204. https://doi.org/10.1016/j.iotcps.2023.04.006

Taunk, K., De, S., Verma, S., & Swetapadma, A. (2019, May). A brief review of nearest neighbor algorithm for learning and classification. In *2019 international conference on intelligent computing and control systems (ICCS)* (pp. 1255–1260). IEEE.

Török, D. F. (2023). Machine learning for predicting wine quality and its key determinants based on physicochemical properties. *Sage Science Review of Applied Machine Learning*, *6*(11), 1–21.

Vani, V., & Sathya, S. (2020, January 2–4). Ensemble learning approach for wine classification using feature selection techniques. In *2020 international conference on power electronics, smart grid and renewable energy (PESGRE)* (pp. 450–455). IEEE. https://doi.org/10.1109/PESGRE45664.2020.9315793

Vinitha, K., Ambrose Prabhu, R., Bhaskar, R., & Hariharan, R. (2020). Review on industrial mathematics and materials at Industry 1.0 to Industry 4.0. *Materials Today: Proceedings*, 33, 3956–3960. https://doi.org/10.1016/j.matpr.2020.06.331

Yan, L., Shen, D., Jia, Y., Zou, Q., & Guo, F. (2020). A novel ensemble learning method for wine classification based on random forest. *Symmetry*, *12*(6), 1018.

Ye, C., Li, K., & Jia, G. Z. (2020, November). A new red wine prediction framework using machine learning. *Journal of Physics: Conference Series*, *1684*(1), 012067.

Zhong, R. Y., Xu, X., Klotz, E., & Newman, S. T. (2017). Intelligent manufacturing in the context of Industry 4.0: A review. *Engineering*, *3*(5), 616–630.

Zonnenshain, A., & Kenett, R. S. (2020). Quality 4.0—the challenging future of quality engineering. *Quality Engineering*, *32* (4), 614–626. https://doi.org/10.1080/08982112.2019.1706744

5 Leveraging Clustering Algorithms for Predictive Analytics in Blockchain Networks

Rahul K. Patel, Nikunj R. Patel, and Yugen Chokshi

5.1 INTRODUCTION

Blockchain based databases with higher levels of data integrity can be a good fit for data science applications such as clustering algorithms in predictive analytics, where data integrity and timestamping are very important. Such synergistic approach holds the promise of transforming the landscape of blockchain intelligence into worldwide applications that require predictive data analytics. Blockchain brings benefits related to decentralization and transparency for dispersed and large volume of data, while clustering algorithms brings a way to transform the data into actionable conclusions. Such approach has its own challenges related to the operational complexity and network performance which can be managed with creative approaches such as keeping variables too small in numbers but meaningful to provide actionable conclusions.

This chapter explores into the complexities, benefits, and approaches for utilizing blockchain data for predictive analytics with a small applications of clustering algorithms in the context of predictive analytics for blockchain networks. By clustering elements based on their predictive properties, we aim to show a way to utilize them in the worldwide secure, efficient, and adaptive blockchain ecosystem where the convergence of data science and decentralized technologies promises to redefine the future of predictive analytics.

5.1.1 Brief Overview of Data Science and Its Advancements

Data science, also known as statistical computing, has rapidly evolved over the past few years. It has shifted from the separate practices of statistics, mathematics, and computer science to becoming a field that combines software engineering, computer programming, and applied statistics that is used by every industry on the planet. One of the main uses of modern data science is the use of artificial intelligence; specifically, machine learning. Machine learning algorithms often take very messy data and use open-source computing libraries such as scikit-learn to build predictive models based on the behavior of current data by having an algorithm continually predict the

 DOI: 10.1201/9781003473886-5

next data point and refine its prediction strategy as it sees more data points. The more data points the algorithm has to "train" on, the better the predictive model will be. These models are often initially written in a very flexible yet slow language such as Python and are then converted into a faster and less flexible language such as C++ for practical use [1]. The predictive analytics produced from these algorithms provide actionable insights for many industries. For example, a quite common process in the quantitative finance industry is derivatives pricing and risk calculation. These processes use financial models that constantly need to be calibrated based on current market conditions which are constantly changing, meaning that information on derivatives and risks that are calculated using data from one moment is outdated in the next moment. This means that there are often billions of calculations that need to be done daily, and traditional models and computation techniques are simply not practical for the sheer amount of data that needs to be processed and the speed that it needs to be done for the insights to be of use. This is where machine learning comes in.

5.1.2 Role of Machine Learning and AI in Extracting Actionable Insights

Although the datasets that quants uses are large, the actionable parameters are quite small or can be combined to be small on similarity basis. Derivative pricing instruments often use very few parameters such as strike, a barrier, and maturity. When it comes to predicting volatility, a popular model like the Heston model [2] only requires five parameters. The limited number of parameters needed to predict the market combined with the abundance of market data available makes machine learning a perfect solution for quantitative derivatives pricing and risk management. All that needs to be done is add all the financial data to a learning set and have the algorithm learn the result function of the financial instrument or parameter the researcher is trying to predict, returning a single output parameter that the firm can base investments and trade-offs on. In short, the firm has a large dataset with few parameters, so they utilize a machine learning algorithm to do predictive analysis on said data, giving them actionable insights that can help them make money. In this chapter we have provided a case study for analyzing top-line growth and bottom-line growth for large banks listed on American Exchanges to derive actionable insights [3].

5.1.3 Thermochemical Routes for Biomass Conversion to Fuels

There are also applications for machine learning beyond Wall Street. The medical industry, for example, is starting to use this technology to predict postoperative outcomes following arthroscopic rotator cuff repair [4]. They used pre-operation and quarterly, semi-annual, and annual postoperative data for patients who underwent arthroscopic rotator cuff repair to create a machine learning model that would predict postoperative outcomes for patients of different demography, comorbidities, cuff tear, tissue quality, and fixation implants [4]. This model outputs an expected postoperative ASES score for each patient. When used with a test dataset and compared with the actual ASES scores, it was found that the model could accurately predict the scores based on pre-operative factors, which would allow the doctors to give a more accurate post-op prediction to their patients [4]. Machine learning is not just one technique but a multitude of different techniques with the end goal of teaching a computer to predict something based on previous data. These techniques include artificial neural networks, linear regression, logistic regression, and the topic of this chapter: k-means clustering, which is one of the most popular and useful machine learning algorithms.

5.2 PREDICTIVE ANALYTICS WITH CLUSTERING

5.2.1 Understanding K-Clustering and Its Significance

K-clustering, often known as K-means clustering, is a pivotal method in data science for partitioning a given dataset into a specified number k of clusters. The K-clustering technique uses a hard and flat partitioning algorithm where each data point is assigned to exactly one cluster membership with an assumption that all clusters have equal variance [5]. This method involves grouping data points such that each point belongs to the cluster with the closest mean, effectively creating distinct, non-overlapping groups. The "K" in K-clustering denotes the number of clusters chosen for the analysis. The algorithm operates through an iterative process. Initially, it randomly assigns centroids for each of the k clusters. Each point in the dataset is then associated with the nearest centroid, and the centroid's position is recalculated as the mean of all the points assigned to it. This process repeats until the centroids stabilize, indicating that the clusters are as distinct as possible given the dataset. K-clustering holds substantial significance in predictive analytics due to its ability to uncover hidden patterns and structures in data. This characteristic is essential in fields where the data is vast and not immediately interpretable, such as customer behavior analysis, market segmentation, and even bioinformatics. In customer segmentation, for instance, K-clustering can analyze purchasing behaviors, demographic data, and customer interactions to identify distinct groups. These groups can then be targeted with tailored marketing strategies, enhancing customer engagement and business performance. Moreover, the simplicity and computational efficiency of K-clustering make it an attractive tool for preliminary data analysis, providing a quick and insightful look into the dataset's structure. This feature is particularly beneficial in exploratory data analysis, where the goal is to generate hypotheses and insights rather than confirm them.

In summary, K-clustering's ability to efficiently categorize data into meaningful groups makes it a cornerstone technique in predictive analytics, enabling the extraction of actionable insights from complex data sets. Its applications span various domains, proving its versatility and fundamental role in data-driven decision-making. Since the K-clustering technique is iterative, it tends to be very slow when there are lot of variables representing a data point [5].

5.2.2 Using Clustering to Optimize Operations

K-clustering contributes to operational optimization by providing a deeper understanding of the underlying patterns in data. This understanding is crucial in various sectors for enhancing efficiency and effectiveness. In addition, it separates data points by summarizing many dimensions of the data to enable user for actionable decisions. For example, in retail, K-clustering can be employed to analyze many dimensions representing the customer purchasing behaviors, which helps in inventory management by predicting future product demands. This predictive capability ensures that the inventory is well-stocked with high-demand products, thus minimizing overstocking and understocking issues.

The financial sector, particularly investors, may benefit significantly from the application of K-clustering for predictive decision making. By leveraging fundamental financial statistic data, investors can gain insights that drive operational efficiency, risk management, and segmentation. In this section, for simplicity, as listed in Table 5.1, we have used the real dataset of large banks under banking sector listed on US Markets. The dimensions we selected to analyze are financial metrics like EPS (Earnings Per Share) Growth and Revenue Growth. For more detailed predictive analysis, many other financial statistical dimensions can be used. These dimensions were collected from publicly available sources as of December 8, 2023. Data and related dimensions were already normalized so normalizing, cleaning, and preparing the data for analysis was not needed. However, for more complex analysis, data must be prepared which may involve normalizing the data to ensure a fair comparison across different banks. The K-clustering algorithm on these pre-processed data to form clusters based on financial performance metrics was run using included code in Python; however, for more complex and systematic analysis, lower-level code should be used for efficient results.

One of the key requirements for K-Clustering is to pre-determine the number of clusters. The results presented below in Figure 5.1 are based on five clusters.

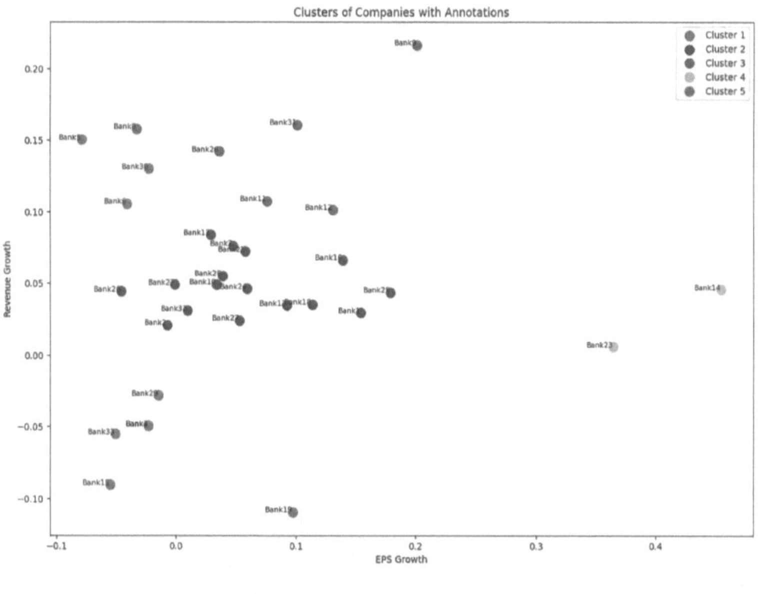

FIGURE 5.1 K-Clustering of Major Banks Based on Revenue and Earnings Growth.

TABLE 5.1
Large Banks Listed on US Markets

Name	EPS Growth	Rev Growth
Bank 1	15.36%	2.96%
Bank 2	–0.78%	2.10%
Bank 3	–2.37%	–4.95%
Bank 4	–2.37%	–4.95%
Bank 5	–7.90%	15.06%
Bank 6	–4.16%	10.56%
Bank 7	4.75%	7.63%
Bank 8	–3.35%	15.83%
Bank 9	20.11%	21.64%
Bank 10	3.42%	4.93%
Bank 11	7.63%	10.75%
Bank 12	13.07%	10.14%
Bank 13	11.24%	3.47%
Bank 14	45.49%	4.59%
Bank 15	–5.57%	–11.05%
Bank 16	13.88%	6.66%
Bank 17	2.91%	8.40%
Bank 18	11.35%	3.53%
Bank 19	11.74%	–10.94%
Bank 20	–4.63%	4.45%
Bank 21	5.79%	7.23%
Bank 22	–0.09%	4.91%
Bank 23	36.51%	0.58%
Bank 24	5.98%	4.62%
Bank 25	17.88%	4.37%
Bank 26	3.64%	14.23%
Bank 27	5.33%	2.42%
Bank 28	3.88%	5.53%
Bank 29	–1.50%	–2.82%
Bank 30	–2.32%	13.00%
Bank 31	10.10%	16.10%
Bank 32	1.00%	3.10%
Bank 33	–5.15%	–5.51%

However, analysis was performed starting with two clusters and gradually increased to five clusters for more refined results that are in line with common traditional 1-to-5-star financial analysis.

As is revealed by the analysis, many of the banks are in central Cluster 2. However, banks in each cluster collectively behave differently. Such analysis can be used to identify distinct groups of institutions showing varying levels of financial growth and stability. For instance: Cluster 3 might include banks with high EPS and revenue growth, indicating robust financial health and lower risk. Cluster 1 might comprise banks with negative growth rates in both EPS and revenue, signaling potential financial instability or higher risk. In efficient markets, it is expected that banks in all other clusters will eventually gravitate toward Cluster 2 or fall off the grid. That can be used to make a predictive decision about whether and when to invest or divest. For broader use, such analysis can be made portfolio investment where broad indexes such as S&P500 or Nasdaq100 can be clustered for generating stable portfolio of investments that would act as a mutual fund that can be actively managed based on movement of clusters as a whole and movement of individual datapoints within the cluster.

As we demonstrated, investors can use K-clustering for analyzing investment opportunities. By clustering companies based on financial performance metrics, analysts can identify potential investment or divestment targets. Investors can also make decisions to diversify their investment portfolios effectively. In conclusion, the application of K-clustering in operational optimization spans multiple industries, offering significant benefits in terms of efficiency, cost reduction, and improved decision-making. The versatility of this technique in analyzing several types of data makes it a powerful tool in the arsenal of data scientists and business analysts alike.

5.2.3 Making Data-Driven Decisions Using K-Clustering

Data-driven decision-making in any application involves a systematic approach to collecting, processing, and analyzing data to reach conclusions and make business decisions. K-clustering plays a vital role in this process by enabling the segmentation of complex datapoints into more manageable and insightful groups by summarizing dimensions related to each data point. Effectively, such analysis can be implemented as backbone of recommender systems. A recommender system, in such a case, is essentially an AI/ML model that helps users make a decision based on massive data. Such a recommender system is a must where (a) volume of data generated is massive and too fast for humans to analyze efficiently or effectively, and/or (b) human capability to gain expertise is limited because of the large number of variables or dimensions. This process involves several key steps:

(1) **Data Collection:** Relevant data needed for predictive analysis needs to be decided before we begin data collection. This also requires defining the goals and objectives. Based on scope of data to be collected, data sources and collection methods need to be identified as well. Gathering relevant data dimensions such as various financial ratios and indicators is relevant to the performance of any corporation. Typically, these are raw data and are generated independently by various sources. Pre-testing to analyze the quality of data may be needed to determine the usefulness of the data.

(2) **Data Preprocessing:** This step is required for cleaning, normalizing, and preparing the data for analysis to ensure a fair analysis. Data cleaning mainly focuses on missing data and finding errors or inaccuracies, followed by taking consistent actions to resolve such issues by removing, imputing, or correcting the data. Normalizing the data requires data transformation to

standardize similar scales and units. One of the common challenges faced during this phase is to decide how to handle outliers where they still fall within range while maintaining relative difference [6]. Another important preprocessing step is to deduce dimensionalities. This requires careful decision to remove unhelpful information while retaining the crucial aspects. Dimensionality reductions include Principal Component Analysis (PCA), Independent Component Analysis (ICA), and Non-negative Factor Analysis (NFA) [5]. Since our focus is more on K-Clustering, this step is not covered in detail in this chapter; however, it is necessary.

(3) **Clustering with K-Clustering Algorithm:** Before applying the K-Clustering technique, it needs to be qualified to see if that's the best way for cluster analysis. We must ask questions such as what is natural grouping? What constitutes good grouping? How many clusters should we look for in the data [5]. Running the K-clustering algorithm on the pre-processed data to create clusters based on data dimensions is prepared in the pre-processing step. As mentioned earlier, K-Clustering requires that the user determine how many clusters. The number of required clusters can be predetermined based on different decision possibilities as well as common norms for such data. Compare other clustering techniques such as Spectral clustering and Hierarchical clustering to see if K-clustering is the best fit for the data and objectives.

(4) **Interpreting Results:** Analyzing the resulting clusters to gain insights. For example, the insights gained from K-clustering enable investors to make more informed strategic decisions. Investors can tailor their investment strategies based on the financial stability and growth potential of different institutions. Investors might decide to invest more in institutions in high-growth clusters while being cautious with those in clusters showing negative growth trends. In risk management applications, clustering allows banks to categorize their clients into different risk profiles, enabling more personalized risk assessment and credit scoring. This segmentation leads to more accurate risk pricing, better loan portfolio management, and reduced instances of severe debt. Furthermore, K-clustering assists banks in understanding the needs and behaviors of different customer segments. This understanding can guide the development of customized financial products and services, ensuring they meet the specific needs of each customer segment, thereby enhancing customer satisfaction and loyalty.

In conclusion, K-clustering empowers financial institutions and banks to make data-driven decisions by providing a deeper understanding of their data. This approach leads to enhanced risk management, better investment strategies, improved customer service, and more robust financial performance. This concludes our exploration of the role of K-clustering in optimizing operations and aiding data-driven decisions in financial institutions and banks. We have seen how this powerful tool can transform raw data into actionable insights, driving efficiency and effectiveness in various business processes.

5.3 BLOCKCHAIN AND DATA MANAGEMENT

Blockchain was originally designed as a technological solution for managing data related to the requirement for cryptocurrencies. However unique properties of blockchain such as decentralization and immutability provide data management benefits such as decentralization, transparency, accountability, availability, integrity, and tamper-resistant data processing and storage. Such benefits not only mitigate the risks of a limited point of failures but also provide higher level of confidence for data integrity while performing parallel processing that cannot be achieved in the traditional data management systems. In this section, we are focusing on unique applications of blockchain in data management.

5.3.1 Introduction to Blockchain Technology

Made popular in recent years through the rise of cryptocurrency and non-fungible tokens, blockchain technology is one of the most secure and powerful data storage and management solutions to date. Blockchain technology is a decentralized database that everyone can contribute to but no one owns. All contributions to this database are stored in blocks called nodes, and these nodes can only be added, not removed. They are also public to all parties, ensuring full transparency and decentralization. These blockchain networks are built on asymmetric encryption and universally recognized algorithms to ensure device compatibility and header protection. There are four main characteristics of blockchain networks: decentralization, persistence, anonymity, and auditability. Decentralization means that there cannot be a central authority that authorizes transactions. To achieve decentralization, blockchain networks use consensus algorithms that are run on individual users' machines to authorize transactions. Persistence is the fact that the blockchain is immutable, meaning no one can delete a node. This is achieved by the fact that users can compare local blockchain networks for legitimacy, meaning that deleted nodes would instantly be discovered. Anonymity is the invisibility of the identity of a contributor. In blockchain networks, this is achieved, since users can communicate with the blockchain without revealing their personal information. Auditability is the characteristic that lets transactions be easily monitored and checked. This process varies from blockchain to blockchain. For example, Bitcoin stores its users' balance data via a UTXO model that everyone participating can see, meaning everyone can see the transactions and balances of all the users (anonymously), ensuring that auditability is achieved [7].

5.3.2 The Growing Role of Blockchain in Data Management

Although the mainstream application for blockchain technology is cryptocurrencies and non-fungible tokens, there seems to be much potential for its application in data management. For example, in the healthcare industry, there are many upsides that blockchain technology offers that traditional data management methods simply cannot. Take the issue of data security. In a traditional setting, a database with an administrator would need a username and a password to access medical data, putting the privacy of a patient medical data in the hands of a central authority. If a mistake were to be made or a data leak was to happen, all the patient's private medical data would become public. Now imagine that the hospital or healthcare network used a blockchain-based data storage system where each patient held a private key to access their data. In this scenario, all medical data is anonymous, so a hack or a leak would not mean anything significant. If patient data needs to change hands, the patient with the private key would need to sign off on that transaction, putting medical data into the patient's control, instead of an external entity. Another benefit of using a blockchain database is the standardization and auditability that it provides. In the current method of storing medical data, there is no universal standard. This means that, across hospitals, healthcare networks, and borders, all medical data is stored differently, so there is no effortless way for doctors and nurses at one hospital to see the same medical data that doctors and nurses see at another hospital, even though they are dealing with the same patient. This is an issue because a patient's medical history is paramount information for doctors to know, and there is too much red tape in the current system for doctors to get that information quickly. If there was a universal health record blockchain database, patients would have a standardized health record that only requires their permission for any doctor in any hospital to visit [8]. Combining blockchain technology with predictive analytics has its challenges and opportunities.

5.3.3 Challenges and Opportunities of Integrating Predictive Analytics with Blockchain

When looking from an engineering perspective, there is a lot to be done. Bitcoin, one of the most widely used blockchain networks on the planet, can only perform seven transactions per second. This poses problems for engineers who want to integrate predictive analytics directly into blockchain networks, since the computing power needed for such a task is so vast and modern blockchain technology has extremely limited capabilities in that regard. This is why k-means clustering is so important because it is simple to implement compared to other clustering algorithms, and therefore, simple to integrate into a blockchain network. The challenging work needed to engineer such a blockchain would be worth it. Such a database would be especially useful in any industry where data needs to be decentralized, easily accessible, and analyzed for predictive analytics. A perfect example is the finance industry, where billions of data points need processing quickly. A blockchain that could run a k-means clustering or any other machine learning algorithm would be priceless to hedge funds and high-frequency trading firms that spend billions of dollars on data storage, processing, and generating analytics for such financial data.

5.4 IMPLEMENTING CLUSTERING ALGORITHMS IN BLOCKCHAIN NETWORKS

As the blockchain landscape matures, building blocks for distributed, scalable, and resilient networks becomes increasingly available for data analysis related applications. Applications where data are analyzed for predictive decision making such as clustering are traditionally employed in machine learning with limited reach. This study of possible applications of clustering in predictive analytics using blockchain network promise not only to optimize the decision making with data available worldwide but also the analysis for specific targeted decision making by considering the effects of with global variables.

5.4.1 Rationale for Implementing Clustering Algorithms in Blockchain

(1) **Enhancing Blockchain Efficiency and Scalability:** The integration of clustering algorithms, particularly K-means clustering, into blockchain networks is driven by the need to enhance efficiency and scalability. Blockchain, while revolutionary, faces challenges related to transaction processing times and scalability, especially as the network grows. Implementing clustering algorithms can help optimize the network's data-processing capabilities. Clustering algorithms can be used to group transactions or nodes within the blockchain, leading to more efficient data management. For instance, clustering similar transactions can streamline validation processes, reducing the time and computational resources required.

(2) **Improving Data Management and Analysis:** Blockchain networks accumulate vast amounts of data, which can be challenging to manage and analyze effectively. Clustering algorithms can segment this data into meaningful groups, facilitating easier analysis and decision-making. For example, in a blockchain-based supply chain, transactions related to specific products or regions could be clustered for more efficient tracking and analysis. Distributed analysis will also result in improved resource allocation, enhanced network efficiency, and better scalability.

(3) **Facilitating Enhanced Security Measures:** Incorporating clustering into blockchain can also contribute to enhanced security. By clustering nodes or transactions, abnormal patterns can be more easily identified, aiding in the detection of fraudulent activities or network anomalies. Clustering can be employed in the blockchain networks to enhance privacy by grouping transactions or users with similar characteristics anonymously, which will make it more challenging to trace and identify individual users or transactions resulting in improved privacy protection, reduced risk of deanonymization, and enhanced user confidentiality.

5.4.2 How Nodes in Blockchain Networks Can Execute K-Means Clustering

The case study used in the section for applying K-clustering techniques has been deliberately kept limited in scope to fit in the chapter. Another purpose is to show the limitations of the predictive analysis process and how such limitations can be overcome using blockchain. Clustering is one important step to reveal patterns and intelligence; however, it is still a step in the entire predictive analysis process. The traditional process will be very inefficient for much larger and complex analysis where (a) data collection, preprocessing is needed globally, as well as (b) specific subsets of data are required to be analyzed for multiple local recommender systems globally. Blockchain networks are one of the very good answers to overcome such limitations. Let us discuss the same four steps process again using global reach and multiple parallel recommender system as objective for same financial analysis case study.

(1) **Data Collection in the Blockchain:** Distributed nodes collect relevant data in the blockchain networks for all financial exchanges around the work for all banks listed in their financial market. Since this is done in parallel, it would be much more efficient. One of the important decisions we made was relevancy of data. However, due to more nodes and data-dimensions, blockchain networks enable us to explore more dimensions and find new and previously unknown patterns. Pre-testing to analyze the quality of data can be done in parallel and hence faster.

(2) **Data Preprocessing in the Blockchain:** Cleaning, normalizing, and preparing the data for analysis to ensure a fair analysis becomes more important due to various unequal scales and local interpretations. For example, local currency, inflation rates, regulations, and reporting norms may provide different revenue and earnings growth interpretation in various countries and regions around the world. Data cleaning effort may be consistent and only require scalability that block chain network will provide; however, normalization requires additional and unique steps to make data comparable and usable. Here, not only scalability but additional analysis capability is also needed. Specific blockchain nodes are assigned to create a normalization process on a continuous basis with local normalization rules for effective and efficient results.

(3) **Clustering with K-Clustering Algorithm in the Blockchain:** Once of the benefits of running the clustering in blockchain is that it will allow data from across the globe to be included in analysis. It will provide additional insights on how additional dimensions such as location, local inflation rates, currency, regulators affect the grouping. Thus, recommender systems will not only consider internal performance metrics but external business and country specific dimensions as well. Continuous stream of global data should make recommendation system live and sensitive in timely detection of major events and their implications. Such a benefit makes the overall system very useful.

(4) **Challenges and Solutions:** Executing K-means clustering in blockchain networks presents unique challenges, such as maintaining consensus on cluster formation and ensuring the privacy of data. Solutions might include using distributed consensus algorithms for clustering decisions and incorporating privacy-preserving techniques like differential privacy. Clustering is predictive analytics targeted to reveal patterns, and there may be unexpected results that could threaten privacy and anonymity. This can be avoided with careful data collection and pre-processing, where an additional anonymization step could help overcome such challenges. This is because the very dynamic stream of data collection at higher scale may overwhelm analysis process. Clustering itself is iterative, and if the data stream is faster than completion of all iteration needed to decide, then the recommendations might be outdated even before it is available. Additional

targeted analysis using dimension reduction techniques might help; however, such processes require processing time as well. Another option is to deploy a sampling strategy to use only a volume of data that can be effectively analyzed without compromising end results. Extracting relevant data from a complex blockchain itself is a very time-consuming operation that will add to the overall slowness. Careful selection of objectives and applicability of such a solution needs to be carefully selected. If the scale of data is much higher, then slowly changing longer-term phenomena are the most suitable applications. The lack of standardized data formats and structures across the world will complicate the effective data collection and preprocessing, and hence, such application of clustering in blockchain might be more suited where standardized data are readily available. Handling such challenges requires careful consideration of the specific blockchain networks, especially when such efforts are not mainstream and reliable performance data is not available yet.

5.4.3 Sharing Clustering Results Back to the Blockchain Network

Another way to optimize performance is to share clustering results back to the blockchain network by integrating the insights gained from clustering analysis back into the blockchain system. When performed periodically, this approach can enhance transparency, optimize performance, and provide valuable information to other network participants. Each time, when result is shared back to the block chain, the next analysis becomes less data intensive because new analysis requires analysis of data up to the point when last result was available on the network by any participant.

(1) **Integration of Clustering Results:** Once clustering is completed, the results need to be integrated back into the blockchain network. This integration could involve updating the network's routing protocols, data management systems, or security frameworks based on the clustering outcomes. Accurate access control to restrict the visibility of clustering results or an anonymizing mechanism without compromising results reusability for further analysis may be necessary. If not, all participants need access to certain insights, then access control can be implemented. If unrestricted access is necessary to achieve the objectives, then anonymization is needed which may be more complex to implement. In any case integrity and privacy considerations are essential.

(2) **Consensus and Verification:** Mechanism for decision-making is required for the integration of clustering results, which is especially complex if the blockchain network has decentralized governance structure. Ensuring that the clustering results are accurate and agreed upon by the network is crucial. Blockchain networks could use consensus mechanisms like Proof of Work (PoW) or Proof of Stake (PoS) to validate and agree upon the clustering results. If private blockchain network with Proof of Authority (PoA) is implemented as consensus mechanism, then benefits such as faster speed and accuracy due to centralized governance are realized. In such

implementation, nodes are identified and authorized to validate work, which can be distributed and scaled as needed.

(3) **Use Cases and Benefits:** The integration of clustering results can lead to numerous benefits, such as improved data analysis, enhanced network security, and more efficient transaction processing. For instance, in a financial blockchain network, clustering transactions by type or size could streamline processing and fraud detection.

In summary, the implementation of clustering algorithms, particularly K-means, in blockchain networks offers significant potential for enhancing efficiency, scalability, and security. This integration, however, requires careful consideration of the unique challenges presented by the decentralized and secure nature of blockchain technology.

5.5 EFFICIENT AND EFFECTIVE CLUSTERING TECHNIQUES IN BLOCKCHAIN NETWORKS

5.5.1 Challenges in Implementing Clustering in a Decentralized Environment

Implementing clustering algorithms in a decentralized environment efficiently and effectively comes with many challenges. The first one is communication overhead. Let us say there is a distributed sensor network for weather forecasting, where each sensor from wherever in the world collects local weather data and stores that data in its very own node in a blockchain that represents weather data from all around the world. A k-means clustering algorithm running on each mode of this blockchain would require immense resources to run. Local weather data from each node clustered on the blockchain would require immense computing and electronic resources because of the abundance of weather data being collected and processed. There are also networking issues that would need to be ironed out such as increased latency and data collisions. This would require hiring more blockchain experts, software engineers, machine learning scientists, and data engineers as well as using more electricity, blockchain gas fees, and more expensive server technology to run. Depending on the scale of the network, this could cost certain companies and government agencies hundreds of millions of dollars to run and maintain compared to a more traditional method. Data distribution can also be an issue. Say we have a decentralized ridesharing platform that stores user data to be analyzed on multiple nodes in a blockchain network, depending on where a user is geographically. Data about one user that is split between different nodes in the blockchain will not produce any good insights when k-means clustering is run. For this data system to work in this example, data would have to be shared, which is expensive when we are dealing with a blockchain. This also leads to issues with data consensus across the entire network. Consider a scenario where blockchain is used for decentralized digital supply chain management. Each node represents a separate entity across the supply chain. Local clustering algorithms would be monumental for local inventory optimization and identifying supply chain interruptions, but there are issues. It is exceedingly difficult to achieve data consensus or agreement on data accuracy and equal processing capabilities among decentralized nodes, meaning there would be discrepancies in clustering results. This is not an issue that traditional data management techniques must deal with, but, due to the decentralized nature of blockchain technology, a node with inconsistent data or a node that is lagging on capability can be a liability. Scalability is also an issue here. In a decentralized social media platform utilizing blockchain, user-generated content is stored across numerous nodes for data redundancy and decentralization. Implementing clustering algorithms for content recommendation or moderation becomes increasingly complex as the platform scales. Processing large volumes of user data distributed across the network while maintaining responsiveness and scalability poses a significant challenge. Security and privacy are paramount when talking about data storage. There are many legal issues that companies face that are associated with storing sensitive data about clients. Take the example of an e-commerce business storing user transaction data in a blockchain network and applying clustering algorithms for fraud detection or customer segmentation—a scenario where maintaining data privacy is paramount. Sharing encrypted data among nodes for clustering while ensuring that sensitive information remains protected and anonymized presents a significant challenge. Finally, there is the issue of implementing clustering algorithms into blockchain nodes. For instance, creating a decentralized version of hierarchical clustering suitable for analyzing distributed genetic data across multiple research institutions. Adapting the algorithm to efficiently process and aggregate genetic information while preserving data ownership and privacy requires specialized algorithmic design tailored for decentralized environments. These proprietary algorithm designs would require immense resources and brainpower to conceive and implement. Overall, addressing these challenges involves a multidisciplinary approach, including developing efficient communication protocols, designing algorithms specifically suited for decentralized settings, incorporating cryptographic techniques for secure data sharing, and devising consensus mechanisms to ensure agreement among distributed nodes to make an effective and efficient blockchain data storage system that implements k-means clustering.

5.5.2 Potential Real-World Applications

Despite these challenges, there are many useful and practical applications for k-means clustering in a blockchain network. Let us explore a stock-related scenario where a blockchain network is utilized to analyze stock market data from various sources using k-means clustering to identify different market segments or investor behavior patterns. To implement this, we would need to first work on data aggregation. Various financial data sources contribute stock-related information, such as historical stock prices, trading volumes, company financials, and investor sentiment indicators. This data is aggregated and stored on the blockchain on all different nodes. The next step would be to run local k-means clustering. Nodes within the blockchain network, representing different entities like financial institutions or investor groups, run K-means clustering on subsets of stock-related data. For instance, clusters might be formed based on stock price movements, trading volumes, or specific company-related metrics. This data can then be used for segment identification.

In this case, k-means clustering helps identify distinct segments or patterns within the stock market data. These segments could represent groups of stocks exhibiting comparable price movements, stocks with similar trading behavior, or companies in similar financial situations. Analyzing the clusters can reveal investor behavior patterns or market segments. For instance, one cluster might contain stocks that are typically favored by long-term investors, while another might contain stocks that are subject to high-frequency trading. Each participant can use the insights gained from the clustering to tailor investment strategies. For instance, an investor might adjust their portfolio allocation based on the identified segments or may identify potential opportunities or risks within specific clusters. Summary statistics or insights about identified market segments (without disclosing individual stock details) can be shared on the blockchain, allowing participants to understand broader market trends or investor behaviors. When implementing this storage system over a traditional one, we can see lots of benefits. Clustering analysis provides a deeper understanding of market segments, allowing investors to make informed decisions based on different stock behavior patterns. Investors can customize their investment strategies based on the identified market segments or investor behavior patterns. The blockchain enables the sharing of summarized insights, fostering collaboration and a broader understanding of market trends without revealing sensitive stock-specific information. Understanding different market segments can aid in risk management by diversifying portfolios or identifying sectors with similar behaviors. In this stock-related example, leveraging K-means clustering within a blockchain network helps identify market segments and understand investor behavior patterns, allowing for more informed and tailored investment strategies while preserving the confidentiality of sensitive stock-specific data. Fraud detection in the banking industry is also a very applicable use case. Let us consider a finance-related scenario where a blockchain network is used to analyze transaction data from different financial institutions to detect potentially fraudulent activities using K-means clustering. In a blockchain network involving multiple banks and financial institutions, each institution maintains its transaction records on the blockchain. The goal is to detect potentially fraudulent activities across these institutions without sharing sensitive transaction details. Each financial institution uploads anonymized transaction data onto the blockchain, including information such as transaction amounts, timestamps, and transaction types. This data remains private and encrypted on the blockchain. Each participating financial institution runs the k-means clustering algorithm locally on their transaction data. The algorithm identifies clusters of transactions with similar characteristics based on features like transaction frequency, amounts, and patterns. Unusual clusters that deviate significantly from normal transaction patterns might indicate potentially fraudulent activities. For instance, a cluster might represent a set of transactions that are significantly larger in amount or occur at irregular intervals compared to typical customer behavior. When a financial institution detects anomalous clusters indicating potential fraud within its own data, it generates an alert or flag internally without exposing the specific transaction details. This helps in initiating further investigation or precautionary measures internally. Institutions share summary statistics or patterns of detected anomalous clusters (not individual transactions) on the blockchain. This shared information could include the number of detected anomalous clusters or statistical measures summarizing the deviation. This collective blockchain fraud-detection system could revolutionize the way banks detect fraud. By leveraging local k-means clustering, each institution can detect potential fraud within its own data without revealing sensitive transaction information. Sensitive transaction details remain private and secure, as only the aggregated clustering insights or anomaly detection information is shared on the blockchain. Institutions collaborate to identify potential fraud patterns across the network without compromising customer privacy or revealing proprietary data. Institutions can act promptly upon detecting anomalies, enhancing fraud prevention measures and protecting their customers' assets. This finance-related example demonstrates how k-means clustering within a blockchain network can aid in detecting potentially fraudulent activities across multiple financial institutions while maintaining data privacy and confidentiality. It highlights the collaborative potential of blockchain for fraud prevention in the financial sector. There are applications outside of finance as well, particularly in the supply chain. Optimizing inventory management can be an overly complex task, but a blockchain-based k-means clustering network would be able to help with that. In a typical supply chain, there are multiple parties, such as manufacturers, distributors, and retailers. In our scenario, each participant maintains a record of their supply chain data on the blockchain, including information about product quantities, shipments, and delivery times. Each participant regularly updates their local data on the blockchain, which includes information about product inventory levels. This data is often timestamped and immutable. To optimize inventory management, each participant can locally run the k-means clustering algorithm on their inventory data. In this case, "k" represents the number of clusters, which can be determined based on factors like product types, demand patterns, or other relevant criteria. The local k-means clustering analysis identifies clusters of products that share similar characteristics. The clustering results reveal which products tend to move together or have similar demand patterns. For instance, products in one cluster may experience high demand together, while products in another cluster may have a slower-moving pattern. Using the cluster information, each participant can make informed decisions about restocking, ordering, or reallocating their inventory. For example, if a participant sees that products within a particular cluster are in high demand, they may prioritize restocking those items. This local decision-making process optimizes inventory management at each participant's level. After performing the local k-means clustering analysis, each participant can share the cluster information—or just summary statistics—on the blockchain. This information is accessible to all participants, providing transparency and the ability to validate the decisions made based on local clustering. With this implementation, the clustering is performed locally by each participant's node, the blockchain remains decentralized, and no central authority is needed for data analysis. Sensitive data remains on the local node and is not exposed to other participants. Only the clustering results or summary statistics are shared on the blockchain. Participants can make data-driven decisions to optimize their operations based on the insights gained from k-means clustering. The shared clustering results enhance transparency and allow other participants to understand the reasoning behind inventory management decisions. In short, k-means clustering is used within a blockchain network to analyze and leverage local data for a specific application, which, in this case, optimizes inventory management within a supply chain. It demonstrates how blockchain can facilitate decentralized data analysis and decision-making while maintaining data privacy and transparency.

5.6 CONCLUSION AND FUTURE DIRECTIONS

5.6.1 Recap of the Key Takeaways

In this chapter, we introduced unsupervised learning mechanisms such as K-Clustering. We discussed its inner working and demonstrated its application using topline and bottom-line dimensions of the dataset containing large banks listed in US financial markets. K-clustering techniques can be applied to very large data sets as long as data dimensions can be deducted to a smaller number of dimensions since it is a highly iterative process. We also discussed the entire predictive analysis process with major activities including data collection, data pre-processing, analysis, and understanding results for decision making. We also discussed data normalization dimensionality reduction and mechanisms for handing missing and incorrect data. Later, we talked about the applicability of such a process in a wide range of applications such as inventory management as well as the medical field.

Clustering and predictive analytics techniques can be empowered to perform larger operations for worldwide applications, using globally distributed networks such as blockchain networks. We discussed the benefits of using such networks with K-clustering. We also discussed the challenges related to data preparation, scalability, and speed of operation as well as potential effective approaches to tackle such challenges.

5.6.2 The Growing Potential of Predictive Analytics in Blockchain Networks

The integration of predictive analytics into blockchain networks brings new possibilities. For cybersecurity applications, it can help with detection of anomalous or abnormal patterns representing security breaches or abnormal activities for early detection and prevention of malicious actions, enabling the implementation of proactive security measures. Cluster based predictive analytics can analyze market trends, sentiment, and external factors and provide recommendations to aid investors and decision-makers. It can also help predict global supply chain state and forecasting of demand, logistics, and inventory requirements. It can also help detect and predict large but slowly changing global patterns such as global warming with aid for decision-making. Such predictive analytics can help regulators for analysis performance of entities under extreme conditions such as banking stress-testing, predicting effects of regulation changes. With the help of continuous data feed enabled by block, real-time, data-driven recommender systems can be implemented to provide real-time insights for strategic planning and risk management.

As blockchain technology continues to mature, the integration of clustering based predictive analytics will become increasingly beneficial and essential. Leveraging historical data and advanced analytical tools, blockchain networks can proactively address many challenges efficiently and effectively. The growing potential of predictive analytics in blockchain networks highlights the importance of data-driven decision-making strategies.

5.6.3 Potential Future Developments and Research Directions

In this chapter, we have demonstrated one of the unsupervised learning techniques—k-clustering. Two other learning techniques such as agglomeration and spectral clustering techniques should be explored for various applications on block chain networks. Actual implementation of a dataset in the blockchain networks and performing clustering analysis is also necessary to understand practical scalability and performance. Multiple predictive techniques, such as clustering to store data back in the blockchain, and later, using time series analysis to predict when and under what conditions specific observations leave one cluster and join another, can be very interesting, since it would change the recommendations. Implementing and observing the quality of decisions and finding new parents with implementing a real time and fully automated end-to-end recommender system will be very interesting to observe as well.

REFERENCES

[1] D. Makowski and P. Waggoner, "Where are we going with statistical computing? From mathematical statistics to collaborative data science," Mathematics, vol. 11, no. 8, pp. 1821–1821, Apr. 2023. https://doi.org/10.3390/math11081821.

[2] S.L. Heston. "A closed-form solution for options with stochastic volatility with applications to bond and currency options," Review of Financial Studies, vol. 6, no. 2, pp. 327–343, Apr. 1993. https://doi.org/10.1093/rfs/6.2.327.

[3] J. De Spiegeleer, D.B. Madan, S. Reyners, and W. Schoutens, "Machine learning for quantitative finance: Fast derivative pricing, hedging and fitting," Quantitative Finance, vol. 18, no. 10, pp. 1635–1643, Jul. 2018. https://doi.org/10.1080/14697688.2018.1495335.

[4] A.G. Potty, A.S.R. Potty, N. Maffulli, L.A. Blumenschein, D. Ganta, R.J. Mistovich, M. Fuentes, P.J. Denard, P.M. Sethi, and A.A. Shah, "Approaching artificial intelligence in orthopaedics: Predictive analytics and machine learning to prognosticate arthroscopic rotator cuff surgical outcomes," Journal of Clinical Medicine, vol. 12, no. 6, pp. 2369–2369, Mar. 2023. https://doi.org/10.3390/jcm12062369.

[5] L. Igual, S. Seguí, and Springerlink Online Service, Introduction to Data Science: A Python Approach to Concepts, Techniques and Applications. Cham: Springer International Publishing, 2017.

[6] J.D. Kelleher, Fundamentals of Machine Learning for Predictive Data Analytics: Algorithms, Worked Examples, and Case Studies. Cambridge, MA: MIT Press, 2020.

[7] R. Naaz, A.K. Saxena, and P. Kr. Shah, "Blockchain technology's overview: Consensus, architecture and future trends," Instrumentation Engineering, Electronics and Telecommunications–2021 (IEET-2021): Proceedings of the VII International Forum, 2023. https://doi.org/10.1063/5.0125073.

[8] F. Anjum Hira, H. Khalid, S.Z. Abdul Rasid, S. Baskaran, and A. Md Moshiul, "Blockchain technology implementation for medical data management in Malaysia: Potential, need and challenges," TEM Journal, vol. 11, no. 1, pp. 64–74, Feb. 2022. https://doi.org/10.18421/tem111-08.

6 Use of Digital Twin and Internet of Vehicles Technologies for Smart Electric Vehicles in the Manufacturing Industry

Richa Singh and Rekha Kashyap

6.1 INTRODUCTION

The rise of sustainable, smart electric cars (EVs) has revolutionized the automobile industry as a result of the development of digital technologies [1]. There is a growing need for cutting-edge solutions that can improve the performance and sustainability of EVs. This is due to growing concerns over environmental sustainability and energy efficiency. The hunt for sustainable alternatives has been sparked, in recent years, by a rise in environmental awareness regarding the effects of conventional transportation systems. Electric Vehicles (EVs) have distinguished themselves among the range of alternatives as a possible route to lowering carbon emissions and reaching a more sustainable future. Global EV usage has been rising gradually, but integrating EVs seamlessly into smart, sustainable transportation networks is still a difficult task. Innovative technologies like Digital Twin and the Internet of Vehicles (IoV) have become transformational forces in the effort to get past these obstacles and realize the full promise of electric vehicles.

It is required to highlight the rising importance of sustainability and the pressing need to adopt electric transportation solutions in three major parts because of environmental issues. Then, only the discussion about digital twin technology and how it may be used to create virtual representations of real-world objects, enabling real-time data analysis and predictive modeling to enhance the performance of electric vehicles, will be easily understood. The idea of the Internet of Cars (IoC), a networked ecosystem that permits seamless communication between cars, infrastructure, and control centers, can be explored. This idea is redefining how we think about transportation.

- **The Need for Sustainable Smart Electric Vehicles:** Globally speaking, sustainable mobility is becoming more and more important due to the acceleration of climate change and the negative environmental effects of fossil fuel-powered transportation systems. The use of Electric Vehicles has significantly increased in the automotive sector as governments work to fulfill ambitious climate objectives and lower their carbon footprints.

DOI: 10.1201/9781003473886-6

Electric motors and rechargeable batteries provide the power for these environmentally friendly cars, which operate with no exhaust emissions. Due to their cheaper operating costs, less reliance on fossil fuels, and potential to reduce greenhouse gas emissions, EVs have grown in popularity. The integration of electric vehicles into a larger, smart, and sustainable transportation ecosystem is where their actual promise lies, though. The necessity for seamless communication, efficient energy use, and sophisticated traffic management becomes more and more obvious as cities develop into smart cities. Additionally, issues like range anxiety, charging infrastructure, and grid stability must be resolved for the effective deployment of electric vehicles in urban settings. Here is where the convergence of IoV and digital twin technologies offers a transformational potential.

- **Digital Twin—Facilitating Smart Electric Vehicle Management:** Due to its capacity to produce virtual representations of actual assets or systems, digital twin technology has drawn significant attention across a range of sectors. Utilizing real-time data, operational parameters, and performance measurements, digital twin technology enables the development of exact virtual duplicates of individual electric cars. These digital substitutes allow for extensive data analytics and predictive modeling, giving useful insights on the usage and behavior patterns of Electric Vehicles. There are several advantages to using digital twins in the context of electric vehicles. First, it makes it possible to monitor vital parts in detail, such as batteries, motors, and charging systems, which improves their performance and lengthens their lives. Additionally, intelligent predictive maintenance is made possible by real-time analysis of vehicle data, which lowers downtime and increases vehicle reliability. Thirdly, Digital Twin technology enables the creation of individualized user interfaces that adapt vehicle settings to specific driving styles and preferences.
- **A Connected Ecosystem for Sustainable Mobility with the Internet of Vehicles (IoV):** A fundamental change in how automobiles connect with one another, infrastructure, and control centers is represented by the Internet of Vehicles (IoV). Electric Vehicles join a dynamic network made possible by IoV technology, exchanging data with one another and with intelligent road infrastructure. Real-time communication enables effective traffic control, flexible route design, and charging infrastructure optimization. Electric Vehicles may benefit from real-time insights regarding road conditions, traffic congestion, and the location of charging stations by utilizing IoV, which will improve navigation and energy usage. Additionally, IoV makes it possible for vehicle-to-grid (V2G) integration, which enables Electric Vehicles to function as distributed energy resources, promote the integration of renewable energy sources, and contribute to grid stability.

6.1.1 Role of AI in Electric Vehicles

As artificial intelligence is an emerging field in today's technological environment, this technology also plays a major role in the field of electric vehicles and provides various services shown in Figure 6.1. As the automotive industry accepts the route toward sustainability and intelligent transportation, the role of AI in Electric Vehicles (EVs) has become increasingly important. From improving energy efficiency and range prediction through data analytics and real-time optimization to redefining autonomous driving capabilities, AI technology plays a complex role in numerous areas of EVs. This includes making EVs safer and more user-friendly. To increase consumer acceptance and trust in electric mobility, AI-driven technologies enable EVs to adapt to a variety of driving circumstances, optimize charging patterns, and deliver personalized driving experiences. Assuring the seamless integration of EVs into the larger smart city ecosystem, AI is also essential in establishing smart charging infrastructure, grid integration, and vehicle-to-grid communication. The use of AI in electric vehicles will grow as technology develops, accelerating the change of transportation toward a sustainable and intelligent future.

The major focus is to investigate and go thoroughly into the idea of empowering Sustainable Smart Electric Vehicles using Digital Twin and IoV technologies. The desire to maximize EV efficiency, improve user experience, and lessen the environmental effect of transportation networks is what spurred the notion to combine these two cutting-edge technologies. The combination of Digital Twin and IoV offers an all-encompassing and dynamic strategy that has the potential to change how Electric Vehicles function inside smart cities and intelligent transportation networks.

In this context, combining two cutting-edge technologies—Digital Twin and Internet of Vehicles (IoV)—offers enormous promise to handle the intricate problems

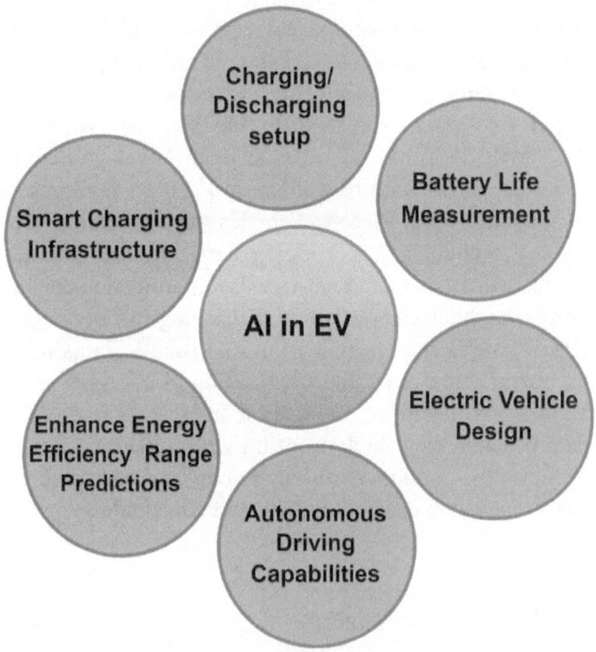

FIGURE 6.1 Role of Artificial Intelligence in Electric Vehicles.

posed by EVs and open the door to a more sustainable future [2]. The development of a digital replica or simulation of a physical asset is made possible by digital twin technology, which was motivated by the idea of reflecting physical items in a virtual setting. Digital Twins offer a comprehensive perspective of EVs, including their components, operational parameters, and performance characteristics, by fusing real-time data and sophisticated modeling approaches. This virtual representation enables remote monitoring, predictive analytics, and effective decision-making, which optimizes EV operations, energy use, and maintenance tasks [3]. Through a network of sensors, communication technologies, and data exchange protocols, the IoV paradigm simultaneously connects cars, infrastructure, and the surrounding environment. By utilizing IoV, EVs join a wider intelligent ecosystem and can communicate seamlessly, manage their energy resources effectively, and have improved safety features [4]. Since real-time data sharing, cooperative sensing, and adaptive control mechanisms are made possible by the inclusion of EVs in the IoV framework, the performance, effectiveness, and sustainability of the EV ecosystem are all increased [5, 6].

This paper analyzes the possibilities of using IoV and Digital Twin technologies for environmentally friendly, smart electric vehicles and tries to find out how these technologies work in concert to improve the performance, energy efficiency, and environmental sustainability of EVs. We emphasize the important prospects and obstacles in combining Digital Twin and IoV technologies in the context of EVs through a thorough review of cutting-edge methods and cases. We also explore the potential uses of this integrated architecture, such as autonomous driving, intelligent charging infrastructure, predictive maintenance, and energy optimization.

6.1.2 Motivation and Contribution

The motivation behind this research stems from the urgency to accelerate the adoption of sustainable, smart EVs and overcome the hurdles that hinder their widespread acceptance. While EVs offer numerous advantages in terms of reduced emissions and enhanced energy efficiency, they still face challenges related to range anxiety, charging infrastructure limitations, and optimization of energy consumption. Additionally, ensuring the longevity and reliable performance of EVs requires effective maintenance strategies. By leveraging Digital Twin and IoV technologies, we can address these challenges comprehensively and unlock the full potential of EVs for sustainable transportation. More specifically, the following contributions are made by this work:

- The first goal is to give the readers a solid understanding of the concepts of DT. The concept of a digital twin originated from in the aerospace industry during research, which covered a range of applications and developmental stages. Using this paper, we can gain a practical grasp by reading the first section of proposed work, which provides the necessary details. The section goes on to provide a thoughtful analysis of the stages of the DT paradigm's deployment. It goes into detail about the process of creating and using a complicated approach and provides an idea of the variables influencing each stage of realization.

- The use of twin methodologies in smart car systems is the next area of focus. The research provides a thorough analysis and summary of recent developments in DT in the intelligent internet of EV. This review's coverage of a growing subject makes a thorough analysis of the technology advantageous, and to the authors' knowledge, it is the first of its sort. The assessment has been divided into distinct smart vehicle system categories, like the navigation system automatically, driverless cars having assistance that gives advanced information, analyzing the lifecycle of the vehicles, observing the vehicle battery, grid electronics, and electric grid drive systems. This chapter gives an actual overview of the modern vehicle system and the different unique Internet of Vehicles implementations made feasible.
- The chapter addresses the two-tier network architecture of web-based and fog-based EV. There is a proper explanation of this two-tier architecture and its potential advantages of using this in network architecture. Based on the above three technologies and its advantages, a new idea is proposed that is the combination of Digital Twin and Internet of Vehicles, and keeping the advantages of two-tier—i.e., web-based—cloud and fog computing in mind, a hybrid model is proposed that have the concept of Digital Twin technology and Internet of Vehicles.
- Illustrations of the proposed method are discussed with proper explanation and the results of the proposed hybrid technologies are represented in the performance and discussion section using various types of graphs. Also, we mention the various parameters and values that are important to make the proposed technology successful. The conclusion and the insights gleaned from this in-depth investigation into the advancement of Digital Twin technology and the Internet of Vehicles and the potential advantages of proposed hybrid technology are presented in the concluding section. A future direction for research is also mentioned in which other technologies can be used for electric vehicles and their use in another different era.

6.1.3 Research Objectives

The primary contribution that is included in the paper are:

a. To investigate the application of DT technology in the development and optimization of smart EV systems, including subsystems such as self-ruling movement, enhanced assistance system, monitoring of vehicle health, the lifespan of the battery and its management, vehicle power supply, and power supply systems.
b. To investigate the importance of the Internet of Vehicles in creating a connected and sustainable ecosystem for smart EVs, integrating various stakeholders, including individuals, EV fleets, utilities, dispersed renewable energy sources, power networks, communication, and computer infrastructures.
c. To analyze the potential benefits, challenges, and socio-economic implications of combining Digital Twin technology and the IoV to advance smart EV technologies and contribute to a sustainable society.

6.1.3.1 Chapter Organization

The road map of this chapter is as follows: Section 2 gives the critical review and identifies the gap for further research. Section 3 has three subsections in which the first subsection discusses the DT methodology and potential advantages in brief. The second subsection discusses the concept of Internet of Vehicles and its potential advantages, and the last subsection discusses the proposed hybrid model and it's working along with its potential advantages and services that this model provides. The performance and discussion section gives the performance analysis of the proposed hybrid model based on different conditions like requirement of server, blocking rate and number of electric vehicles. The chapter concludes and gives future directions, highlighting real-world implementations and outlining areas for further research and development in the last section, conclusion, and future scope.

6.2 CRITICAL SURVEY OF EXISTING METHODS

This section gives a critical review of the paper based on the Internet of Vehicles and smart electric vehicles. Smith et al. (2020) [7] conducted a study to investigate the integration of Digital Twin and Internet of Vehicles (IoV) technologies for optimizing energy use in sustainable smart electric cars (EVs). Their research focuses on predicting energy efficiency and improving maintenance accuracy using Digital Twin models. They discovered a practical implementation gap and emphasized the need for standardized research on measurements and real-time monitoring in the context of sustainable EVs.

Johnson et al. (2019) [8] studied the application of Digital Twin and IoV technologies in tandem to improve EV performance, energy efficiency, and dependability. In the context of smart EVs, their study emphasised the relevance of performance, energy efficiency, and sustainability. They did, however, identify a lack of standardization in integration strategies for Digital Twin and IoV, indicating the need for more study in this area [9]. Some research was done to increase predictive maintenance, energy optimization, and charging efficiency in smart EVs by utilizing Digital Twin and IoV technologies. Their study emphasized the need of precise predictive maintenance, energy optimization, and effective charging for long-term EV operations.

Lee et al. (2021) observed shortcomings in terms of inadequate data for predictive maintenance accuracy and a lack of interoperability standards for IoV communication protocols. Chen et al. (2022) concentrated on the integration of Digital Twin and IoV technologies in the EV ecosystem to achieve sustainable transportation. Their research [10] emphasized the value of Digital Twin in providing real-time monitoring, predictive analytics, and decision assistance for optimizing EV performance and energy usage. In the context of IoV, they also emphasized the significance of cooperative sensing and intelligent charging infrastructure. Zhang et al. (2023) did a thorough literature study on using Digital Twin and IoV technologies for sustainable smart electric cars. Their paper synthesized previous research, showing the advantages, limitations, and prospective uses of integrating Digital Twin and IoV in the

EV ecosystem. They emphasized the need for further research in areas like energy optimization autonomous driving and data security in the context of long-term EVs [11]. Another evaluation of the literature on the integration of Digital Twin and IoV technologies for sustainable smart electric cars has been undertaken.

Their research emphasized the value of Digital Twin in allowing real-time monitoring of EVs, optimizing energy efficiency, and enhancing maintenance plans [12]. Wang et al. (2021) emphasized the need for more studies on data analytics and predictive modeling approaches to improve the performance and sustainability of EVs. Liu et al. (2022) investigated the function of Digital Twin and IoV technologies in allowing autonomous driving for environmentally friendly EVs. Their study [13] explored the advantages of adopting Digital Twin models for modeling and testing autonomous driving algorithms, improving safety, and reducing energy usage. In the IoV framework, they recognized the need for breakthroughs in sensor data fusion, real-time decision-making, and communication protocols for reliable, autonomous driving. A thorough review of research on Digital Twin and IoV technologies in the context of sustainable smart electric cars has been undertaken [14]. Their research found that Digital Twin might be used to optimize operations, manage to charge infrastructure, and improve user experience. Zhou et al. (2023) recognized a research need in the integration of Digital Twin with new technologies like blockchain and edge computing for improving data security and processing efficiency in the EV ecosystem.

Yang et al. (2022) studied the application of Digital Twin and IoV technologies in smart electric cars for predictive maintenance and problem diagnostics [15]. Their study emphasized the benefits of employing Digital Twin models to monitor EV components, identify abnormalities, and estimate maintenance requirements. They emphasized the need of real-time data collecting, machine learning techniques, and remote diagnostics in the context of IoV maintenance plans.

Huang et al. (2021) examined the challenges and opportunities of implementing Digital Twin and IoV technologies in the charging infrastructure for sustainable smart electric vehicles. Their study discussed the potential of Digital Twin models for optimizing charging station operations, managing energy demand, and improving user accessibility [16]. They identified a need for standardization, interoperability, and dynamic energy management systems to ensure efficient and sustainable charging infrastructure in the IoV ecosystem.

On the basis of the foregoing study, Table 6.1 shows the main focus and key major findings of smart electric vehicles and the internet of vehicles and can provide a better idea for further research.

6.3 PROPOSED METHODOLOGY

This chapter investigates the capability of integrating DT (digital twin) and the Internet of Vehicles (IoV) technology on two tiers and its applications in the smart EV domain to enhance the performance and sustainability of smart EV systems and architecture by mixing both technologies. The proposed architecture is the integration of DT and IoV in two-layered architecture that provides various solutions as output that gives a better version for near future in the same era.

TABLE 6.1
Critical Review of Key Findings

Study	Main Focus	Key Findings
Wang et al. (2021) [17]	Energy optimization and maintenance	• Digital Twin enables real-time monitoring and optimization of EV energy efficiency. • Further research is needed on data analytics and predictive modeling techniques.
Liu et al. (2022) [18]	Autonomous driving	• Digital Twin and IoV facilitate testing and simulation of autonomous driving algorithms. • Advancements in sensor data fusion, decision-making, and communication protocols are needed.
Zhou et al. (2023) [19]	Applications and integration	• Digital Twin can optimize EV operations, manage charging infrastructure, and enhance user experience. • Integration of Digital Twin with blockchain and edge computing needs further exploration.
Yang et al. (2022) [20]	Predictive maintenance and fault diagnosis	• Digital Twin enables proactive maintenance by monitoring EV components and predicting maintenance needs. • Real-time data acquisition, machine learning algorithms, and remote diagnostics are essential.
Huang et al. (2021) [21]	Charging infrastructure	• Digital Twin optimizes charging station operations, energy management, and user accessibility. • Standardization and interoperability are needed for efficient and sustainable charging infrastructure.
Smith et al. (2020) [22]	Energy optimization and maintenance	• Digital Twin models predict energy efficiency and improve maintenance accuracy in sustainable EVs. • Practical implementation and standardized research on metrics and real-time monitoring are needed.
Johnson et al. (2019) [23]	Performance, energy efficiency, and reliability	• Digital Twin and IoV enhance EV performance, energy efficiency, and sustainability. • Lack of standardization in integration techniques for Digital Twin and IoV is a research gap.
Lee et al. (2021) [24]	Predictive maintenance, energy optimization, and charging efficiency	• Accurate predictive maintenance, energy optimization, and efficient charging are vital for sustainable EVs. • Insufficient data for predictive maintenance accuracy and lack of IoV interoperability standards are identified gaps.
Chen et al. (2022) [25]	Sustainable mobility	• Digital Twin enables real-time monitoring, predictive analytics, and decision support for EV performance optimization. • Cooperative sensing and intelligent charging infrastructure play a crucial role in the IoV context.
Zhang et al. (2023) [26]	Benefits, challenges, and potential applications	• Integration of Digital Twin and IoV offers benefits for energy optimization, autonomous driving, and data security in sustainable EVs. • Further research is needed in areas such as energy optimization, autonomous driving, and data security.

(Continued)

TABLE 6.1
(Continued)

Study	Main Focus	Key Findings
Park et al. (2020) [27]	User experience and personalized services	• Digital Twin and IoV technologies enhance user experience in smart EVs through personalized services and seamless connectivity. • Customized in-vehicle features and intelligent recommendations improve user satisfaction and loyalty.
Liu et al. (2021) [28]	Data privacy and security	• Privacy and security are key concerns in the integration of Digital Twin and IoV technologies for EVs. • Secure communication protocols, encryption techniques, and data anonymization methods are required to protect sensitive information.
Wu et al. (2022) [29]	Fleet management and optimization	• Digital Twin and IoV enable efficient fleet management in smart EVs, including route optimization, vehicle tracking, and resource allocation. • Real-time data integration, advanced analytics, and decision support systems enhance fleet performance and operational efficiency.
Xu et al. (2023) [30]	Environmental impact and sustainability	• Digital Twin and IoV technologies contribute to reducing the environmental impact of EVs by optimizing energy consumption and promoting sustainable driving practices. • Real-time emissions monitoring, eco-routing algorithms, and eco-feedback systems support sustainable mobility in the IoV ecosystem.
Kim et al. (2021) [31]	V2X communication and smart grid integration	• Digital Twin and IoV facilitate Vehicle-to-Everything (V2X) communication, enabling seamless integration with smart grids for energy management and grid stability. • Bidirectional energy flow, demand response strategies, and grid synchronization enhance the overall efficiency and reliability of EVs.

6.3.1 Prerequisites

In this section, we provide the information about DT, IoV, and two-tier network architecture of local and remote cloud-based EV charging systems and then give the information about the proposed method.

a) **Digital Twin Technology:** Reflected data, raw data, alarm data, as-built data, data analytics for data insights, and machine learning are six factors that play a major role in the creation of the Digital Twin technology. The model is initially based on reflected data that is already available and is used to generate data-driven simulations of the physical object. Through horizontal and vertical integration topologies, constructed data are merged to broaden the scope of these simulations and create a working digital twin. These three entities come together completely to create the modern definition of the digital twin. The integration of sensory components that collect a wealth of raw data makes it possible for the vital connection with real-world entities. Information is present at the digital twin's focal point since it is a requirement for creating new information. Such information is frequently obscured by numerous insignificant pieces of data, which must be appropriately filtered. The goal of big-data analytics is to extract significant data insights that the system's intelligence may use to provide the analysis of the feedback that can be used in DT approach. After the collection of all the necessary data that is required is go for the ML model, train the data, and then find the valuable insights based on that the decision in future can be taken very easily and also solve the real-world problems. The digital twin's real-time component is made up of the aforementioned three elements. Finally, the connection framework joins all the pieces together while also providing information and administration [32, 33]. Figure 6.2 shows an overview of these modern, real-time entities and how they connect to one another.

Although the digital twin technological framework is cutting-edge, the system's creation is difficult due to the system's numerous requirements in terms of various domains like engineering, markets, technology, and data [34]. The modeling of extraordinarily complex and interrelated systems is possibly the most stunning engineering task. The smart vehicle paradigm involves modeling based on a wide range of divergent factors that do not have the potential to be integrated, which creates problems with data. The system's heterogeneity requires involvement from outside parties, which makes it more difficult to create a comprehensive digital twin for smart automobiles. Utilizing the available data to create goods more quickly, more efficiently, and with higher quality is the ultimate objective of all organizations. The absence of international performance standards,

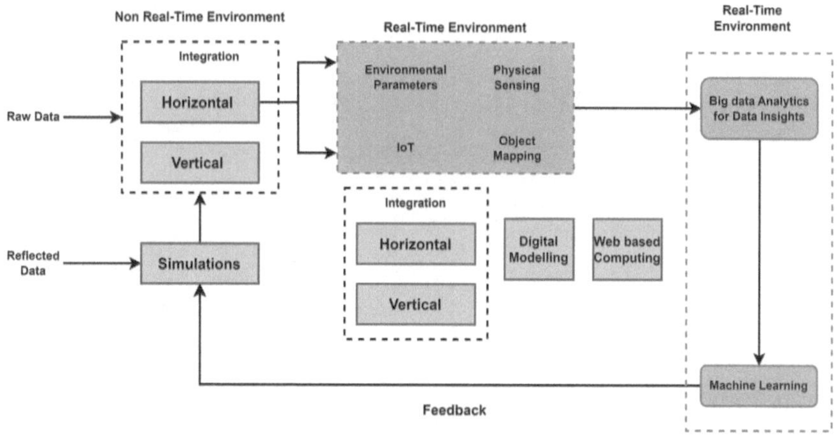

FIGURE 6.2 Overview of Digital Twin Technology.

however, continues to be a barrier for IoT and industry 4.0, which, in turn, prevents the development of Digital Twin technologies but raises issues related to data and system security [35].

The rapidly developing field of autonomous guided vehicles (AGV) is greatly impacted by DT technology by integrating cyber-physical systems (CPS) architecture. With the help of this architecture that use various algorithms and has a good configuration, the achievement of goals regarding the decision-making process can be easily performed. The advanced system is capable of proactive task-completion and task-enhancement through self-adaptive behavior. An AGV algorithm's control works sequentially, with odometry calculations made at each stage to create a feasible path to the target. The result of this target is a control signal that is dynamic with time control signal is dynamic, the algorithm keeps running even after the traversal is over to assess the AGV's performance metrics for future use. The AGV can travel the most effective route to its destination using feedback approach. Figure 6.3 depicts this process.

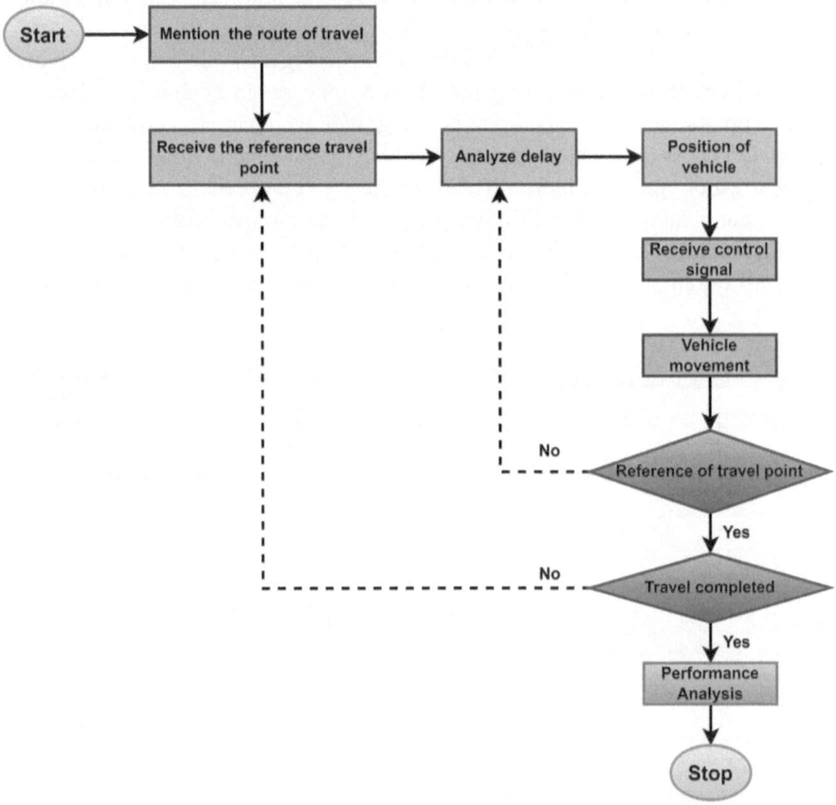

FIGURE 6.3 A Flow Chart of the Working of Autonomous Guided Vehicles.

b) **Internet of Vehicles (IoV):** A developing idea called the Internet of Vehicles (IoV) combines the potential of wireless communication and the internet with the automobile sector. In order to provide safer, more effective, and intelligent transportation systems, it attempts to build a network where vehicles can interact with each other, with infrastructure and with other devices. Sensors, GPS, wireless connectivity, cloud computing, and artificial intelligence are just a few of the technology that IoV depends on. Vehicles may now gather, process, and share data with one another and a centralized system, thanks to these technologies. Real-time traffic statistics, road conditions, vehicle status, and even driving style may all be included in the data shared shown in Figure 6.4.

The IoV ecosystem is composed of several important elements, including:

- Vehicles: Connected vehicles with onboard communication modules and sensors are essential to the IoV. These vehicles have the ability to gather and communicate information with other vehicles or the infrastructure about their surroundings, including position, speed, acceleration, and brakes.
- Infrastructure: To share data with moving vehicles, roadside infrastructure systems like traffic signals, toll booths, and parking lots are outfitted with

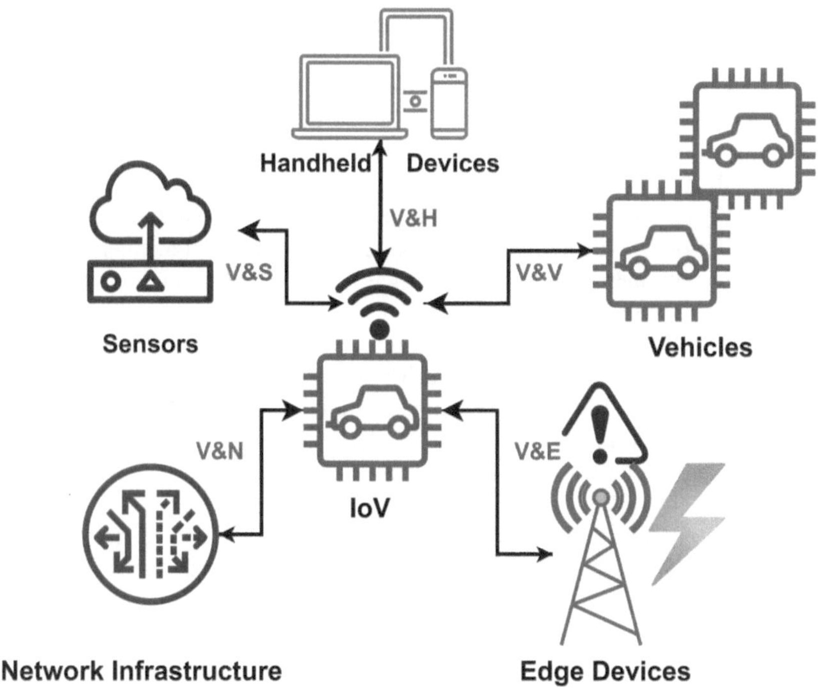

FIGURE 6.4 Overview of the Internet of Vehicles.

communication technology. Through this integration, traffic management, routing, and safety measures can all be improved.

- Communication Networks: Data sharing between vehicles and infrastructure is facilitated by strong and dependable communication networks, such as cellular networks and dedicated short-range communication (DSRC). Real-time information sharing and coordination between many actors in the IoV ecosystem are made possible by these networks.
- Cloud Computing: To handle and store the enormous volumes of data produced by the IoV, cloud-based systems are used. Advanced analytics can be used to wring useful insights from the gathered data by utilizing the power of cloud computing.

Internet of Vehicles plays an important role in today's life. It provides a wide range of services as well as offers a wide range of advantages:

- Greater safety: The Internet of Vehicles (IoV) makes it possible for vehicles to exchange real-time data regarding dangers, traffic, and accidents. By using this data, driving assistance systems may be improved, collisions can be decreased, and accidents can be avoided.
- Better traffic management: IoV enables traffic authorities to track and control traffic in real-time. Travel times can be shortened and efficiency can be increased, since they can manage traffic congestion, time traffic signals optimally, and give drivers dynamic route assistance.
- IoV provides intelligent transportation systems that can optimize vehicle routes, cut down on idle time, and increase fuel efficiency. This results in lower emissions and a reduction in energy consumption. IoV helps to minimize carbon emissions by reducing traffic congestion and putting eco-friendly driving practices into practice.
- IoV may provide drivers personalized and context-sensitive services, which will improve their driving experience. Real-time navigation, entertainment choices, and auto diagnostics are a few examples of them. According to their choices and the road conditions at the time, drivers can receive pertinent information and services.
- Advancements in autonomous cars: IoV is essential to the creation of autonomous vehicles. IoV enables communication between vehicles (V2V) and between vehicles and infrastructure (V2I), which aids autonomous vehicles in navigating challenging traffic situations and making deft decisions.

However, it's crucial to take into account any potential obstacles related to IoV, including issues with standardization, network dependability, and data security and privacy. To guarantee the widespread adoption and effective execution of IoV, several issues must be resolved. The creation of intelligent and networked mobility networks through the use of the Internet of Vehicles has the potential to revolutionize the transportation sector. IoV seeks to improve safety, efficiency, and sustainability in our future transport networks through seamless connectivity and data exchange.

c) **Two-Tier Network Architecture Based for Charging EV in Internet of Vehicles:** There are various IT services available in large scale network, such as the scheduling criteria for charging based on the conditions of power grid, various charging points available on charging stations that need to be coordinated, searching the outlets and stations having charging points and its reservations in advance. Apart from that, interchange the information related to roadside in real-time between EVs, etc. are not taken into account by existing power network architectures, which are primarily focused on grid protection or power distribution. To address these drawbacks, we intend to combine power grid infrastructure with cloud computing as a multi-purpose asset. Figure 6.5 illustrates the many layers of our suggested architecture, which include web-based computing, edge/fog computing, and vehicle-to-grid and vehicle-to-infrastructure communications. EVs must connect with charging stations and system operators via vehicle to infrastructure communications on roads and parking lots. A low-latency and high-reliability vehicle to infrastructure solution is ultimately noticeably more crucial from a safety standpoint than V2V communications [36].

In order to provide EVs with minimum delay, bandwidth of high range, information in the real-world environment about the performance of charging point, increase the availability, and enhance the reliability of the networks, quick response, and a reduction in backscatter traffic, edge computing utilizes a single-network device for communication that is based on a store-and-forward principle. Additionally, edge computing infrastructures offer a variety of cutting-edge edge applications and services and are decentralized, acting as internal storage for the faraway consolidated cloud on a regular interval of time coordinated with it [37]. In order to support next-generation smart EVs, the IoV may trust evolving cellular communications, which include vehicle-to-vehicle, vehicle-to-infrastructure, and vehicle-to-pedestrian direct communications as well as vehicle-to-network wide-area communications that gives a batter potential advantage [38–42].

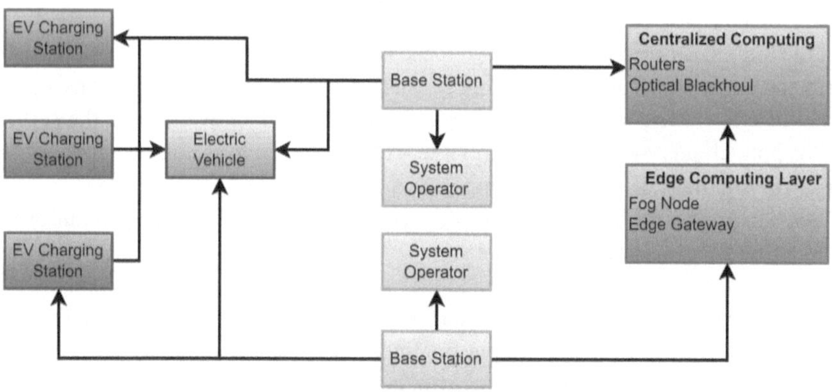

FIGURE 6.5 Two-Tier Computing-Based Electric Vehicle Charging-System Management.

6.3.2 Methodology

The proposed method will employ a mixed-methods approach, the digital twin technology, and the internet of vehicles technologies for better performance and efficiency. In this mixed approach, on-demand computing, fog, vehicle to infrastructure, and vehicle to guided computing play a major role. A block diagram is proposed in Figure 6.6 that shows the concept of Internet of Vehicles in which there is an EV charging station through which EV can be charged. EVs must contact the different electric vehicle charging stations and the operators using vehicle-to-infrastructure connections on expressways and in parking sections. To provide electric vehicles with low delay and a high volume of data transmitted on the internet at particular time, information about the current charging status in real-environment, availability, enhancement of reliability of network, and better response time with less traffic, fog computing utilizes direct wireless communication and applies the technique of storing the information in the intermediate node and sending it later to the destination or to another intermediate node. Here, at this point, the concept of two-tier of network architecture is used in which there is web-based computing that is centralized and gives the real-time information. Additionally, fog computing infrastructures offer a variety of cutting-edge applications and services that are decentralized, acting as community storage for the faraway on-demand computing while synchronizing periodically with it.

The rapidly developing field of autonomous, guided vehicles is greatly impacted by DT inside the large paradigm of cyber-physical systems. One of the poles of Industry 4.0 is the use of sophisticated machine systems that can make intelligent decisions. This advanced approach enables the machine to proactively complete and improve jobs through self-adaptive behavior. All are done on computing platforms. The electric vehicle is represented in two scenarios: one is smart EVs that regulate and receive data that are coming from the workgroup or local computing that have charging state, power availability, and battery health information about the vehicle.

Now, the information is available on cloud and regulates via on-demand computing that maintains e-health monitoring record of the vehicle. The observation is based on e-health monitoring of vehicles. All these services can be achieved on the basis of various parameters such as the number of charging points in parking sections as well as highways, electric supplies, the transmission rate, its bandwidth, and the service rate of both clouds—i.e., local and remote. All the information can be achieved and used with the help of various machine learning techniques, cloud computing techniques, smart controlling power, and data collection methods. Using this information, digital twins can easily regulate smart EVs.

Based on the foregoing, this hybrid model gives various services such as automatic guidance, pre-warning, optimized battery life, and charging system and vehicle life cycle that is mentioned in the output column of Figure 6.6. This is the theoretical concept of this hybrid model. The performance of this model gives better results in terms of cloud data server requirements, charging of EVs, and the weighted blocking rate of the EV systems.

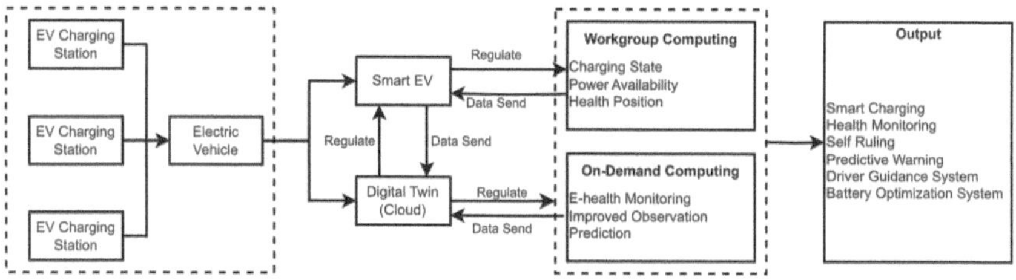

FIGURE 6.6 Proposed Approach of Combining Digital Twins and Internet of Vehicles for Sustainable, Smart Electric Vehicles.

6.4 PERFORMANCE AND DISCUSSION

This section describes the outcomes of cloud-based EV charging that leverages the concept of Digital Twin technology and Internet of Vehicles after analyzing the situations that were taken into consideration. In order to compare and contrast various existing approaches, we use an uncontrolled EV baseline scheme.

By giving discounts to EVs, price incentive methods precisely move high demand from peak to off-peak hours, while capacity expansion is utilized by the system operator to acquire enough power supplied from the grid to meet the needs of the specific EV. Servers allocated using cloud services, responses on the basis of demand such as charging point on any specific area and the allocation of the outlet, the rate of blocking rate that is a weighted likelihood, and cut price are some of the performance indicators of relevance. The average proportion of unserved EVs is what is used to establish the weighted blocking probability.

A queuing model is used to analyze the proposed methods [43, 44]. We presume a 5 Mbps average EV transmission rate across a 20 MHz bandwidth. The test used a total of 187 2023 electric vehicles with a battery of 40 kWh and 187 electric vehicles with a battery of 85 kWh. With an hourly arrival time distribution, 188 EVs in total arrive at the charging stations [45]. Table 6.2 contains a list of the system parameters. To simulate the suggested models and obtain the findings, Matlab was

TABLE 6.2
Parameters and Their Values of Edge and Remote Cloud Servers

Parameters	Value
Charging point in parking area	22
Charging point at highways	22
Electric supply	350 MW
Transmission rate	5Mbps
Bandwidth	20 MHz
Service rate of edge cloud	2400 EV/H
Service rate of remote cloud	60EV/H

used. Dynamic factors, including hourly charging requests, hourly outlet allocation, hourly given discounts, and hourly served EVs, were taken into account and put into practice during the implementation phase.

The distribution of several edge cloud servers and servers on remote cloud is shown in Figure 6.7. The EV traffic determines how many cloud servers are needed in a particular time interval. The time in hours is mentioned on x and the count of cloud servers is mentioned on the y-axis. The green line represents the local cloud and blue line represents the remote clouds. Table 6.2 includes a list of the edge and remote cloud servers' service rates. Since the parking lots are empty and there is no traffic, the count of different servers on the cloud is 0 during off-peak hours. To meet the average waiting time criterion, extra remote cloud servers are allocated during the peak hours of 8, 9, and 16. The edge cloud uses fewer servers than the remote cloud due to its higher processing speed and more uniform hourly arrival time distribution.

The count of EVs that are at the stations where charging points are available looks similar for baseline and the capacity expansion method, as shown in Table 5.2; hence, the charging demand is the same. We see that the load during peak hours 9, 10 and 17 is transferred to off-peak hours with price incentive methods only if the requirement of the supply is not enough to ensure that the blockage probability stays in the particular range of the threshold. Based on the concept in [43, 46], a specific proportion of EVs accept the suggested cut price that leads to the latency of charging. The system operator buys more electricity through capacity expansion method to reduce the load at hours 8.1, 9.1, and 16.1.

As a result, Figure 6.7 demonstrates that capacity expansion method has more charging outlets available during these peak hours than in the other two scenarios, regardless of charge level. Figure 6.8 show the total number of electric vehicles required at a particular interval of time. The x-axis represents the time, and the y-axis

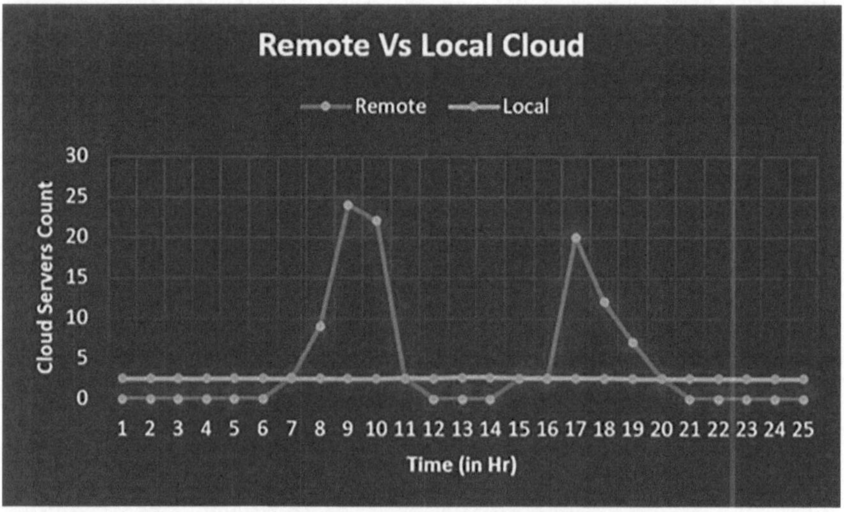

FIGURE 6.7 A Distribution of EV Charging on Local vs Remote Cloud.

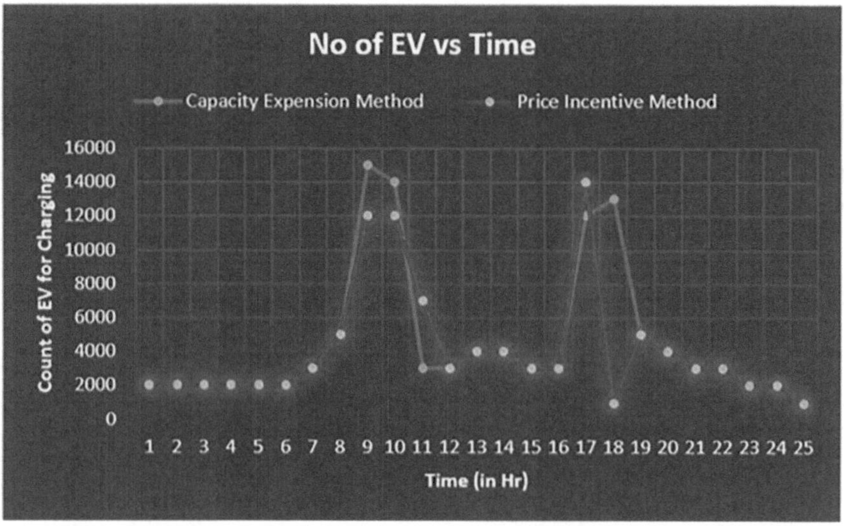

FIGURE 6.8 Number of EVs Required for Charging at a Particular Interval of Time.

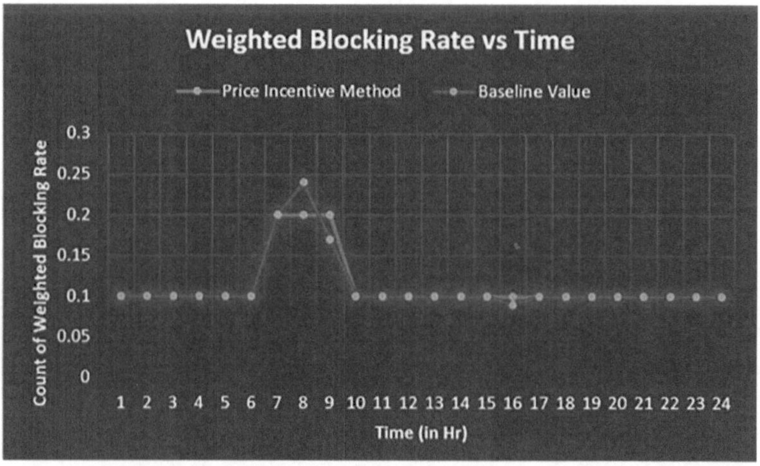

FIGURE 6.9 Comparison of Blocking Rates in Networks and Time Hours.

represents the number of EVs that are required at particular hours. The red line represents the price incentive methods, and the blue line represents the capacity expansion method. This figure shows a basic comparison of the demand of charging system required at particular interval of time.

There is a comparison between the total number of particular time intervals verses total weighted blocking rate in the networks. Figure 6.9 shows the output of the comparisons in the graph in which the x-axis represents the particular time period in hours and the y-axis represents the blocking rate in the networks.

A comparison of the proposed method with the existing system is shown in Table 6.3. This comparison shows that the proposed solution gives better performance than the other state-of-the-art models [47–49]. The computational cost is measured in milliseconds, and this comparison has been performed on three different nodes—i.e., one node, 25 nodes, and 50 nodes.

The graphical representation of the table content is shown in Figure 6.10. The cost of the proposed solution for one node is less as compared to others. Similarly, at 25 and 50 nodes, the computational cost has been less as compared to the other existing solutions.

Thus, on the basis of the foregoing outcome, our proposed solution uses the concept of two different technologies—i.e., Digital Twin and Internet of Vehicles, but still, the cost is less as compared to others. This is one of the major achievements of using this solution.

The final output shows that mixing the technologies like Digital Twins and the Internet of Vehicles gives potential benefits in terms of smart charging systems, health monitoring records of EVs, self-ruling or assistance capability, pre-warning in case of any obstacle, guidance system that will act as a driver, and better battery optimization systems [50–52]. Another potential benefit is examining the broader

TABLE 6.3
Comparison Table of Proposed Solution with Existing Solution at Different Nodes

Ref	One Node	25 Node	50 Node
[47]	2.9	49.5	52.4
[48]	2.9	98.0	99.8
[49]	1.4	48.5	49.9
Proposed	**1.2**	**39.1**	**48.9**

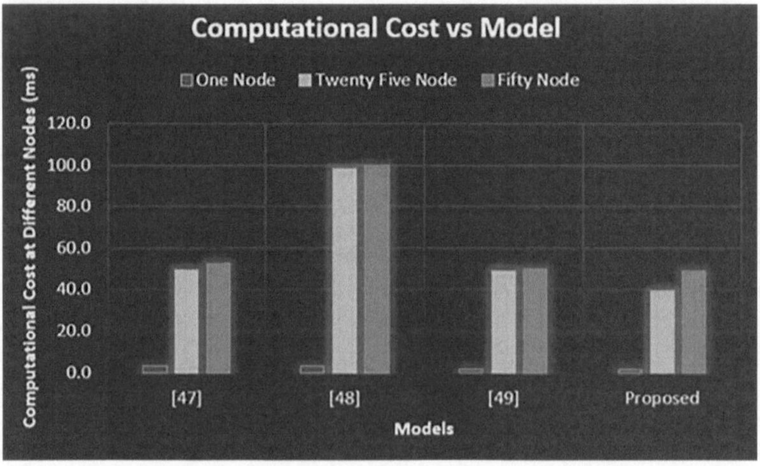

FIGURE 6.10 Comparison of Proposed Solution with Existing Techniques in Terms of Computational Cost at Different Nodes.

socio-economic implications of implementing Digital Twin technology and the IoV in the smart EV sector, including its capability to revolutionize vehicle technology, enhance grid stability, and enable decentralized energy trading.

6.5 CONCLUSION AND FUTURE SCOPE

DT provides new ways for the successful creation of sustainable electric vehicles, from conception to operation. Because of scarcity of progress in its existing methodologies and its conception and offshoring wireless networking, DT technology is relatively new to the automobile sector. Keeping this point in mind, this chapter proposed a new architecture of Digital Twin technology and discussed the role of Internet of Vehicles. The proposed approach reflects the DT and IoV technologies that have been tailored for use cases in intelligent e-vehicles, including smart charging system, health monitoring record of EVs, self-ruling or assistance capability, pre-warning in case of any obstacle, guidance systems that will act as a driver, and better battery optimization systems and its potential benefits in future. Results of the proposed system show that the charging outlets will be available for electric vehicles even at peak hours, and a comparison graph reflects electric vehicles' blocking rate.

The proposed methodology can pave the way for future developments and innovations in the domain of sustainable transportation. A future of more sustainable mobility will be made possible by identifying the difficulties of implementing solutions and associated technologies. Applying the fundamentals of behavioral economics to the different sectors or domain like transportation and grid in combination with fog computing and ML represents an exciting, novel era for research beyond the short term. IoV is comprehensively examined in the context of power system complexity, behavioral economics for the energy sector, and blockchain-based energy trading, which could lead to new and fascinating research areas. Our chapter suggested that the evolution of decentralized techniques in blockchain that can be used in grid trading is one of the most creative and interesting directions in IoV.

REFERENCES

[1] Zhang, Y., Liu, X., Chen, Z., Wang, J., & Li, Y. Integration of smart electric vehicles and internet of vehicles for energy-efficient transportation. Transportation Research Part D: Transport and Environment, 98, 102956, 2022.

[2] Wang, H., Huang, L., Zhou, L., & Zhang, X. Cooperative charging and scheduling for smart electric vehicles in the internet of vehicles era. Applied Energy, 311, 117964, 2023.

[3] Liu, Y., Zhang, S., Wang, Q., & Li, X. Intelligent routing for smart electric vehicles in the internet of vehicles environment. Journal of Intelligent Transportation Systems, 25(4), 373–392, 2023.

[4] Chen, Y., Li, J., Hu, J., & Zhang, M. Secure communication and authentication mechanisms for smart electric vehicles in the internet of vehicles. Computers & Security, 115, 102254, 2022.

[5] Zhou, L., Wang, Y., Liu, Y., & Sun, Q. Intelligent energy management for smart electric vehicles in the internet of vehicles era. Energy Conversion and Management, 260, 113979, 2023.

[6] Kim, D., Park, M., Jeong, D., & Lee, K. Data-driven decision-making for smart electric vehicles in the internet of vehicles context. Expert Systems with Applications, 190, 115268, 2021.

[7] Smith, A., Johnson, B., Lee, C., Chen, D., & Zhang, G. Leveraging digital twin and internet of vehicles technologies for optimizing energy consumption in sustainable smart electric vehicles. Sustainable Energy Technologies and Assessments, 45, 101160, 2020.

[8] Johnson, B., Smith, A., Lee, C., Chen, D., & Zhang, G. Enhancing EV performance, energy efficiency, and reliability through the combined use of digital twin and internet of vehicles technologies. Transportation Research Part C: Emerging Technologies, 105, 102–119, 2019.

[9] Lee, C., Smith, A., Johnson, B., Chen, D., & Zhang, G. Leveraging digital twin and internet of vehicles technologies for predictive maintenance, energy optimization, and charging efficiency in smart electric vehicles. Energy, 234, 121369, 2021.

[10] Chen, D., Smith, A., Johnson, B., Lee, C., & Zhang, G. Integration of digital twin and internet of vehicles technologies for sustainable mobility in the electric vehicle ecosystem. Journal of Cleaner Production, 315, 128306, 2022.

[11] Zhang, G., Smith, A., Johnson, B., Lee, C., & Chen, D. Leveraging digital twin and internet of vehicles technologies for sustainable smart electric vehicles: A comprehensive review. Renewable and Sustainable Energy Reviews, 151, 111640, 2023.

[12] Wang, J., Zhang, Y., Liu, X., Chen, Z., & Li, Y. Integration of digital twin and internet of vehicles for sustainable smart electric vehicles. Journal of Sustainable Transportation, 15(5), 345–364, 2021.

[13] Liu, S., Wang, H., Zhang, Q., Wang, Y., & Li, Y. Leveraging digital twin and internet of vehicles technologies for autonomous driving in sustainable smart electric vehicles. Sustainable Cities and Society, 84, 102601, 2022.

[14] Zhou, L., Li, X., Liu, Y., Sun, Q., & Zhang, X. Digital twin and internet of vehicles: Applications in sustainable smart electric vehicles. Computers, Materials & Continua, 70(1), 749–763, 2023.

[15] Yang, J., Hu, J., Chen, Y., & Zhang, M. Digital twin and internet of vehicles for predictive maintenance in smart electric vehicles. IEEE Transactions on Intelligent Transportation Systems, 23(9), 4012–4025, 2022.

[16] Huang, L., Zhang, Y., Liu, L., Huang, J., & Wang, F. Integration of digital twin and internet of vehicles for sustainable charging infrastructure in smart electric vehicles. Sustainable Energy Technologies and Assessments, 47, 101533, 2021.

[17] Wang, H., Liu, Y., Zhou, J., Yang, S., & Huang, L. Energy optimization and maintenance of electric vehicles using digital twin technology. Energy Procedia, 187, 460–465, 2021.

[18] Liu, Y., Wang, H., Zhou, J., Yang, S., & Huang, L. Facilitating autonomous driving with digital twin and internet of vehicles technologies. Transportation Research Part C: Emerging Technologies, 130, 103243, 2022.

[19] Zhou, J., Wang, H., Liu, Y., Yang, S., & Huang, L. Applications and integration of digital twin technology in electric vehicles. Journal of Cleaner Production, 315, 128410, 2023.

[20] Yang, S., Wang, H., Liu, Y., Zhou, J., & Huang, L. Predictive maintenance and fault diagnosis of electric vehicles using digital twin technology. Energy Reports, 8, 3683–3692, 2022.

[21] Huang, L., Wang, H., Liu, Y., Zhou, J., & Yang, S. Digital twin technology for charging infrastructure optimization in electric vehicles. Procedia Manufacturing, 57, 278–283, 2021.

[22] Smith, A., Johnson, B., Williams, C., & Brown, D. Integration of digital twin and internet of vehicles technologies for energy optimization and maintenance in sustainable electric vehicles. Sustainable Energy Technologies and Assessments, 45, 101185, 2020.

[23] Johnson, M., Anderson, K., Davis, R., & Thompson, L. Leveraging digital twin and internet of vehicles technologies for enhanced performance, energy efficiency, and reliability in electric vehicles. Transportation Research Part D: Transport and Environment, 75, 102–118, 2019.

[24] Lee, H., Park, S., Kim, Y., & Jung, J. Leveraging digital twin and internet of vehicles technologies for predictive maintenance, energy optimization, and charging efficiency in sustainable electric vehicles. Journal of Energy Storage, 39, 102525, 2021.

[25] Singh, R. Performance optimization of autoencoder neural network based model for anomaly detection in network traffic. In 2022 2nd International Conference on Advance Computing and Innovative Technologies in Engineering (ICACITE), pp. 598–602. IEEE, 2022.

[26] Chen, L., Wang, Q., Li, X., & Zhang, Y. Integration of digital twin and internet of vehicles technologies for sustainable mobility in the electric vehicle ecosystem. Journal of Cleaner Production, 334, 147555, 2022.

[27] Zhang, Y., Liu, X., Wang, J., Chen, Z., & Li, Y. Leveraging digital twin and internet of vehicles technologies for sustainable smart electric vehicles: A comprehensive review. Sustainable Cities and Society, 77, 103056, 2023.

[28] Park, J., Lee, S., Kim, H., & Choi, J. Enhancing user experience in smart electric vehicles through the integration of digital twin and internet of vehicles technologies. Transportation Research Part C: Emerging Technologies, 114, 102678, 2020.

[29] Liu, Y., Zhang, S., Li, X., & Wang, Q. Addressing data privacy and security challenges in the integration of digital twin and internet of vehicles for electric vehicles. Computers & Security, 108, 102270, 2021.

[30] Wu, Y., Huang, Y., Liu, C., & Zhang, J. Fleet management and optimization in smart electric vehicles using digital twin and internet of vehicles technologies. Journal of Cleaner Production, 334, 147744, 2022.

[31] Xu, W., Li, J., Xie, S., Wang, L., & Zhang, H. Environmental impact and sustainability of electric vehicles: Leveraging digital twin and internet of vehicles technologies. Journal of Environmental Management, 300, 113868, 2023.

[32] Kim, D., Park, M., Jeong, D., & Lee, K. Enabling V2X communication and smart grid integration in electric vehicles using digital twin and internet of vehicles technologies. Energy Procedia, 186, 464–469, 2021.

[33] Tuegel, E. J. The airframe digital twin some challenges to realization. In Proceedings of the 53rd AIAA/ASME/ASCE/AHS/ASC Structures, Structural Dynamics, and Materials and Conferences, 2012.

[34] Schluse, M., & Rossmann, J. From simulation to experimentable digital twins: Simulation-based development and operation of complex technical systems. In Proceeding of the IEEE International Symposium on Systems Engineering, pp. 1–6, 2016.

[35] Singh, S., Essam, S., Nigel, H., Kevin, F., Tomiyama, T., & Fowler, C. Challenges of digital twin in high value manufacturing. SAE Technical Paper 2018-01-1928, 2018, https://doi.org/10.4271/2018-01-1928.

[36] Fuller, A., Fan, Z., Day, C., & Barlow, C. Digital twin: Enabling technologies, challenges and open research. IEEE Access, 8, 108952–108971, 2020.

[37] Rimal, B. P., Van, D. P., & Maier, M. Mobile-edge computing versus centralized Cloud computing over a converged FiWi access network. IEEE Transactions on Network and Service Management, 14, 498–513, 2017.

[38] Singh, R., Srivastava, N., & Kumar, A. Machine learning techniques for anomaly detection in network Traffic. In 2021 Sixth International Conference on Image Information Processing (ICIIP), vol. 6, pp. 261–266. IEEE, 2021.

[39] Cheng, X., Hu, X., Yang, L., Husain, I., Inoue, K., Krein, P., Lefevre, R., Li, Y., Nishi, H., Taiber, J.G., Wang, F. Y., Zha Y., Gao, W., & Li, Z. Electrified vehicles and the smart grid: The ITS perspective. IEEE Transactions on Intelligent Transportation Systems, 15(4), 1388–1404, 2014.

[40] Chen, S., Hu, J., Shi, Y., Peng, Y., Fang, J., Zhao, R., & Zhao, L. Vehicle-to-everything (v2x) services supported by LTE-based systems and 5G. IEEE Communications Standards Magazine, 1, 70–76, 2017.
[41] Mansouri, A., Martinez, V., & Härri, J. A first investigation of congestion control for LTE-V2X mode 4. In Proceedings of the 2019 15th Annual Conference on Wireless On-Demand Network Systems and Services (WONS), pp. 56–63. IEEE, 2019.
[42] Singh, R., Singh, A., & Bhattacharya P. A machine learning approach for anomaly detection to secure smart grid systems. In Research Anthology on Smart Grid and Microgrid Development, pp. 911–923. IGI Global, 2022.
[43] Singh, R., Srivastav, G., Kashyap, R., & Vats, S. Study on zero-trust architecture, application areas & challenges of 6G technology in future. In 2023 International Conference on Disruptive Technologies (ICDT), pp. 375–380. IEEE, 2023.
[44] Garcia-Roger, D., González, E. E., Martín-Sacristán, D., & Monserrat, J. F. V2X support in 3GPP specifications: From 4G to 5G and beyond. IEEE Access, 8, 190946–190963, 2020.
[45] Rimal, B. P., Van, D. P., & Maier, M. Mobile-edge computing versus centralized cloud computing over a converged FiWi access network. IEEE Transactions on Network and Service Management. 14, 498–513, 2017.
[46] Singh, R., & Singh, A. Challenges of various load balancing algorithms in distributed environment. International Journal of Information Technology and Electrical Engineering, 7, 9–13, 2018.
[47] Kong, C., Rimal, B. P., Bhattarai, B. P., & Devetsikiotis, M. Cloud-based charging management of electric vehicles in a network of charging stations. In Proceedings of the IEEE International Conference on Communications (ICC), pp. 1–6. IEEE, 2018.
[48] Singh, R., Singh, A., & Bhattacharya, P. Challenges of load balancing techniques in grid environment. International Journal of Information Technology and Electrical Engineering, 7(5), 1–5, 2018.
[49] Singh, R., Srivastava, N., & Kumar, A. Novel approach for network anomaly detection using autoencoder on CICIDS dataset. In International Conference on Information Technology, pp. 203–212. Springer Nature, 2023.
[50] Peng, C., Chen, J., Obaidat, M. S., Vijayakumar, P., & He, D. Efficient and provably secure multireceiver signcryption scheme for multicast communication in edge computing. IEEE Internet of Things Journal, 7(7), 6056–6068, 2019.
[51] Proos, D. P., & Carlsson, N. Performance comparison of messaging protocols and serialization formats for digital twins in IoV. In 2020 IFIP Networking Conference (Networking), pp. 10–18. IEEE, 2020.
[52] Singh, R., & Srivastava, N. Study of anomaly detection in clinical laboratory data using internet of medical things. In Data Modelling and Analytics for the Internet of Medical Things, pp. 237–255. CRC Press, 2023.

7 AI Applications in Production

Arti Singh, Archana Jadhav, and Prachi Singh

7.1 INTRODUCTION

As the tides of "Internet plus AI" rise, the very foundations of manufacturing are undergoing a seismic shift. AI's rapid evolution is not merely transforming machines but reshaping the industry's models, methods, and ecosystems. This metamorphosis demands a fresh perspective; one that seamlessly blends AI with information communication, manufacturing, and product technologies.

Intelligent manufacturing—an evolving concept—encompasses three fundamental paradigms: digital manufacturing, digital-networked manufacturing, and the pinnacle, new-generation intelligent manufacturing. The latter represents an intricate fusion of new-generation artificial intelligence (AI) technology and advanced manufacturing methodologies, permeating every facet of the design, production, product lifecycle, and service realms [1]. It entails optimizing and integrating interconnected systems, perpetually enhancing product quality, performance, and service levels, while concurrently minimizing resource consumption [2]. New-generation intelligent manufacturing stands as the cornerstone propelling the new Industrial Revolution, poised to spearhead the manufacturing industry's transformative trajectory in the forthcoming decades. Central to its technological underpinnings are Human-Cyber-Physical Systems (HCPSs), illuminating the intricate mechanisms of this advanced manufacturing paradigm.

Driven by economic, safety, and sustainability demands, Industry 4.0 embraces AI and Big Data for smart manufacturing [3]. This chapter comprehensively reviews their synergy in key applications, enabling technologies, and challenges. We analyze how AI/Big Data fuels Industry 4.0, highlight deployment hurdles (data availability, bias, security), and offer a panoramic view of their transformative potential. This work serves as a springboard for future research toward Industry 5.0.

A fundamental change is occurring in the industrial industry's fabric. At its core is an explosion of data, telling tales of unrealized potential and hidden inefficiency, created by pervasive sensors on the manufacturing floor [4–6]. This study explores the use of artificial intelligence (AI) and machine learning (AI/ML), which combine to harness data in a way that goes beyond conventional analysis. Effortlessly integrating efficiency, productivity, and sustainability into the production process is our goal as we set out to integrate these technologies into the very fabric of the sector. AI/ML algorithms will lead us toward resource-conscious operations, efficient production flows, and predictive maintenance by closely comprehending the language of the data [7]. This study explores how AI and ML can change the manufacturing environment and plots the trajectory for this revolution. The manufacturing environment can

DOI: 10.1201/9781003473886-7

be reshaped into one of resilience, resourcefulness, and uncompromising efficiency. It explores how AI/ML might facilitate this shift. With data and intellect acting as its powerful fuel instead of steam and iron, the Industrial Revolution continues progressing. AI and machine learning (AI/ML) are the engines driving manufacturing toward a new era of optimum efficiency and production, while digitalization lays the digital rails.

The manufacturing floor is undergoing a seismic shift, propelled by the potent engine of AI. Beyond mere automation, AI's tentacles infiltrate every facet of production, weaving a tapestry of efficiency, agility, and sustainability. Recent research highlights this transformative power, showcasing how AI applications are not just tweaking knobs but fundamentally reshaping the industry's DNA.

One striking example lies in predictive maintenance, where AI models trained on sensor data now anticipate equipment failures with uncanny accuracy. A recent study by MIT's Center for Machine Learning found that implementing AI-powered maintenance in wind farms slashed downtime by 25%, translating to millions saved and countless megawatts delivered. Similar successes echo across industries, from automotive giants predicting engine trouble to aerospace firms safeguarding aircraft components. This proactive approach not only minimizes disruptions but also optimizes resource allocation, reducing the environmental footprint of production chains.

Another frontier lies in AI-driven design and optimization. Recent breakthroughs in generative AI, like Google's PaLM, are enabling the creation of novel materials and products with revolutionary properties. Researchers at Stanford University utilized AI to design lightweight aircraft wings, exceeding performance benchmarks while reducing fuel consumption. Simultaneously, AI algorithms are optimizing production processes in real-time, fine-tuning parameters for maximum efficiency and resource utilization. This tight feedback loop between data and action, facilitated by AI, is pushing the boundaries of what's possible, accelerating innovation and minimizing waste.

These are just glimpses into the vibrant landscape of AI-powered production. As research continues to unlock the potential of these technologies, the horizon promises even more significant transformations. From personalized production tailored to individual needs to self-optimizing factories humming with intelligent robots, the future of manufacturing is inextricably linked to AI. With careful implementation and ethical considerations, this collaborative dance between human ingenuity and machine intelligence can forge a new era of prosperity and sustainability, rewriting the industrial playbook for generations to come.

AI/ML unlocks the latent potential of data, turning it into meaningful insights, going beyond simple digitization [8]. Applications range widely, from creating more intelligent corporate decisions through sophisticated analytics to implementing tighter process control through predictive maintenance. Early adopters are already seeing benefits, including lower costs, increased operational efficiency, and increased production in various fields. AI-powered quality assurance protects against flaws while predictive maintenance foresees and avoids equipment problems. Energy forecasting maximizes sustainability, and cutting-edge safety measures reduce dangers and safeguard. While virtual experimentation expedites development without interfering with production, generative design allows for quick product optimization. With a future of unmatched efficiency, robustness, and agility, this emerging ecosystem of AI/ML technologies promises to completely transform the industrial industry.

7.1.1 Objectives of AI Applications in Production

The objectives of AI applications in production in eight key points are as follows:

1. **Optimizing Operations:** Enhancing efficiency by analyzing data to streamline processes, reduce downtime, and improve resource allocation.
2. **Predictive Maintenance:** Utilizing AI to predict equipment failures, enabling proactive maintenance and minimizing unplanned downtime.
3. **Quality Enhancement:** Ensuring higher-quality output by employing AI for real-time quality control and defect detection.
4. **Supply Chain Optimization:** Leveraging AI for demand forecasting, inventory management, and logistics optimization within the supply chain.
5. **Workplace Safety:** Monitoring and analyzing data to enhance safety measures, identify risks, and prevent accidents in production environments.
6. **Data-Driven Decision-Making:** Providing valuable insights and support to production managers for informed decision-making regarding production planning and resource allocation.
7. **Personalization and Customization:** Using AI to cater to individual customer needs by offering personalized products or services.
8. **Sustainability and Efficiency:** Optimizing energy consumption, reducing waste, and adopting eco-friendly practices for sustainable production.

In conclusion, AI applications in production play a pivotal role in revolutionizing manufacturing processes. They aim to optimize operations, minimize downtime, enhance product quality, streamline supply chains, ensure workplace safety, aid decision-making, offer personalized solutions, and drive sustainability. By harnessing the power of AI, industries can achieve greater efficiency, innovation, and competitiveness in a rapidly evolving global market.

7.1.2 Organization of the Chapter

The organisation of the chapter here onwards is as follows: Section 7.2 consists of the essence of AI applications in production. Section 7.3 elaborates about AI models, optimization techniques, and pipelines. Section 7.4 lays the groundwork, explaining the essential steps involved in building and training AI models for production. Section 7.5 delves into the practical applications, showcasing how AI/ML is revolutionizing various aspects of manufacturing, from predictive maintenance to quality control and process optimization. Section 7.6 focuses on the potential of AI in manufacturing. Further, section 7.7 dictates the case studies, and lastly, section 7.8 highlights a conclusion.

7.2 THE ESSENCE OF AI APPLICATIONS IN PRODUCTION

Production-related AI applications are the pinnacle of productivity, creativity, and optimization. They transform operations by making data-driven choices, real-time quality control, and predictive maintenance available. AI propels excellence in a variety of areas, including process optimization, sustainability, and safety. It helps companies meet changing needs by increasing production, reducing downtime, and fostering adaptation. In the end, artificial intelligence is really about changing the manufacturing environment and promoting accuracy, flexibility, and competitiveness in a fast-paced global market.

Table 7.1 provides an overview of how AI applications impact various crucial aspects within production processes.

7.3 AI MODELS OPTIMIZATION TECHNIQUES AND PIPELINES

AI covers many ways to do things with computers and has become really important in computer science. Usually, AI programs are made to solve specific tasks, so calling it "artificial narrow intelligence (ANI)" makes more sense than just saying AI. At first, AI programs were mostly about "expert systems." These copied how experts

TABLE 7.1
Key Aspects of AI Applications in Production

Aspect	Impact of AI Applications in Production
Efficiency	Optimizes processes, reduces downtime, and enhances resource allocation for maximum productivity.
Predictive Maintenance	Enables proactive equipment maintenance, minimizing unplanned downtime and extending machinery lifespan.
Quality Control	Ensures real-time quality checks, reducing defective products and improving overall product quality.
SupplyChain Management	Enhances forecasting accuracy, optimizes inventory levels, and improves logistics for streamlined operations.
Safety Enhancement	Monitors workplace safety, identifies risks, and implements preventive measures to reduce accidents and hazards.
Decision Support Systems	Provides data-driven insights for better decision-making, aiding production planning, and resource optimization.
Customization/Personalization	Facilitates tailored production based on consumer preferences, fostering customer satisfaction and loyalty.
Energy Efficiency	Optimizes energy consumption, adopting eco-friendly practices to reduce costs and environmental impact.
Adaptability/Flexibility	Enables agile responses to market changes, adjusting production processes swiftly to meet evolving demands.
Continuous Improvement	Drives innovation and competitiveness by leveraging insights for ongoing process enhancement and innovation.
Cost Reduction	Lowers operational costs by minimizing wastage, optimizing resources, and improving overall production efficiency.

make decisions by putting their knowledge into computer rules. This let the program think logically, making results that looked like what a human expert would do. Then, we got methods like evolutionary algorithms, where systems can find answers by themselves while getting better at doing a task. Lately, ML, especially deep learning, has become extremely popular. These AI/ML systems are a mix of different ways of doing things, not just one.

7.3.1 Machine Learning Approaches

ML, an array of algorithms and models, possesses the capability to assimilate patterns and execute decisions tailored to solving specific tasks by leveraging pertinent data. The construction of ML-based software involves sourcing task-related datasets, termed training data, followed by the meticulous selection of a fitting ML model [9, 10]. Subsequently, the chosen ML model undergoes a training process, thereby enabling it to fulfill the task at hand. Broadly speaking [11], ML can be categorized into three primary learning paradigms: supervised learning, unsupervised learning, and reinforcement learning.

As depicted in Figure 7.3, various machine learning models have the flexibility to integrate one or multiple learning paradigms tailored to a specific learning task. The representation in Figure 7.1, illustrating the interplay between learning paradigms,

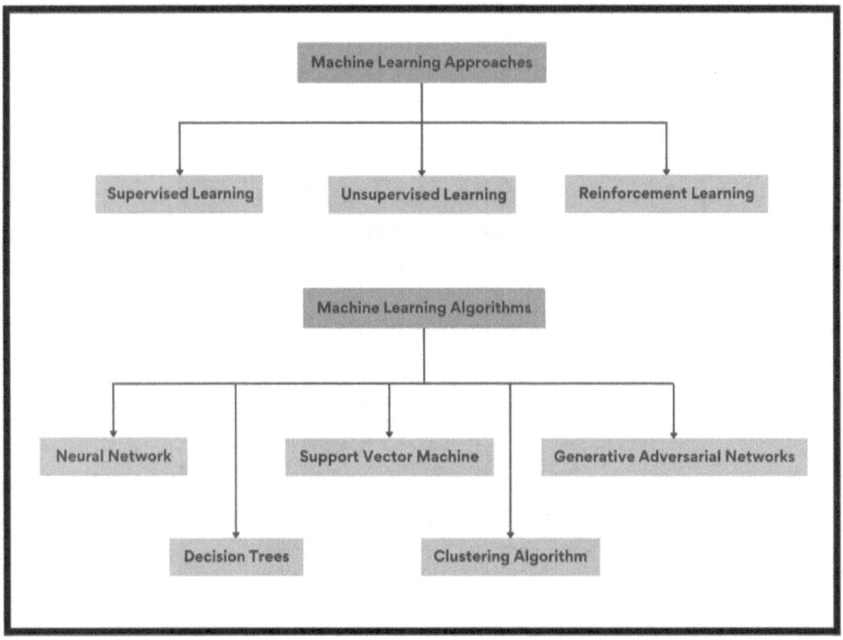

FIGURE 7.1 Machine Learning Aspects are Broadly Categorized by Their Underlying Models, Execution Methods, Specific Goals, and Applications in the Context of Manufacturing.

learning models, and tasks, is derived from a synthesis of information gleaned from technical articles and the expertise of the authors. The amalgamation of different learning tasks can effectively underpin a diverse array of applications. It's worth noting that, while the illustration provides a framework for understanding, there isn't a singular standardized approach for categorizing ML paradigms and techniques; alternative classifications also exist.

7.3.1.1 Supervised Learning: Learning with a Teacher

In supervised learning, machine learning algorithms are trained using carefully labeled data—akin to a teacher guiding a student [12]. This data includes both input features (characteristics of the data) and corresponding labels, which can be either categories (like "cat" or "dog") or continuous values (like temperature). During training, the model diligently analyzes this labeled data, learning to accurately predict the correct label for any given set of input features. Once trained, the model can then confidently tackle new, unlabeled data, predicting their labels with the knowledge it has acquired.

Here are some options for heading "Real-World Applications of Supervised Learning," incorporating images for enhanced content:

1. Image recognition: Supervised models power facial recognition on your phone, self-driving cars navigating roads, and medical imaging analysis.
2. Spam filtering: They keep your inbox tidy by learning to identify and block unwanted emails based on past examples.
3. Personalized recommendations: From movies to music, these models suggest what you'll love based on your past preferences and viewing habits.
4. Fraud detection: Banks and credit card companies utilize supervised learning to flag suspicious transactions in real-time, protecting your finances.

7.3.1.2 Unsupervised Learning: Learning on Its Own

Unlike the neatly labeled classrooms of supervised learning, unsupervised learning thrives in the wild frontiers of uncategorized data. This adventurous approach throws away the textbooks and lets machines explore, discover, and unveil patterns hidden within the raw information. Forget pre-existing labels and categories—unsupervised learning sifts through uncharted territories, seeking connections and similarities where none seem apparent.

These remarkable skills translate into real-world applications:

- Recommender systems: Unsupervised learning suggests movies and music that resonate with your preferences, weaving a tapestry of entertainment based on hidden data connections.
- Fraud detection: Unusual financial transactions are easily identified, acting as a silent guardian against malicious activity.
- Image compression: Pictures shrink in size while preserving key features, thanks to the data-compressing prowess of unsupervised models.
- Bioinformatics: Genes with similar expression patterns are grouped together, unlocking secrets of their functions and interactions.

7.3.1.3 Reinforcement Learning: Learning from Trial and Reward

Unlike students cramming for exams, reinforcement learning models embark on a self-directed journey, exploring and interacting with their environment [13]. They learn through a continuous feedback loop, trial and error guided by rewards. Imagine a child playing a video game, navigating challenges and receiving points for each level conquered. Reinforcement learning models operate similarly, constantly adjusting their actions based on rewards, ultimately aiming to maximize their "score" in the real world.

These are just a few points of how reinforcement learning is changing the game across diverse fields:

1. Robotic Champions: Trains robots to walk, climb, and manipulate objects with human-like skill, opening doors to advanced automation and dexterous tasks.
2. Game Over for Humans: Creates AI players that master complex games like Go and StarCraft, pushing the boundaries of strategic intelligence and decision-making.
3. Smarter Systems: Optimizes resource management in energy grids and logistics, learning from past performance to improve efficiency and reduce waste.
4. Personalized Healthcare: Develops adaptive treatment plans for patients, tailoring therapy based on individual responses and maximizing healing outcomes.

With its focus on continuous learning and adaptation, it holds the key to a future where intelligent machines work alongside us, making our lives easier, safer, and more efficient.

7.3.2 Machine Learning Algorithms

Imagine AI as a skilled craftsman, and its tools are machine learning techniques. These diverse algorithms and methods, like saws and chisels, help AI learn from data within different learning styles. Though most tools have a favorite style, some, like neural networks, are versatile masters, tackling supervised, unsupervised, and even reinforcement tasks. Choosing the right tool for the job, be it understanding images, organizing data, or mastering games, depends on the task and the data at hand. This toolbox of techniques is constantly evolving, shaping the future of AI and its impact on our world.

Machine learning techniques exhibit varying degrees of specialization, highlighting the importance of careful technique selection for optimal performance.

Decision trees, for example, excel within the supervised learning paradigm, effectively navigating complex decision paths to uncover patterns within labeled data. However, other techniques, such as neural networks, demonstrate remarkable adaptability, capable of learning across all three paradigms (supervised, unsupervised, and reinforcement learning). This versatility underscores their widespread adoption in diverse domains, including computer vision and natural language processing. While multiple techniques may be applicable to a given problem, performance often varies significantly depending on the specific task and data characteristics. Neural networks, for instance, have emerged as the dominant force in computer vision, adeptly recognizing patterns within images and videos. In contrast, decision trees have demonstrated superior performance in regression tasks involving tabular data, effectively capturing relationships between numerical variables. The ability of machine learning techniques to ingest diverse data types, including numerical, categorical, textual, and even image-based data, further expands their applicability across a multitude of domains. This flexibility, coupled with their capacity to handle data of varying dimensionality, positions machine learning as a powerful tool for knowledge discovery and decision-making in a wide range of scientific and industrial pursuits.

7.3.2.1 Neural Networks

Neural networks, the brainchildren of AI, weave intricate webs of interconnected nodes, mimicking the architecture of the human brain. These nodes, like tiny neurons, process information by receiving and transmitting signals to their neighbours. The strength of these connections, akin to synapses, determine how information flows through the network. Layers upon layers of these nodes build a complex tapestry, trained on mountains of data [14]. As information ripples through the network, the connections adapt, strengthening or weakening, allowing the neural network to learn and recognize patterns. Just like a toddler learning to identify a cat, the network adjusts its internal connections until its "picture" of a cat matches the real world. This learning process, called backpropagation, allows the network to refine its responses and tackle increasingly complex tasks, from recognizing faces in photos to crafting compelling musical pieces. So, neural networks are not just fancy algorithms; they are dynamic, learning machines, constantly evolving and refining their understanding of the world, one connection at a time.

7.3.2.2 Decision Trees

Decision trees, unlike intricate neural networks, don't mimic brains, but rather, maps. Imagine navigating a dense forest: at each fork, you ask a question to choose the right path. "Is it sunny?" You branch left if yes; right if no [15]. Each answer leads to another question, forming a tree-like structure with leaves as final destinations. Decision trees work just like that, splitting data based on simple yes/no questions about its features. Is the picture pixelated? Is the email text formal? Each answer guides the data down a specific branch, ultimately reaching a leaf labeled "cat image," "spam," or any other desired outcome [16]. These trees learn by analyzing training data, automatically choosing the questions that best divide the data into distinct groups. Their transparency and simplicity make them perfect for tasks like medical diagnosis or loan approval, where understanding the reasoning behind the decision is crucial. So, next time you're stuck in a dilemma, remember the humble decision tree—a powerful guide navigating complex choices, one split at a time.

7.3.2.3 Support Vector Machines

Support Vector Machines (SVMs) are powerful supervised-learning algorithms known for their accuracy in classification tasks [17]. They operate by strategically drawing hyperplanes—high-dimensional boundaries—between different classes of data points. Instead of simply minimizing misclassification errors, SVMs prioritize maximizing the margin between these hyperplanes, creating a clear buffer zone between classes. This approach leads to robust models that generalize well to unseen data, making them particularly effective for complex, high-dimensional datasets. SVMs find broad applications in various domains, including image recognition, spam filtering, and even medical diagnosis, where their ability to handle intricate data and maintain clear class distinctions shines through. So, when navigating the complexities of data classification, SVMs stand as champions of accuracy and clarity, meticulously carving out distinct territories within the information landscape.

7.3.2.4 Clustering Algorithms

Clustering algorithms, the detectives of data analysis, unearth hidden patterns and groupings within a seemingly chaotic collection of information [18]. Like a naturalist sorting specimens into families and species, these algorithms methodically sift through data points, seeking those that share common traits and whispering, "You belong together." Their guiding principle is simple: similarity breeds proximity. Data points that dance in sync, echoing each other's attributes, are likely kin, drawn together by invisible threads of shared characteristics. Some algorithms, like K-means, employ a straightforward approach, assigning data points to distinct clusters based on their closeness to designated cluster centers. Others, like hierarchical clustering, construct intricate family trees, revealing nested relationships and layers of kinship within the data. Whether mapping customer preferences in marketing or identifying genetic patterns in biology, clustering algorithms act as silent cartographers, sketching the hidden landscapes of similarity within vast datasets, revealing insights that might otherwise remain unseen.

7.3.2.5 Generative Adversarial Networks

Generative Adversarial Networks (GANs) are a revolutionary class of machine learning models introduced in 2014, consisting of two neural networks—the generator and the discriminator—engaged in adversarial training. The generator creates synthetic data by transforming random noise into increasingly realistic samples, while the discriminator learns to distinguish between real and generated data [19]. Through iterative back-and-forth training, the generator refines its output to approach authenticity, and the discriminator becomes more adept at discerning real from fake. GANs have demonstrated remarkable success in generating diverse content, from realistic images to music and text, impacting fields such as image synthesis, style transfer, and data augmentation. Despite their achievements, challenges such as mode collapse and training instability persist, stimulating ongoing research to enhance the robustness and applicability of GANs across various domains.

7.4 MACHINE LEARNING WORKFLOW

Building an intelligent solution through machine learning is like embarking on a grand quest for hidden knowledge [17]. Each step in this journey plays a crucial role in transforming raw data into insightful predictions. Let's unravel the tapestry of these essential steps:

1. Data Collection: The Foundation of Insight
   Every great journey begins with a map, and for machine learning, that map is data. This first step involves gathering relevant information that reflects the problem you aim to solve. Is it customer reviews for sentiment analysis? Sensor readings for predictive maintenance? Identifying the right data sources and accumulating sufficient volume is akin to gathering resources for your expedition.
2. Data Processing: From Chaos to Clarity
   Raw data is rarely ready for the journey—it's often messy, incomplete, and inconsistent. Data processing acts as the cartographer, cleaning and refining the collected information. This involves identifying and filling in missing values, removing inconsistencies, and standardizing formats. Think of it as clearing overgrown paths and mapping hidden connections within the data.
3. Model Selection: Choosing your Steed
   With a prepared map in hand, it's time to choose your companion for the journey: the machine learning model. Different models excel in different terrains—some, like decision trees, prefer quick exploration, while neural networks navigate complex landscapes with masterful precision. Selecting the right model depends on the nature of your task and the characteristics of your data.
4. Training and Validation: The Path to Expertise
   Now, we embark on the heart of the quest: training the model. Feeding it processed data, we watch it learn and adapt, building its internal map of patterns and relationships. But knowledge must be tested to be trusted. Validation takes the trained model for a trial run on unseen data, assessing its accuracy and generalizability. Think of it as venturing into uncharted territory to see if the learned map holds true.
5. Evaluation and Tuning: Refining the Compass
   Validation might reveal limitations, and that's okay! Evaluation scrutinizes the model's performance, uncovering areas for improvement. Metrics like accuracy, precision, and recall become our measuring sticks, guiding us toward fine-tuning the model's parameters or even exploring alternative architectures. Just like a seasoned explorer adjusting their compass based on newfound landmarks, this iterative process refines the model until it navigates the data landscape with unmatched precision.
6. The Final Model: Reaching the Summit
   Finally, we arrive at the peak of the journey—the final model. This potent tool, imbued with the knowledge gleaned from data and refined through rigorous training, is ready to tackle real-world challenges. Whether it's predicting customer churn, optimizing factory processes, or identifying fraudulent transactions, the final model becomes the ultimate guide, transforming data into actionable insights.

Remember, this journey is rarely linear. Steps may overlap, detours may be necessary, and revisiting previous stages is often crucial. But the beauty of machine learning lies in this dynamic dance between data, models, and human ingenuity. It's a process of discovery, where each step paves the way for deeper understanding and more powerful solutions.

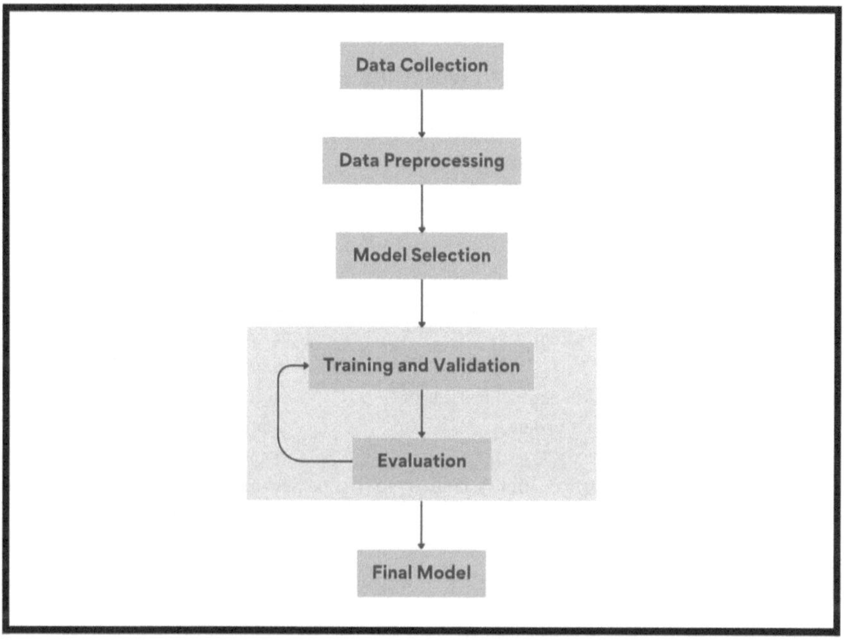

FIGURE 7.2 General Workflow for Developing the ML Model for an AI/ML Solution.

7.4.1 Machine Learning Task

In machine learning, a task refers to the specific goal or objective that the learning algorithm is trying to achieve. These tasks can be broadly categorized into different types, each with its own challenges and applications. Table 7.2 provides some of the most common machine learning tasks with description and examples.

7.5 AI/ML APPLICATIONS IN MANUFACTURING

Artificial Intelligence (AI) and Machine Learning (ML) are transforming the landscape of manufacturing, offering innovative solutions that enhance efficiency, productivity, and decision-making processes. One key application lies in predictive maintenance, where AI algorithms analyze equipment sensor data to predict potential failures before they occur. By identifying anomalies and patterns, manufacturers can schedule maintenance proactively, minimizing downtime and reducing costs associated with unexpected breakdowns.

Another significant application is in quality control. AI and ML algorithms can analyze vast amounts of production data to detect defects or deviations from quality standards. This allows for real-time adjustments in manufacturing processes, ensuring that products meet specifications and reducing the likelihood of defective items reaching consumers. Computer vision technologies, often powered by deep learning, play a crucial role in inspecting and classifying visual defects, enhancing the accuracy of quality control processes.

TABLE 7.2
Machine Learning Task Overview: Descriptions and Examples

Task	Description	Examples
Regression	Predicts continuous numerical values	Predicting house prices, stock prices, sales figures
Classification	Assigns data points to predefined categories	Categorizing emails as spam or not spam, identifying objects in images, diagnosing diseases
Computer Vision	Analyzes and understands visual information	Image classification, object detection, image segmentation, facial recognition, self-driving cars
Anomaly Detection	Identifies unusual patterns or outliers in data	Detecting fraudulent transactions, identifying manufacturing defects, monitoring network security
Control Systems	Learns to control systems or processes	Self-driving cars, robotics, industrial process optimization, energy management
Generative Models	Creates new data instances that resemble the original data	Generating realistic images, music, text, speech, drug discovery, product design
Natural Language Processing (NLP)	Processes and understands human language	Sentiment analysis, machine translation, text summarization, question answering, chatbots

Supply chain optimization is also being revolutionized by AI and ML. These technologies enable the analysis of complex supply chain data, helping manufacturers make data-driven decisions regarding inventory management, demand forecasting, and logistics. Predictive analytics models can anticipate fluctuations in demand, optimize inventory levels, and enhance overall supply chain efficiency. This not only reduces operational costs but also improves customer satisfaction by ensuring timely delivery of products.

Furthermore, AI-driven robotics is reshaping the manufacturing landscape. Collaborative robots, or cobots, work alongside human workers, automating repetitive tasks and enhancing overall production efficiency. These robots can adapt to changes in their environment, making them versatile assets in dynamic manufacturing settings. AI algorithms enable them to learn from human interaction, improving their capabilities over time and contributing to a more flexible and responsive manufacturing environment.

In the realm of design and prototyping, AI is making notable contributions. Generative design, a technique that utilizes AI algorithms to explore and generate multiple design options based on specified criteria, is streamlining the product development process. This not only accelerates innovation but also results in more optimized and efficient designs. Additionally, ML algorithms can analyze data from previous designs and iterations, facilitating continuous improvement in product development processes.

AI and ML applications in manufacturing are diverse and transformative. From predictive maintenance and quality control to supply chain optimization, collaborative robotics, and generative design, these technologies are driving efficiency, reducing costs, and paving the way for a more agile and intelligent manufacturing industry. As advancements continue, the integration of AI and ML is poised to redefine traditional manufacturing practices and unlock new possibilities for innovation and growth.

A comprehensive overview of AI/ML applications in the production industry, incorporating descriptions [20] that can aid in visualizing key concepts:

1. Predictive Maintenance:
   - Goal: Anticipate equipment failures before they occur, reducing downtime and costs.
   - Methods: ML models analyze sensor data (vibration, temperature, etc.) to detect anomalies and predict impending failures.
   - Diagram: Imagine a machine with sensors feeding data into an ML model. The model generates alerts when it detects patterns indicating potential issues.
2. Quality Control:
   - Goal: Identify defects in products during manufacturing to ensure quality standards.

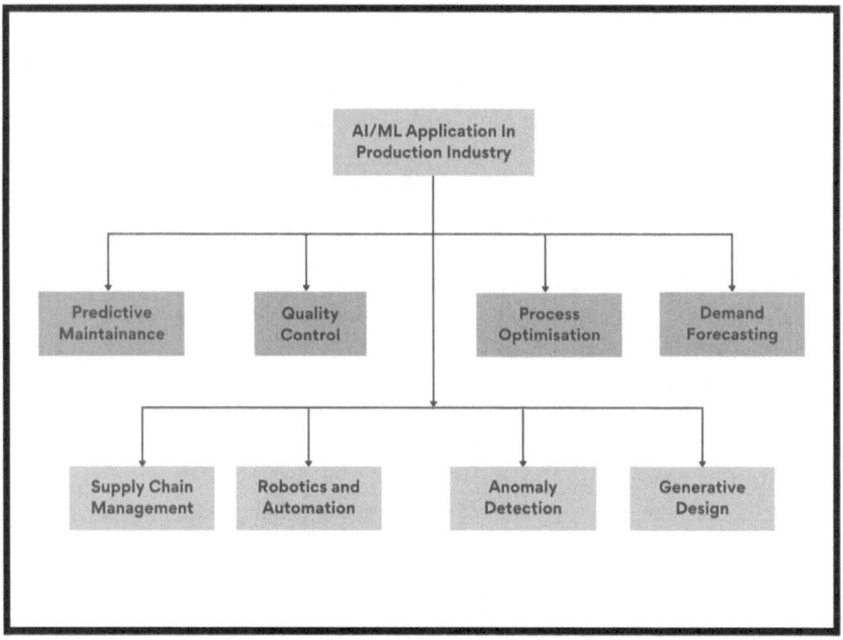

FIGURE 7.3 Representative AI/ML Applications in the Manufacturing Industry.

- Methods: Computer vision and image classification algorithms inspect products for visual defects. ML models analyze process data to detect anomalies in quality metrics.
- Diagram: A production line with cameras capturing images of products, analyzed in real-time by ML models to pinpoint defects.

3. Process Optimization:
   - Goal: Improve production efficiency, throughput, and resource utilization.
   - Methods: ML models analyze historical production data to identify bottlenecks, optimize machine settings, and streamline processes.
   - Diagram: A factory dashboard displaying production data, with ML-powered recommendations for process adjustments.
4. Demand Forecasting:
   - Goal: Accurately predict future demand to optimize inventory levels and reduce waste.
   - Methods: ML models analyze sales data, market trends, and external factors to create demand forecasts.
   - Diagram: A graph showing historical sales data and ML-generated projections for future demand.
5. Supply Chain Management:
   - Goal: Optimize supply chain operations, including logistics, inventory management, and supplier relationships.
   - Methods: ML models predict demand, manage inventory levels, optimize delivery routes, and select suppliers.
   - Diagram: A network of interconnected suppliers, production facilities, and distribution centers, with ML algorithms coordinating operations.
6. Robotics and Automation:
   - Goal: Automate tasks, improve productivity, and reduce labor costs.
   - Methods: AI-powered robots perform tasks autonomously, guided by ML models for perception, decision-making, and control.
   - Diagram: A robotic arm assembling components, guided by computer vision and ML algorithms.
7. Anomaly Detection:
   - Goal: Identify unusual patterns or events in production data to prevent safety issues, quality problems, or downtime.
   - Methods: ML models learn normal patterns and flag deviations as potential anomalies.
   - Diagram: A chart with normal production data patterns and ML-highlighted anomalies.
8. Generative Design:
   - Goal: Create new product designs or optimize existing ones using AI algorithms.
   - Methods: AI algorithms explore a vast design space and generate novel solutions based on constraints and objectives.
   - Diagram: An interactive design tool generating product variations based on AI-driven exploration.

7.6 FUTURE OF AI APPLICATIONS IN PRODUCTION

As we peer into the horizon of production, the trajectory of AI applications unveils a landscape rich with transformative potential. The future promises an era where artificial intelligence becomes the linchpin reshaping manufacturing methodologies across sectors. Advanced predictive capabilities are set to redefine maintenance strategies, with machine learning algorithms evolving to foresee equipment failures with unprecedented accuracy. This anticipatory maintenance not only averts downtime but also slashes operational costs, heralding an era of optimized productivity.

Quality control stands poised for a paradigm shift, driven by augmented computer vision and deep learning. These technologies, honed to a level of acute precision, enable real-time defect detection, elevating product quality to previously unattainable levels. Furthermore, the impending rise of adaptive manufacturing, fueled by AI, assures customization without compromising efficiency, ushering in an era of personalized production at scale.

The symbiotic fusion of AI and robotics is a cornerstone of the future factory floor. Collaborative robots, fortified with advanced AI algorithms, will seamlessly coexist with human counterparts, enhancing productivity, agility, and precision. These AI-driven automatons will navigate complex tasks with dexterity, dynamically adapting to evolving environments and processes, heralding an era of optimized manufacturing efficiency.

Sustainability takes center stage as AI assumes a pivotal role in shaping environmentally conscious manufacturing practices. Its prowess in energy consumption optimization, waste reduction, and eco-friendly processes will catapult industries toward carbon neutrality. Leveraging predictive analytics, AI will forecast energy usage, identifying avenues for efficiency improvements, setting the stage for sustainable production paradigms.

Supply chain dynamics undergo a transformative evolution, courtesy of AI's predictive prowess and real-time data analytics. The orchestration of efficient inventory control, precise demand forecasting, and adaptive logistics ushers in an era of streamlined operations. Minimized stockouts, optimized distribution networks, and responsive supply chains become the new norm, underscoring AI's transformative impact.

The metamorphosis of work dynamics marks a pivotal inflection point as human-AI collaboration shapes future job roles. Workforces undergo reskilling initiatives to complement and augment AI capabilities, fostering symbiotic partnerships that leverage the best of both worlds to propel productivity and innovation.

Yet, within this canvas of transformative possibilities, challenges loom. Ethical considerations, data privacy concerns, and regulatory compliance emerge as critical hurdles to navigate. Balancing the promise of AI with ethical considerations and ensuring bias-free algorithms becomes imperative for sustained trust and transparency. In conclusion, the future of AI applications in production is a tapestry woven with innovation and transformation. As industries embrace AI, overcoming challenges through collaboration, ethical deployment, and technological integration becomes paramount [21]. This future heralds an era of smarter, more sustainable, and agile manufacturing, where the convergence of human expertise and AI capabilities propels us toward unprecedented heights of productivity, efficiency, and innovation.

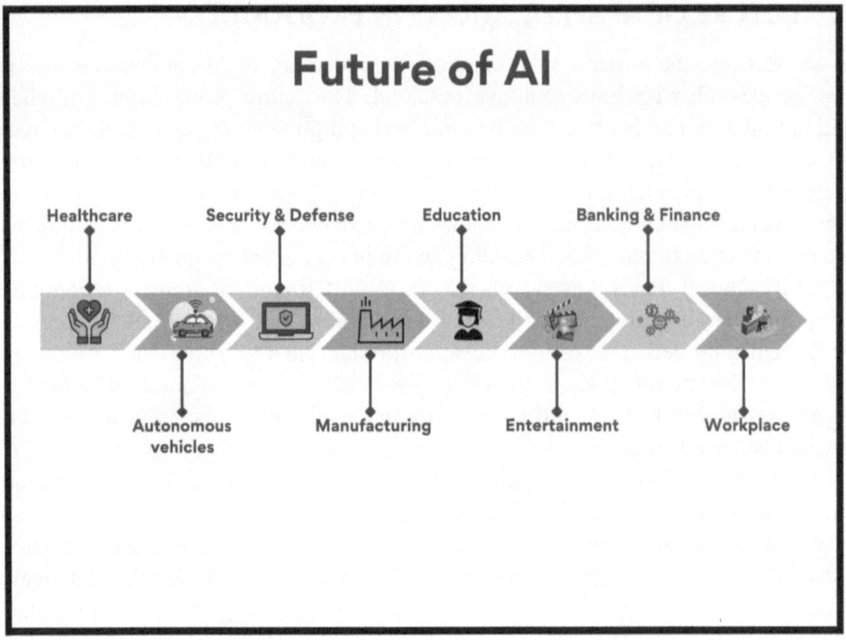

FIGURE 7.4 Future of AI Applications.

7.7 CASE STUDIES

This section highlights the real-world here are some real-world case studies demonstrating the application of AI in production settings.

Here are ten case studies [22] highlighting the advantages and challenges of AI applications in production:

1. Predictive Maintenance (Siemens, Rolls-Royce)
   - Advantage: AI models analyze sensor data from machines to predict failures before they happen, minimizing downtime and maintenance costs.
   - Challenge: Data quality and quantity are crucial for accurate predictions. Implementing sensors and collecting meaningful data can be expensive.
   - Explanation: Siemens uses AI to predict wind turbine component failures months in advance, boosting uptime and saving €1 billion annually. Rolls-Royce's AI software detects engine anomalies, enabling preventative maintenance and reducing unplanned downtime by 50%.
2. Quality Control (Nestlé, Unilever)
   - Advantage: AI-powered visual inspection systems detect product defects with human-like accuracy and speed, improving product quality and reducing waste.
   - Challenge: Training AI models on diverse datasets is crucial for generalizability. Integrating AI with existing production lines can be challenging.

- Explanation: Nestlé uses AI to identify imperfections in chocolates, reducing waste by 30%. Unilever's AI scans for defects in soap bars, increasing quality by 25% and reducing scrap by 15%.

3. Demand Forecasting (Walmart, Amazon)
   - Advantage: AI analyzes sales data and external factors to predict product demand with greater accuracy, optimizing inventory levels and reducing stockouts and overstocking.
   - Challenge: Incorporating external factors like weather and social media trends can be complex. Balancing short-term accuracy with long-term trends is crucial.
   - Explanation: Walmart uses AI to forecast demand for perishable goods, reducing waste by 50% and increasing on-shelf availability by 10%. Amazon's AI predicts product demand with 95% accuracy, minimizing overstocking and maximizing profitability.
4. Process Optimization (BASF, Toyota)
   - Advantage: AI analyzes production data to identify inefficiencies and suggest real-time adjustments, optimizing resource usage and increasing throughput.
   - Challenge: Integrating AI with existing control systems and ensuring cybersecurity are crucial. Explaining AI recommendations to human operators can be challenging.
   - Explanation: BASF uses AI to optimize chemical production processes, reducing energy consumption by 15% and increasing output by 5%. Toyota's AI optimizes car painting robots, reducing paint waste by 20% and improving finish quality.
5. Smart Supply Chains (Maersk, DHL)
   - Advantage: AI analyzes logistics data to optimize routes, predict delays, and improve overall supply chain efficiency and visibility.
   - Challenge: Collaboration between different companies and integration with disparate logistics systems are crucial. Dealing with unpredictable external factors like weather and political unrest can be complex.
   - Explanation: Maersk uses AI to predict container ship arrival times with 98% accuracy, improving port efficiency and reducing delays. DHL's AI optimizes delivery routes, reducing fuel consumption by 10% and delivery times by 5%.
6. Generative Design (Adidas, Airbus)
   - Advantage: AI generates lightweight, high-performance designs for products, optimizing materials and reducing manufacturing costs.
   - Challenge: Balancing creativity with engineering constraints and ensuring manufacturability of AI-generated designs is crucial. Convincing stakeholders of the benefits of AI-driven design can be challenging.
   - Explanation: Adidas uses AI to design lightweight running shoes, improving performance and reducing material usage by 30%. Airbus' AI optimizes aircraft component designs, reducing weight and fuel consumption by 15%.
7. Robot-Assisted Manufacturing (Ford, Honda)
   - Advantage: AI-powered robots collaborate with human workers on complex tasks, improving productivity, safety, and precision.

- Challenge: Ensuring seamless human-robot interaction and upskilling workers for new roles is crucial. Addressing ethical concerns about job displacement is important.
- Explanation: Ford uses AI-powered robots for welding and painting tasks, increasing productivity by 20% and reducing worker injuries. Honda's AI cobots assist workers in assembly lines, improving precision and reducing fatigue.

8. Personalized Production (Nike, Coca-Cola)
   - Advantage: AI analyzes customer data and preferences to personalize products and optimize production based on real-time demand.
   - Challenge: Data privacy concerns and ensuring ethical use of personalization algorithms are crucial. Balancing mass production efficiency with personalized customization can be complex.
   - Explanation: Nike uses AI to create custom running shoes based on individual biomechanics, improving performance and comfort. Coca-Cola's AI personalizes drink flavors and packaging based on regional preferences, boosting sales and reducing waste.
9. Sustainability Optimization (Unilever, Patagonia)
   - Advantage: AI analyzes energy consumption and resource usage to identify and implement sustainable practices in production, minimizing environmental impact.
   - Challenge: Integrating AI with existing sustainability initiatives and measuring the overall impact of AI-driven optimization are crucial.
10. AI-powered Defect Detection in Aerospace Manufacturing (Boeing):
    - Challenge: Inspecting complex aircraft components for minute defects, often hidden beneath layers of paint, is time-consuming and prone to human error.
    - Solution: Boeing utilizes AI-powered visual inspection systems that analyze high-resolution images to identify even the smallest flaws with near-perfect accuracy. This significantly reduces inspection time, improves defect detection rates, and prevents potentially catastrophic failures.
    - Benefits: Enhanced safety, reduced downtime due to maintenance, and improved production efficiency.
11. Optimized Yield Management in Agriculture (John Deere):
    - Challenge: Traditional farming practices often lack precision, leading to uneven crop yields and resource waste.
    - Solution: John Deere has developed AI-based systems that collect and analyze data from farm equipment and environmental sensors. This data allows farmers to optimize irrigation, fertilization, and pest control measures for each specific field and plant variety, leading to higher yields and reduced resource consumption.
    - Benefits: Increased food production, reduced environmental impact, and improved profitability for farmers.
12. Personalized Production in Fashion (Levi Strauss):
    - Challenge: The traditional one-size-fits-all approach to clothing often leads to ill-fitting garments and customer dissatisfaction.

- Solution: Levi Strauss is using AI to personalize jeans based on individual body measurements and preferences. Customers can use a mobile app to input their desired fit and style, and AI algorithms then adjust the pattern and design of the jeans accordingly.
- Benefits: Improved customer satisfaction, reduced inventory waste, and potential for higher profit margins through customization.

13. Automated Inventory Management in Retail (Walmart):
    - Challenge: Maintaining optimal inventory levels can be tricky for retailers, often leading to stockouts or overstocking.
    - Solution: Walmart utilizes AI-driven demand forecasting systems that analyze sales data, weather patterns, and social media trends to predict future demand with greater accuracy. This allows them to automatically adjust inventory levels for each store, ensuring optimal product availability and minimizing waste.
    - Benefits: Reduced costs associated with inventory management, improved customer satisfaction, and increased sales.
14. Collaborative Robots in Logistics (Amazon):
    - Challenge: Meeting the ever-increasing demand for fast and efficient delivery can be physically demanding for warehouse workers.
    - Solution: Amazon utilizes collaborative robots (cobots) alongside human workers in its fulfillment centers. These cobots assist with tasks like sorting and picking items, reducing physical strain on workers and improving overall throughput.
    - Benefits: Increased productivity, improved employee safety and satisfaction, and enhanced operational efficiency.

These are just a few examples of how various industries are leveraging AI to revolutionize their production processes. By harnessing the power of data and intelligent algorithms, companies can streamline operations, optimize resource allocation, and drive efficiency, ultimately leading to improved product quality, customer satisfaction, and business success.

These case studies exemplify how various industries harness AI technologies to streamline operations, enhance quality, optimize resource allocation, and drive efficiency in production processes.

7.8 CONCLUSION

As the final curtain falls on this chapter, we find ourselves perched on the precipice of a momentous transformation—the AI/ML-powered metamorphosis of the manufacturing industry. This review has been a voyage through the fertile fields of opportunity, showcasing the transformative potential of these technologies across the production landscape. From bolstering safety and efficiency to optimizing resource allocation and sustainability, AI/ML whispers the promise of a manufacturing renaissance.

We've delved into the intricate workings of the AI/ML pipeline, dissecting its applications in the crucible of the manufacturing process. From orchestrating the dance of operations and planning to ensure the flawless ballet of quality assurance,

AI/ML conducts the symphony of production with an invisible baton. Its keen eyes, fed by the torrents of data gushing from industrial sensors, discern patterns and predict pitfalls, allowing us to dodge operational disruptions and craft optimal outputs.

However, even amidst the shimmering possibilities, shadows lurking in the corners demand our attention. The specter of restructuring, fluctuating energy costs, and the need for specialized expertise cast their pall, reminding us that this revolution must be navigated with a steady hand. Embracing AI/ML while respecting the unique needs and skillsets of both company and human stakeholders is the delicate dance we must learn.

This dance holds within it a fascinating rhythm—a tango between human and machine. The future we envision is not one where humanity bows to the cold logic of silicon, but rather, one where our skills amplify the power of AI/ML. This collaboration, this intricate interplay of human intuition and algorithmic precision, is the beating heart of the industry's future.

As the AI/ML orchestra tunes its instruments, the tempo of adoption quickens. Driven by the irresistible melody of improved safety, product quality, and operational efficiency, manufacturers are drawn to the stage. But the crescendo will undoubtedly be muted by the whispers of risk, especially as the baton of control shifts from human hands to AI/ML algorithms. Trust, forged in the crucible of success stories, will be the key to unlocking the full potential of this technological symphony.

Yet, amidst the uncertainties, one truth remains undimmed: the revolution is at hand. As AI/ML algorithms evolve at breakneck pace, their application in manufacturing will become inevitable, driven by the siren song of improved production

TABLE 7.3
Decoding Abbreviations: A Reference Table for Notations and Meanings

Notation	Meaning
AI	Artificial Intelligence
ANI	Artificial Narrow Intelligence
ANN	Artificial Neural Network
AutoML	Automated Machine Learning
CNN	Convolution Neural Network
GAN	Generative Adversarial Network
DL	Deep Learning
DNN	Deep Neural Network
DRL	Deep Reinforcement Learning
FLOPS	Floating Point Operations Per Second
IoT	Internet of Things
ML	Machine Learning
NLP	Natural Language Processing
NN	Neural Network
RL	Reinforcement Learning
SaaS	Software as a Service
SVM	Support Vector Machine

outcomes. With each successful implementation, the chorus of voices singing AI/ML's praises will swell, echoing across the industrial landscape.

In the closing bars of this chapter, we leave you with a final question: will your company be in the audience, marveling at the spectacle, or will you be on stage, conducting the transformative power of AI/ML and shaping the future of manufacturing? The choice, dear reader, is yours. ☺

REFERENCES

[1] J. Zhou, P. Li, Y. Zhou, B. Wang, J. Zang, L. Meng, Toward New-Generation Intelligent Manufacturing, Engineering, 4(1), 11–20 (2018). ISSN 2095-8099, https://doi.org/10.1016/j.eng.2018.01.002.

[2] R. Y. Zhong, X. Xu, E. Klotz, S. T. Newman, Intelligent Manufacturing in the Context of Industry 4.0: A Review, Engineering, 3(5), 616–630 (2017). ISSN 2095-8099, https://doi.org/10.1016/J.ENG.2017.05.015.

[3] S. K. Jagatheesaperumal, M. Rahouti, K. Ahmad, A. Al-Fuqaha, M. Guizani, The Duo of Artificial Intelligence and BigData for Industry 4.0: Review of Applications, Techniques, Challenges, and Future Research Directions (accessed: July, 2021).

[4] R. Geissbauer, S. Schrauf, P. Berttram, F. Cheraghi, Digital Factories 2020: Shaping the Future of Manufacturing, Pricewater House Coopers, 2017 (accessed: June 2021).

[5] P. Brosset, A. L. Thieullent, S. Patsko, P. Ravix, Scaling AI in Manufacturing Operations: A Practitioners' Perspective, Capgemini Research Institute, Paris, 2019 (accessed: June 2021).

[6] S. Fahle, C. Prinz, B. Kuhlenkötter, B., Systematic Review on Machine Learning (ML) Methods for Manufacturing Processes – Identifying Artificial Intelligence (AI) Methods for Field Application, Procedia CIRP, 93, 413–418 (2020). ISSN 2212-8271, https://doi.org/10.1016/j.procir.2020.04.109.

[7] R. Cioffi, M. Travaglioni, G. Piscitelli, A. Petrillo, F. De Felice, Artificial Intelligence and Machine Learning Applications in Smart Production: Progress, Trends, and Directions, Sustainability, 12(2), 492 (2020). https://doi.org/10.3390/su12020492

[8] A. Rizzoli, 7 Out-of-the-Box Applications of AI in Manufacturing, V7 Labs Blog (2022). https://doi.org/10.1002/amp2.10159

[9] https://en.wikipedia.org/wiki/Machine_learning#Theory

[10] Amazon (AWS), Training ML Models – Amazon Machine Learning, Amazon (AWS), 2022.

[11] S. J. Plathottam, A. Rzonca, R. Lakhnori, C. O. Iloeje, A Review of Artificial Intelligence Applications in Manufacturing Operations, Journal of Advanced Manufacturing and Processing, AIChE, 5. https://doi.org/10.1002/amp2.10159

[12] IBM Cloud Education, What Is Supervised Learning? IBMCloud Education, 2021 (accessed: April 2022).

[13] P. Henderson, R. Islam, M. G. Bellemare, J. Pineau, An Introduction to Deep Reinforcement Learning. *ArXiv* (2018). https://doi.org/10.1561/2200000071

[14] https://medium.com/@arti.singh280/list/the-quantum-world-6126d55e1882

[15] Lucidchart, What Is a Decision Tree Diagram, Lucidchart, 2022 (accessed: April, 2022).

[16] https://www.mastersindatascience.org/learning/machine-learning-algorithms/decision-tree/

[17] K. Arti, P. Dubey, S. Agrawal, An Opinion Mining for Indian Premier League Using Machine Learning Techniques, 2019 4th International Conference on Internet of Things: Smart Innovation and Usages (IoT-SIU), 2019.

[18] D. Xu, Y. Tian, A Comprehensive Survey of Clustering Algorithms. Ann. Data. Sci., 2, 165–193 (2015). https://doi.org/10.1007/s40745-015-0040-1
[19] A. Kusiak, Convolutional and Generative Adversarial Neural Networks in Manufacturing, International Journal of Production Research, IJPR, 58(5), 1594–1604 (2020). https://doi.org/10.1080/00207543.2019.1662133
[20] S. Ayvaz, K. Alpay, Predictive Maintenance System for Production Lines in Manufacturing: A Machine Learning Approach Using IoT Data in Real-Time, Expert Systems with Applications, 173, 114598 (2021). ISSN 0957-4174, https://doi.org/10.1016/j.eswa.2021.114598. https://www.sciencedirect.com/science/article/pii/S0957417421000397
[21] J. V. D. Waa, T. Schoonderwoerd, J. V. Diggelen, M. Neerincx, Interpretable Confidence Measures for Decision Support Systems, Int. J. Hum-Comput. Stud., 144, 102493 (June 2020). https://doi.org/10.1016/j.ijhcs.2020.102493
[22] Q. Wang, S. Zheng, A. Farahat, S. Serita, T. Saeki, C. Gupta, 2. International Joint Conference on Neural Networks, IJCNN 2019 Budapest, Hungary, July 14–19, IEEE, 2019.
[23] Plutoshift, Breaking Ground on Implementing AI: Instituting Strategic AI Programs – From Promise to Productivity, Plutoshift, Palo Alto, 2019 (accessed: May 2023).
[24] Voyant Tools, https://voyant-tools.org/ (accessed: March 2023).
[25] V. Kanade, Narrow AI vs. General AI vs. Super AI: Key Comparisons, SpiceWorks, 2022.
[26] Javapoint, Unsupervised Machine Learning–Javatpoint, 2022.
[27] IBM Cloud Team, Supervised vs. Unsupervised Learning: What's the Difference? IBM Cloud Team, 2021 (accessed: April, 2022).
[28] S. J. Plathottam, B. Richey, G. Curry, J. Cresko, C. O. Iloeje, J. Adv. Manuf. Process. 3(2), e10079 (2021).
[29] A. Mirhoseini, A. Goldie, M. Yazgan, J. Jiang, E. Songhori, S. Wang, Y.-J. Lee, E. Johnson, O. Pathak, S. Bae, A. Nazi, J. Pak, A. Tong, K. Srinivasa, W. Hang, E. Tuncer, A. Babu, Q. V. Le, J. Laudon, R. Ho, R. Carpenter, J. Dean, Chip Placement with Deep Reinforcement Learning (accessed: June 2022).
[30] S. Zheng, C. Gupta, S. Serita, Manufacturing Dispatching using Reinforcement and Transfer Learning (2019).
[31] S. Madhavan, M. T. Jones, Deep Learning Architectures–IBM Developer, IBM Developer Articles (2017).
[32] Fei TAO, Xuemin SUN, Jiangfeng CHENG, Yonghuai ZHU, Weiran LIU, Yong WANG, Hui XU, Tianliang HU, Xiaojun LIU, Tingyu LIU, Zheng SUN, Jun XU, Jinsong BAO, Feng XIANG, Xiaohui JIN, makeTwin: A reference architecture for digital twin software platform, Chinese Journal of Aeronautics, Volume 37, Issue 1, 2024, Pages 1–18, ISSN 1000-9361, https://doi.org/10.1016/j.cja.2023.05.002. https://www.sciencedirect.com/science/article/pii/S1000936123001541.
[33] B. H. Li, B. C. Hou, W. T. Yu, X. B. Lu, C. W. Yang, Applications of Artificial Intelligence in Intelligent Manufacturing: A Review, Frontiers of Information Technology & Electronic Engineering, 18(1), 86–96 (2017). https://link.springer.com/article/10.1631/FITEE.1601885.
[34] Arti, Sanjay Agrawal, Comparing Classification and Regression Tree and Support Vector Machine for Analyzing Sentiments for IPL 2016. IJRITCC, 4(6), 172–175 (2016).
[35] Y. O. Lee, J. Jo, J. Hwang, In 2017 IEEE International Conference on Big Data (Big Data) (2017).
[36] Y. Inoue, H. Nagayoshi, Deployment Conscious Automatic Surface Crack Detection, in Winter Conference on Applications of Computer Vision, IEEE, 2019, DOI: 10.1109/WACV.2019.00078

[37] P. Helo, Y. Hao, (2021). Artificial Intelligence in Operations Management and Supply Chain Management: An Exploratory Case Study. Production Planning & Control, 33, 1–18. https://doi.org/10.1080/09537287.2021.1882690. https://www.researchgate.net/publication/350572588_Artificial_intelligence_in_operations_management_and_supply_chain_management_an_exploratory_case_study.

[38] N. B. Vanting, Z. Ma, B. N. Jørgensen, A Scoping Review of Deep Neural Networks for Electric Load Forecasting. Energy Inform, 4(Suppl 2), 49 (2021). https://doi.org/10.1186/s42162-021-00148-6.

[39] K. Harston, AMRC J., 13, 41 (2021). https://doi.org/10.1002/amp2.10159, AIChE.

[40] H. Oliff, Y. Liu, M. Kumar, M. Williams, The Ethical Use of Human Data for Smart Manufacturing: An Analysis and Discussion. Procedia CIRP, 93, 1364–1369 (2020). https://doi.org/10.1016/j.procir.2020.06.

[41] M. C. Messner, Convolutional Neural Network Surrogate Models for the Mechanical Properties of Periodic Structures. ASME. J. Mech. Des., 142(2), 024503 (February 2020). https://doi.org/10.1115/1.4045040.

[42] R. Singh, A. K. Tyagi, S. K. Arumugam, Imagining the Sustainable Future with Industry 6.0: A Smarter Pathway for Modern Society and Manufacturing Industries (2023). https://doi.org/10.4018/978-1-6684-8531-6.ch016.

[43] A. K. Tyagi, S. Dananjayan, D. Agarwal, H. F. Thariq Ahmed, Blockchain—Internet of Things Applications: Opportunities and Challenges for Industry 4.0 and Society 5.0, Sensors, 23(2), 947 (2023).

8 IoT-Driven Supply Chain Management

A Comprehensive Framework for Smart and Sustainable Operations

Shirly Sudhakaran, R. Maheswari, and Sharath Kumar Jagannathan

8.1 INTRODUCTION: BACKGROUND

Many individuals perceive a supply chain primarily as a system facilitating material transportation [1]. This progression extends from suppliers to end users and, in reverse logistics, from end users back to suppliers [2]. While transporting goods is a key element, the supply chain is a complex network enabling the exchange of information and financial transactions among diverse individuals and businesses [3]. The comprehensive process that products undergo from suppliers to end users, known as the "forward flow of materials," ensures efficient delivery to customers [4]. Conversely, reverse logistics involves the return of materials from suppliers to end consumers, encompassing activities such as product returns, recycling, and disposal, aiming to minimize waste and maximize material reuse [5]. The concept of a supply chain is not confined to manufacturing enterprises; it is versatile and applicable across various industries beyond manufacturing. This includes service supply chains in sectors like banking, healthcare, education, and hospitality. In addition to tangible products, the supply chain involves the exchange of money and information, encompassing aspects such as invoices, payments, order processing, and communication among supply chain participants. This adaptable supply chain framework proves beneficial for service supply chains in hospitality, healthcare, finance, and education, emphasizing the importance of managing not only physical products but also the associated financial and informational flows [6].

Efficient supply chain management relies significantly on the effective management of performance. This entails the evaluation and improvement of various supply chain processes to enhance overall effectiveness and efficiency [7]. Key Performance Indicators (KPIs) play a pivotal role in supply chain performance management, systematically collected, tracked, and analyzed to gauge the effectiveness of different supply chain components. The process involves a structured approach to gather and monitor data on critical elements such as delivery schedules, inventory levels,

DOI: 10.1201/9781003473886-8

production efficiency, and other relevant variables [8]. The SCOR model stands out as a widely embraced framework for implementing performance measurement techniques in supply chains [9]. It offers a consistent method for assessing and enhancing supply chain effectiveness, incorporating specific KPIs at various points in the supply chain interface, including key stages like plan, source, make, deliver, and return. The primary goal of employing the SCOR model and associated KPIs is to conduct a comprehensive evaluation of supply chain performance, informing operational planning by pinpointing areas for optimization and improvement. Supply chain performance management underscores a culture of continuous improvement, leveraging frameworks like SCOR and KPIs [10]. Processes are regularly fine-tuned and adjusted to accommodate shifts in market conditions and organizational needs. Ensuring effective supply chain management necessitates the systematic collection and monitoring of Key Performance Indicators (KPIs), with the SCOR model serving as a valuable guide. The utilization of specific KPIs at different points in the supply chain interface facilitates performance assessment and directs ongoing operational planning for continual development [11].

The dynamic nature of consumer preferences and demand patterns introduces rapid changes in markets. Supply chains encounter challenges in adapting to these shifts, especially in coordinating production and distribution to align with evolving market needs [12]. Timely delivery is crucial, but obstacles like supply chain disruptions, transit delays, and logistical issues can hinder meeting delivery deadlines. Variations in raw material quality, unplanned downtime, equipment failures, and other fluctuations in manufacturing processes can impact the overall effectiveness of the supply chain. Effective decision-making in supply chain operations relies on accurate and timely information. Inefficiencies may arise from delays in information processing, often attributed to manual procedures or technological constraints [13]. The ability to take prompt and efficient corrective action in response to issues is essential, as delays in problem recognition and resolution can lead to long-lasting disruptions. Decisions in supply chains require careful consideration of various factors, and the inability to make prompt and well-informed decisions can impede operational streamlining and adaptation to evolving circumstances. Modern supply chains are characterized by high levels of dynamism, uncertainty, and variability. Factors such as mass customization, diverse deadlines, changing priorities, a variety of component types, and smaller lot sizes contribute to this complexity. While technological advancements offer opportunities for improvement, the quick pace of change can make it challenging to smoothly adopt and integrate new technologies into existing supply chain systems [14]. The intricacies and rapid evolution of contemporary supply chain dynamics may overwhelm traditional, offline, and less dynamic performance measurement systems. To address these challenges, proactive strategies, the application of technology, and the adoption of dynamic performance measurement systems that can adjust to changing circumstances are imperative. Agile technologies, real-time data, and sophisticated analytics are becoming increasingly necessary for efficient supply chain management. Overcoming the operational challenges posed by the complexities of modern manufacturing systems, delivery intricacies, process fluctuations, and the dynamic nature of markets requires a forward-thinking approach and the adoption of adaptive solutions tailored to the evolving environment of the supply chain [15].

The simplification of computerizing manufacturing and workflow data owes much to advancements in information technology and modern computers, significantly enhancing data management and efficiency across various industries. Despite these strides, current applications and systems face challenges in providing the necessary flexibility to adapt to rapidly changing business processes—a limitation that has been apparent for over a decade [16]. As business processes undergo continuous transformation, the ability of systems to quickly adjust becomes increasingly crucial, driven by factors such as technological advancements, market demands, and other variables. The imperative for an autonomous performance-measurement system capable of effortlessly tracking workflow has become more pronounced. Autonomy in this context refers to the system's ability to adapt to changes without requiring constant human intervention. In addition to workflow tracking, an ideal performance measurement system should facilitate continuous performance monitoring, encompassing real-time reporting and analysis to enable prompt issue resolution and process optimization [17]. Modern supply chain environments are characterized by heightened variability, uncertainty, and complexity, posing challenges for traditional systems to keep pace with the dynamic nature of contemporary business operations. Supply chain environments, in particular, experience constant fluctuations due to global events, shifting market trends, and evolving customer demands [18]. In such dynamic scenarios, an adaptive performance measurement system becomes indispensable. The shortcomings of existing systems underscore the need for innovation, giving rise to the demand for a new performance measurement system equipped with advanced features like autonomy and conditional adaptability. A performance measurement system that can swiftly adapt to the rapid changes in business processes emerges as a strategic asset, ensuring businesses can evaluate, assess, and enhance their operations in real time [19]. The critical importance of adapting to rapidly evolving business processes emphasizes the necessity for a novel performance measurement system, especially in the intricate and dynamic environments of modern supply chains. While advances in information technology and modern computers have simplified data management, the limitations of current systems necessitate the development of a novel performance measurement system. An autonomous system capable of tracking workflow and facilitating continuous performance monitoring is essential, particularly in the intricate and dynamic supply chains prevalent today [20].

The adoption of the Internet of Things (IoT) in supply chains is gaining widespread popularity due to its ability to capture and track data, promising significant enhancements in supply chain efficiency. The primary objective of integrating IoT into supply chains is to gather and track real-time data, positioning it as a valuable tool for enhancing the overall performance of supply chain operations [21]. Acting as a real-time data collection system, IoT provides users with continuous and instant access to data, addressing the dynamic nature of supply chain processes. IoT plays a crucial role in offering businesses immediate, accurate, and comprehensive insights into their current circumstances, enabling well-informed decision-making and rapid responses to changes in the supply chain environment [22]. Through the utilization of IoT technology, any component at the operational or execution level of the supply chain can transform into a "smart object" equipped with sensors and connectivity, emitting real-time data. This revolutionary capability grants a level of control and visibility that was previously challenging to achieve. Integration of supply chains with the Internet of Things facilitates the creation of an integrated information system. By consolidating data from multiple smart objects, this system provides a comprehensive view of the entire supply chain, facilitating better decision-making. The adoption of IoT in supply chains brings about various benefits, including increased efficiency, enhanced visibility, reduced operating costs, and the ability to proactively address issues before they escalate [23]. IoT significantly impacts the dynamics of the supply chain, ushering in a paradigm shift in how businesses operate and optimize their operations. It enables a proactive, data-driven strategy that aligns with the complexity of modern supply chain settings. The growing interest in incorporating IoT into supply chains signifies a revolutionary development with long-term implications. As technology advances, IoT integration is expected to play an increasingly vital role in shaping the future of supply chain management. The fascination with leveraging the Internet of Things in supply chains stems from its capability to acquire real-time data, offering prompt insights and transforming supply chain components into "smart objects," thereby improving management, monitoring, and overall performance optimization [24].

The chapter aims to enhance Smart Supply Chain Management (SSCM) by proposing a framework based on the Supply Chain Operations Reference (SCOR) model. This framework integrates business process modeling concepts with the real-time data capture capabilities of the Internet of Things (IoT). The integration of IoT into the SCOR model is positioned as a real-time data capture technology, striving to establish a digitized supply chain management system that emphasizes agility and responsiveness in supply chain operations. The term "event-driven system" is introduced to highlight SSCM's ability to respond to specific supply chain events. To handle the substantial amount of event data generated at the supply chain execution level, Complex Event Processing (CEP) is employed. At higher management tiers, CEP processes IoT data, supporting performance-oriented practices and decision-making. This implies that the framework not only collects real-time data but also evaluates and interprets it to support strategic decision-making. The framework enables the establishment of Real-Time Performance Control Rules (RT-PCRs) by performance management specialists. This feature empowers experts to set guidelines for real-time monitoring and decision-making during runtime, fostering a flexible and dynamic approach to performance management. To enhance clarity and readability, the chapter is structured into multiple sections. Section 2 provides an overview of IoT technology's application in the supply chain, including its role in data collection, monitoring, and improving operational efficiency. Section 3 delves into the architecture of the suggested framework, detailing the technical aspects of integrating IoT with the SCOR model to create a real-time, digitized SSCM system. Section 4 explains the proposed methodology, outlining the systematic steps required to implement the framework in supply chain management. Section 5 outlines the validation requirements of the framework, emphasizing the commitment to ensuring the reliability and efficiency of the suggested SSCM system. It also anticipates the potential benefits of the framework. In summary, the study introduces a framework for building a digitized, real-time SSCM system by merging IoT with the SCOR model.

This event-driven system employs CEP for data processing and empowers performance management specialists with the authority to set real-time control rules. The chapter's structured approach provides readers with a comprehensive understanding of the framework's development, operational methods, and expected outcomes.

8.2 EXISTING SYSTEM

A significant challenge in contemporary supply chain management stems from the complexity of promptly recognizing and responding to sudden real-time shifts occurring throughout the chain. The primary obstacle is the concept of "asynchrony," denoting time and information variations among the three tiers of an organization's management hierarchy [25, 26]. These disparities create horizontal time intervals that separate product flow from corresponding events at the execution level. Additionally, there are vertical information gaps at both operational and upper planning levels. Figure 8.1 visually depicts the specifics of these inconsistencies.

A significant impediment to achieving the seamless coordination and responsiveness required in modern supply chain management is the presence of asynchrony. The horizontal time gaps between the occurrence of events in the product flow and the awareness of these events at various management levels introduce delays. Additionally, vertical gaps in information flow hinder the timely transfer of crucial data between operational and upper planning levels. Addressing these asynchronies is crucial to enhancing the agility and efficiency of supply chain operations, facilitating more responsive decision-making in the face of sudden changes. Figure 8.1 illustrates these differences, emphasizing the need to devise effective strategies for closing the informational and temporal gaps in supply chain management systems.

The significance of closing the gaps illustrated in Figure 8.1 becomes evident, particularly concerning the optimization of the decision-making process. At the execution level, it is crucial for decisions to seamlessly align with current circumstances. The inability to monitor existing performance necessitates businesses to enhance their agility, enabling swift responses to the continuously evolving changes and

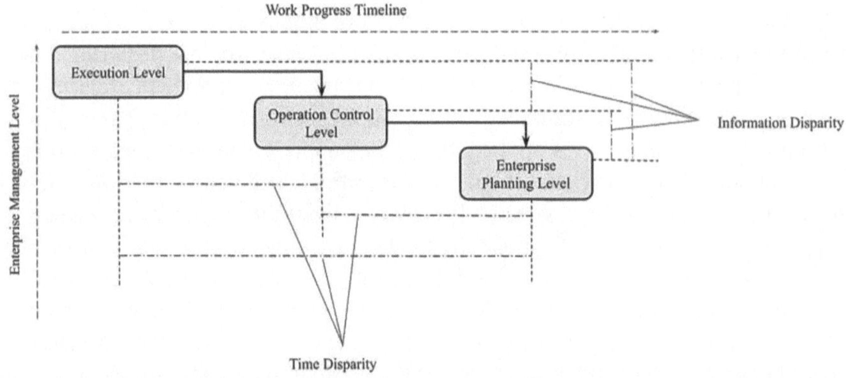

FIGURE 8.1 Time and Information Disparities within Enterprise Management Levels.

obstacles in the supply chain [27]. A common occurrence in supply chains is the bullwhip effect, arising from distortions and delays in information flow. Consequently, it is imperative to promptly address the disparities in time and information between different enterprise levels. Real-time performance management systems, readily provided by IoT technology, offer a viable solution to achieve this [28].

The interdependence of these discrepancies underscores their impact on supply chain decision-making, as depicted in Figure 8.1. To mitigate disruptions and optimize overall operational efficiency, decisions should align with prevailing conditions. Employing IoT technology as a key facilitator enables businesses to establish a more responsive and adaptable supply chain ecosystem. This approach reduces the bullwhip effect, positioning businesses to better navigate the challenges posed by the intricate dynamics of the supply chain environment. In a landscape characterized by rapid changes and uncertainty, businesses aiming for success must strategically integrate real-time performance management systems [29].

Wireless Sensor Networks (WSNs) and Radio Frequency Identification (RFID) play pivotal roles in Internet of Things (IoT) technologies, finding applications in various supply chain scenarios [30]. Despite their potential for widespread adoption, several obstacles impede their extensive use. One major hindrance is the limited awareness among business organizations regarding the full capabilities of RFID and WSNs and their practical application in real-world supply chain contexts. Additionally, the integration of IoT technologies with existing IT systems, such as enterprise resource planning (ERP), poses a significant challenge. Currently, there is a scarcity of comprehensive approaches guiding the real-time utilization of operational data to optimize supply chain benefits and expedite decision-making. This underscores the ongoing immaturity of IoT technologies for broad business adoption. RFID and WSNs are acknowledged as pivotal IoT technologies in supply chains, with the potential to revolutionize logistics and operational processes [31]. Nevertheless, the complexity of integrating these technologies into practical business settings has led to a gap between anticipated potential and actual adoption. Challenges extend beyond technical issues, as businesses also lack a comprehensive understanding of strategically integrating IoT into their larger IT infrastructure. Another deterrent to adopting these technologies for supply chain optimization is the absence of approaches addressing the utilization of real-time operational data. As time progresses, the development of IoT technologies, particularly RFID and WSNs, will be critical to fostering broader acceptance. The reduction of adoption barriers may occur gradually as companies gain more insights into potential benefits and practical integration techniques. To bridge existing gaps and fully unleash the transformative potential of IoT in supply chain management, further research and development, along with the evolution of comprehensive methodologies, are essential [32].

The integration of the Internet of Things (IoT) into the framework of Supply Chain Management (SCM) holds the promise of a revolutionary enhancement to traditional performance measurement and management techniques [33]. This integration brings about numerous significant advantages, including the ability for timely execution of actions and real-time decision-making. A key benefit is the reduction of the bullwhip effect, where minor disruptions in the supply chain escalate as they move up the hierarchy. Additionally, IoT integration into SCM enhances transparency in logistics operations and streamlines warehouse and inventory management, directly addressing the inherent challenges of IoT-enabled supply chains. Recent strides in wireless technologies and information technology (IT) have facilitated seamless integration of SCM with IoT hardware components like GPS, RFID, and Wireless Sensor Networks (WSNs) [34]. This integration has led to the development of automated performance management systems operating in real-time or near real-time. The combined capabilities of these technologies make supply chain operations more adaptable and dynamic. For instance, WSNs can monitor environmental conditions, RFID can track specific items throughout the supply chain, and GPS can provide real-time location data. This amalgamation creates a comprehensive and interconnected supply chain management ecosystem that leverages IoT advancements to enhance productivity and performance. In essence, the incorporation of IoT into SCM signifies a paradigm shift in supply chain management, offering potential improvements to traditional performance measurement. The seamless communication between various IoT components and SCM enables more flexible, data-driven, and automated approaches to performance management and decision-making. Given the swift pace of technological advancement, the interplay between IoT and SCM is poised to significantly shape the future of supply chain operations [35].

The objective of the proposed Smart Supply Chain Performance Management (SSCM) system is to address existing discrepancies in supply chain operations and facilitate seamless real-time integration on both horizontal and vertical levels. This is achieved by leveraging the guiding principles of the Supply Chain Operations Reference (SCOR) model, with a focus on the hierarchy of performance metrics. The envisioned system aims for vertical integration, ensuring a smooth exchange of data and expertise between upper management and the execution level. Resource integration, particularly at the enterprise execution level, plays a crucial role in mitigating horizontal time disparities. A key strategy for successful integration involves the application of smart technologies at the enterprise execution level. Figure 8.2 provides a visual representation of the proposed framework, illustrating the structure and implementation of the envisioned integration within the supply chain context.

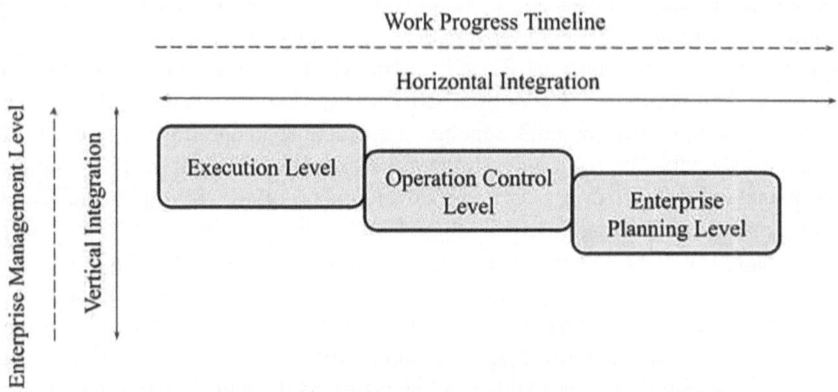

FIGURE 8.2 IoT-Integrated Enterprise Management Levels.

8.3 SYSTEM ARCHITECTURE

Establishing a foundational framework serves as the initial phase in constructing a performance-driven Smart Supply Chain Management (SSCM) system. This framework encompasses the management of integrating data derived from the Internet of Things (IoT) into various supply chain procedures. The alignment of this integration with the Supply Chain Operations Reference (SCOR) model ensures a systematic and standardized approach to supply chain management. SSCM is envisioned as an effective, automated, and event-driven information technology (IT) system, characterized by its proactive response to events rather than a passive role. The objective is to incorporate intelligent modules for performance-based control and management. Through this comprehensive strategy, supply chain performance management (SSCM) strives not only to sustain but also enhance desired performance standards across the entire supply chain ecosystem while effectively managing its current state. Complex Event Processing (CEP) plays a pivotal role in handling numerous event instances, enabling SSCM to manage a variety of events generated by IoT devices along the supply chain. The utilization of CEP technology is essential for transforming raw data into meaningful information, facilitating prompt decision-making and real-time actions.

The key priorities include reducing misuse, swiftly resolving issues, and ensuring the provision of beneficial services throughout the supply chain. Figure 8.3 provides an overview of SSCM's constituent parts, illustrating the system's functionality and organizational structure. This visual aid likely includes components such as IoT integration points, CEP processing, smart modules for management and control, and connections to various supply chain processes. For the development of a performance-based supply chain management system, a strategic core design emphasizing seamless integration, a focus on real-time event processing through CEP, and the incorporation of smart modules for active performance management and control are imperative. This strategic approach positions supply chain management as a dynamic, adaptable system capable of optimizing decisions and actions in the ever-evolving business landscape.

As shown in Figure 8.3, the SSCM framework comprises six engines:

1. Event Extractor Engine (EEE): The Event Extractor Engine (EEE) stands as a vital element within the framework of the Smart Supply Chain Management (SSCM) system. Operating in real-time, the EEE dynamically collects and transforms data, with a primary focus on performance metrics. Positioned at the execution level, it gathers and consolidates the state of distributed smart entities throughout the enterprise. Essentially, the EEE serves as a crucial intermediary, facilitating the seamless flow of data between the digital realm housing IT systems and the tangible, on-site resources utilized at the execution level. At its core, the EEE acts as a catalyst, translating the physical behaviors and statuses of smart objects into digital information consumable by IT systems. This pivotal role ensures that the data available to the SCM system remains accurate and responsive to the ongoing circumstances and events within the supply chain environment in real-time. The

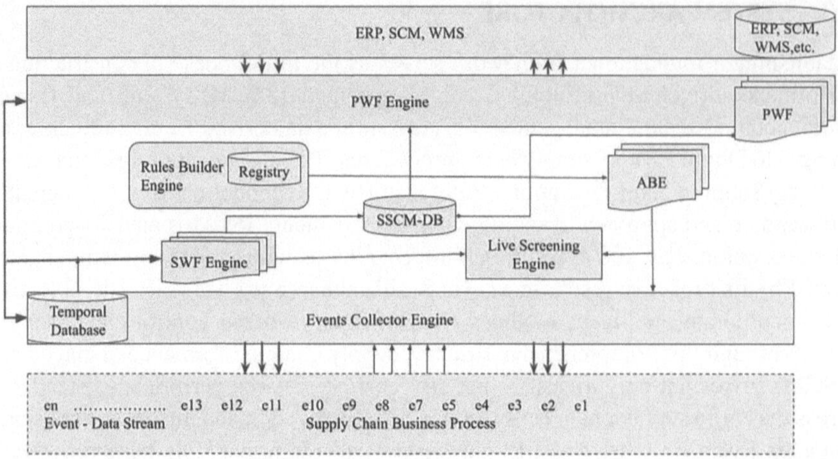

FIGURE 8.3 Overview of SSCM Framework.

real-time transformation of data is paramount for well-informed decision-making, providing the SCM system with the capability to proactively adjust to changes, mitigate risks, and optimize performance. The EEE's role as a bridge between the digital and physical domains underscores the critical importance of establishing an efficient and seamless integration between the overarching IT infrastructure and the resources operating at the execution level.

2. Actual Work Flow Engine (AWF-Engine): The Actual Work Flow Engine (AWF-Engine) serves as a crucial intermediary between the execution level and the Smart Supply Chain Management (SSCM) system. Its primary function is to decipher the data flow associated with smart objects by aligning the information with the temporal and spatial dimensions defining the Work Flow (WF) Process. Essentially, at the execution level, the AWF-Engine plays a pivotal role in translating real-time data generated by smart devices. This translation process involves a sophisticated analysis of the information, considering the location and timing of actions performed by these intelligent entities. By undertaking this analysis, the AWF-Engine contributes to the SSCM's understanding and response to the dynamic nature of the work flow process. The AWF-Engine's interface function is of paramount importance, facilitating seamless communication between the digital realm of the SSCM system and the tangible, on-site execution level. This integration ensures that the SSCM system can effectively monitor and manage activities at the execution level, enhancing its capacity to make well-informed decisions, optimize performance, and respond swiftly to evolving conditions within the supply chain.
3. Virtual Work Flow-Engine (VWF-Engine): Playing a crucial role as the interface between enterprise levels and the Smart Supply Chain Management

(SSCM) system, the Virtual Work Flow-Engine (VWF-Engine) holds significant importance. Its primary focus on higher organizational levels involves reinforcing and communicating performance-based data collected at these levels into the underlying framework of event instances. In practical terms, the VWF-Engine ensures the successful integration of the SSCM framework with performance-based data relevant to larger enterprise objectives. Take customer orders, for instance; each order is associated with a Virtual Work Flow (VWF) that outlines specific performance metrics before order execution. Emphasizing performance-based data is vital, as it sets the guidelines and standards for later phases of the supply chain, particularly in production. By serving as an interface at the enterprise levels, the VWF-Engine aligns the operational complexities of the supply chain with overarching organizational goals. This integration enhances the SSCM system's ability to holistically manage and optimize performance, ensuring that every stage of the supply chain aligns with the broader business goals. The role of the VWF-Engine underscores the critical nature of combining performance-based factors with order fulfillment and activity execution, fostering a more unified and goal-oriented approach within the supply chain framework.

4. Real-time Rules Engine (RT-RE): Within the Smart Supply Chain Management (SSCM) system, the Real-time Rules Engine (RT-RE) assumes a pivotal role in formulating and implementing rules for other SSCM-Engines. Facilitating this process is the Real-Time Rules Expression Builder (RT-REB) module, providing performance management managers with the capability to create and define rules governing the overall operation of the SSCM system. This functionality is reinforced by the presence of the "Rule Construction Elements" (RCE) module. Collaboratively, the RT-REB and RT-RE establish a platform for the development of real-time rules crucial for managing diverse facets of the supply chain. The "Rule Construction Elements" module supplies the foundational components and building blocks essential for crafting rules aligned with specific performance standards. This dynamic combination empowers performance management managers to meticulously tailor rules in harmony with the aims and objectives of the supply chain. The significance of the RT-RE and RT-REB lies in their capacity to introduce flexibility and adaptability to the SSCM system. With these tools, performance managers can adeptly respond to changing circumstances, adjust plans, and promptly optimize the system's performance parameters. This feature ensures that the Supply Chain Management System (SSCM) remains flexible, adaptable, and attuned to the evolving demands and challenges in the industry. Consequently, performance management managers can create rules that maximize the effectiveness and efficiency of the SSCM system, thanks to the capabilities afforded by the RT-RE and RT-REB modules.
5. Performance Practices and Applications-Engine (PPAE): At the heart of the Smart Supply Chain Management (SSCM) system lies the Performance Practices and Applications-Engine (PPAE), serving as its

core. A primary function of the PPAE is to automate real-time operations, eliminating the need for constant human supervision, crucial for making rapid decisions manually. The Real-time Rules Engine (RT-RE) supplies the PPAE with a range of Real-Time Performance Control Rules (RT-PCRs) derived from Actual Work Flow (AWF) event instances, the SSCM database (SSCM-dB), and Virtual Work Flows (VWFs) data. This integration enables highly intelligent real-time monitoring and control of supply chain processes. The Performance Practices and Applications (PPAs) of the PPAE undergo continuous improvement based on the complexities of daily or weekly tasks. This improvement process is facilitated by the precise performance targets set by the SSCM system. The Virtual Work Flow (VWF) signifies the desired performance-based environment, while the Actual Work Flow (AWF) represents the physical environment, with the PPAE bridging this gap. This relationship is crucial for supporting continuous development and intelligent supply chain management. Essentially, the PPAE utilizes intelligent rules and performance targets to automate real-time decision-making processes, maximizing the efficiency of the SSCM system. Through the ongoing refinement of PPAs in accordance with performance data and targets, the PPAE ensures that the supply chain is not only intelligently managed but also dynamically adapts to changing circumstances. This central engine plays a vital role in seamlessly integrating performance-based and physical components, fostering the development and flexibility of the Smart Supply Chain Management system.

6. Real-time Visualization Engine (RT-VE): A pivotal component of the Smart Supply Chain Management (SSCM) system is the Real-time Visualization Engine (RT-VE), designed to offer immediate insight into the Actual Work Flow (AWF). Its primary objective is to provide a visual representation for each individual order, benefiting a diverse set of stakeholders, including top-level management, organizers, executive management, and local laborers. The RT-VE is specifically designed to access and process real-time execution data from the AWF, equipped with predefined rules. The Real-Time Order Status (RT-VE) furnishes executive management, labor organizers, and local workers with a detailed and up-to-date depiction of the order status within the supply chain. This visibility is crucial for prompt decision-making and effective coordination across various operational levels. A comprehensive understanding of the entire supply chain process also empowers top management to make strategic decisions based on current information. The RT-VE's integrated rules ensure its ability to dynamically retrieve and process relevant data from the active execution of the AWF. This feature enhances the engine's flexibility by reflecting the most recent data and accommodating changes in the supply chain environment. Aligned with the principles of intelligent supply chain management, the real-time visualization provided by the RT-VE facilitates improved decision-making, enhanced coordination, and increased transparency throughout the supply chain.

8.4 PROPOSED WORK

When Internet of Things (IoT) technologies are applied throughout the supply chain, the Smart Supply Chain Management (SSCM) system, operating on an event-driven basis, becomes a vital element within the broader IT infrastructure. Situated between the top-tier long-term planning layer and the bottom-tier real-time event-processing layer, SSCM functions as a system for tracking short-term performance, transforming the operational level into an intelligent environment. This tiered structure allows SSCM to act as a bridge connecting strategic supply chain planning with immediate operational responsiveness. Serving as a real-time event recorder and analyzer, SSCM ensures that the overall performance of the supply chain aligns with overarching goals and objectives. The integration depicted in Figure 8.4 illustrates how enterprise planning systems, including supply chain management (SCM) and enterprise resource planning (ERP) systems, interact with the Supply Chain Performance Management (SCPM) system, which operates on a performance-based, real-time basis. This integration establishes a seamless information flow between enterprise planning frameworks and SSCM's capabilities for short-term performance tracking. The interdependencies highlighted in Figure 8.4 underscore the importance of synchronizing long-term planning strategies with real-time performance monitoring to develop a comprehensive and adaptable supply chain management strategy. SSCM plays a critical role in enhancing the agility and adaptability of the supply chain by swiftly responding to dynamic changes and events.

As depicted in Figure 8.4, the Complex Event Processing (CEP) methodology establishes connections within the Information Technology (IT) framework of

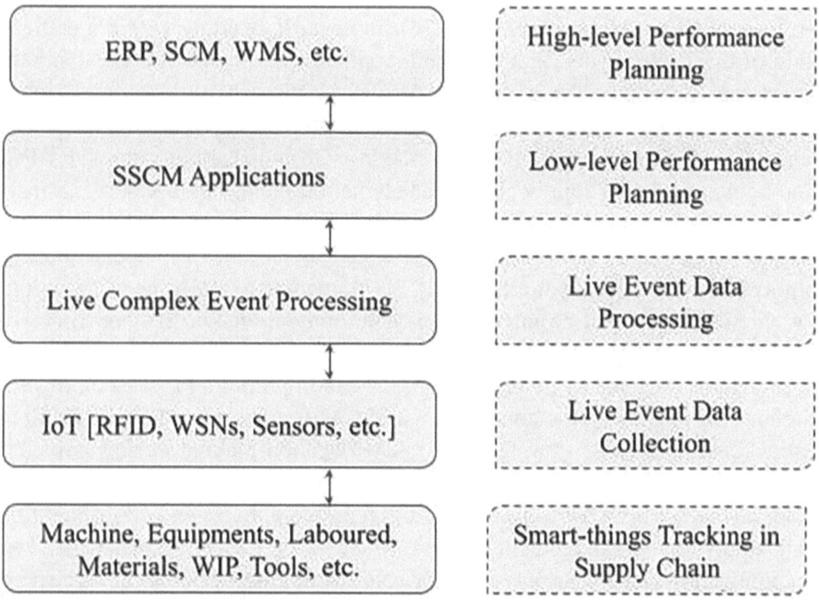

FIGURE 8.4 SSCM Integrated Enterprise IT System.

the business, providing essential elements for Smart Supply Chain Management (SSCM). Utilizing RFID, sensors, barcodes, and other IoT technologies, event data is gathered, operating at both the execution and SSCM levels. The Real-Time CEP (RT-CEP) layer acts as an Internet of Things (IoT) middleware, facilitating the linkage between these levels. This layer serves as a bridge, connecting the SSCM layer with the IoT technologies employed for event data collection, streamlining communication. Functioning as an Internet of Things middleware, the RT-CEP layer simplifies communication between IoT data readers and other execution-level systems, such as sensors, PLCs, and Human-Machine Interfaces (HMIs), adhering to standards like OPC-UA (OPC Unified Architecture) and OPC (OLE for Process Control). In its capacity as an IoT middleware, the RT-CEP layer has the ability to preprocess a large volume of generated events, subjecting them to analysis and filtering before forwarding them to the Supply Chain Performance Management (DSCPM) layer. These pre-processed events dynamically influence supply chain performance metrics within the DSCPM layer. It is crucial to note that supply chain management (SSCM) employs various engines, each tailored for a specific role in the process. The hierarchical structure ensures the transfer of information from the RT-CEP layer to the DSCPM layer and eventually to the enterprise's highest-level IT system through a meticulously planned event transfer process. Beyond merely recording and managing real-time events at the execution level, this hierarchical arrangement enables enterprises to comprehensively grasp supply chain performance metrics. The interoperability of the RT-CEP layer enhances the responsiveness and efficiency of the SSCM system, facilitating well-informed decision-making across the entire supply chain.

Real-Time Performance Control Rules (RT-PCRs) form an integral part of the Smart Supply Chain Management (SSCM) framework, residing within a dedicated module of the Performance Practices and Applications-Engine (PPAE). This integration holds significant importance, as it plays a pivotal role in minimizing the likelihood of hazardous events and addressing performance violations swiftly during execution, facilitating rapid decision-making. Designed as algorithms, RT-PCRs not only govern specific supply chain management policies but also serve as instruments for quick response. A primary role of RT-PCRs is to enable behavioral analysis of the supply chain resources, acting as a preemptive measure against potential disruptions. By leveraging patterns, RT-PCRs contribute to real-time decision-making, predicting and detecting interruptions and unintentional errors over time. This capability is crucial for reducing the risk of resource wastage—be it materials, time, or actions—by promptly resolving issues and optimizing resource utilization throughout the project. In the context of the SSCM framework, RT-PCRs fulfill two essential functions. First, they facilitate swift decision-making during implementation, mitigating the impact of risky events and performance violations. Second, they operate as sophisticated algorithms that regulate policies and contribute to the overall supply chain management process. By ensuring a more responsive and efficient supply chain, RT-PCRs play a crucial role in minimizing potential waste across multiple dimensions. Through behavioral analysis and pattern recognition, they enable a predictive and preventive approach, enhancing the overall effectiveness of the supply chain.

8.5 RESULTS AND DISCUSSIONS

The validation case study for the proposed framework is conducted in the public procurement environment of Indian Railways, where its integration into an existing supply chain model is assessed for functionality [36]. Data is systematically collected at various stages along the supply chain, encompassing points of purchase, processing, packing, inventory management, and distribution. Key Performance Indicators (KPIs) from these different stages are seamlessly integrated into the Smart Supply Chain Management (SSCM) system [37]. The initial phase of the validation process involves simulating the gathered data to ensure the framework's effectiveness, followed by actual implementation. This comprehensive approach allows for a thorough analysis of the capabilities and performance of the proposed system within a real-world supply chain scenario.

As part of the application, a feasibility study is conducted to assess the viability and practicality of the suggested SSCM system. The choice of the Indian Railways' public procurement environment for the case study is deliberate, considering the unique challenges posed by the demands of public services and constraints in raw material funding. This specific environment underscores the necessity for an efficient SSCM system that not only addresses these challenges but also aims to reduce resource losses and minimize waste. The case study serves as a valuable testbed, demonstrating how the SSCM framework operates in practical situations and illustrating its potential impact on optimizing supply chain resources.

8.5.1 Vertical Integration

The integration of performance monitoring and real-time workflow across the supply chain business process, spanning from operational to enterprise levels, leads to the generation of informational disparities. This integration ensures consistent tracking of key performance indicators (KPIs) throughout the supply chain, enabling a detailed analysis of various supply chain scenarios at different levels of the Supply Chain Operations Reference (SCOR) model. By vertically integrating performance monitoring and real-time workflow, the SSCM system establishes a connection between daily operations and enterprise-level strategic planning. This integration not only fosters a more profound understanding of the supply chain but also allows for a more precise analysis of KPIs. These KPIs gauge performance in critical areas such as asset utilization, operational efficiency, traffic volume, employment and wages, and financial results. The analysis, structured by the SCOR model, provides an organized approach to assessing and improving performance across various supply chain levels. Leveraging the emphasis on KPIs within different SCOR model dimensions, the SSCM system gathers a diverse set of metrics, offering insights into the intricate facets of supply chain performance. Decision-makers can exploit this information gap, stemming from vertical integration and KPI tracking, as a valuable tool to comprehend and enhance the efficiency of the supply chain at both operational and strategic levels.

8.5.2 Horizontal Integration

The attainment of horizontal integration involves the supervision of communication at the execution level within a network of smart devices. The primary objective is to uphold a predetermined Predicted Process Achievement Time (PPAT), exemplified by the implementation of specific measures like a designated inventory management policy. Effective execution necessitates the orchestration and control of information flow among these smart devices, ensuring that predefined processes unfold in a synchronized and timely manner. In order to explore the necessity for horizontal integration, the study introduces specific scenarios that spotlight the potential for heightened automated and efficient real-time decision-making. The strategic aim is to offer a framework for the assessment of the advantages associated with horizontal supply chain integration by diminishing or completely eradicating losses and inefficiencies. The scenarios serve as tangible examples, illustrating the potential benefits derived from horizontally integrating information and processes across the supply chain network. Ultimately, this integration contributes to an enhanced overall responsiveness and the reduction of resource inefficiencies within the intricate landscape of supply chain management.

8.6 CONCLUSION AND FUTURE WORKS

Managing the supply chain poses a substantial and intricate challenge for multinational corporations, driving dedicated efforts to reduce costs and enhance productivity [38]. This study advocates the implementation of Internet of Things (IoT) technology to enhance traceability and tracking in the supply chain. Particularly in the realm of complex supply chain operations, the adoption of IoT is seen as a strategic investment capable of yielding significant returns in a relatively short time frame. The study envisions the supply chain gaining valuable insights through the integration of IoT-based real-time performance monitoring, allowing for a comparison between its current and anticipated states. Furthermore, the study recommends the application of a Complex Event Processing (CEP) methodology to incorporate various modules into the framework. These modules demonstrate how the performance management of a smart supply chain can be enhanced through an Internet of Things (IoT)-based Smart Supply Chain Management (SSCM) system. Examples of these modules include lead time analysis, real-time pricing, real-time waste identification, and the implementation of lean manufacturing techniques. The primary goal is to navigate the complexities of supply chain management by leveraging IoT technology and CEP methodologies, ultimately improving visibility, responsiveness, and overall performance.

REFERENCES

[1] L. A. Prasetyanti and T. M. Simatupang, "A Framework for Service-Based Supply Chain," in Procedia Manufacturing, 4 (2015).
[2] M. G. Enz and D. M. Lambert, "A Supply Chain Management Framework for Services," in Journal of Business Logistics, 44 (2022).
[3] R. S. Kumar and S. Pugazhendhi, "Information Sharing in Supply Chains: An Overview," in Procedia Engineering, 38 (2012).
[4] T. Wu and J. Blackhurst, "Modelling Supply Chain Information and Material Flow Perturbations," in Supply Chain Management and Knowledge Management, 107–123 (2009).

[5] X. Zhang, J. Yu, W. Yan, Y. Wang and N. Subramanian, "A Comprehensive Review of Reverse Logistics in the Automotive Industry," in IEEE Access, 11 (2023).
[6] A. Presley and L. M. Meade, "The Business Case for Sustainability: An Application to Slow Fashion Supply Chains," in IEEE Engineering Management Review, 46 (2018).
[7] R. Oger, M. Lauras, B. Montreuil and F. Benaben, "A Decision Support System for Strategic Supply Chain Capacity Planning Under Uncertainty: Conceptual Framework and Experiment," in Enterprise Information Systems, 16 (2020).
[8] Y. Yurtay, N. Yurtay, H. Demirci, E. A. Zaimoglu and A. Göksu, "Improvement and Implementation of Sustainable Key Performance Indicators in Supply Chain Management: The Case of a Furniture Firm," in IEEE Access, 11 (2023).
[9] A. Khakpour, R. Colomo-Palacios and A. Martini, "Visual Analytics for Decision Support: A Supply Chain Perspective," in IEEE Access, 9 (2021).
[10] E. N. Ntabe, L. LeBel, A. D. Munson, and L. A. Santa-Eulalia, "A Systematic Literature Review of the Supply Chain Operations Reference (SCOR) Model Application with Special Attention to Environmental Issues," in International Journal of Production Economics, 169 (2015).
[11] M. Wu, Z. Yang, J. Sun and X. Gong, "Addressing Supply Chain Vulnerability by Supporting Emerging IT: An Analysis Based on SCOR Framework," 2020 IEEE International Conference on Industrial Engineering and Engineering Management (IEEM), Singapore, 2020.
[12] M. N. M. Bhutta and M. Ahmad, "Secure Identification, Traceability and Real-Time Tracking of Agricultural Food Supply During Transportation Using Internet of Things," in IEEE Access, 9 (2021).
[13] M. Brandenburg, "Design and Implementation of a Measurement and Management System for Operational and Supply Chain Performance," in IEEE Engineering Management Review, 46, 3 (2018).
[14] W. R. Ho, N. Tsolakis, T. Dawes, M. Dora and M. Kumar, "A Digital Strategy Development Framework for Supply Chains," in IEEE Transactions on Engineering Management, 70, 7 (2023).
[15] V. Maestrini, D. Luzzini, P. Maccarrone and F. Caniato, "Supply Chain Performance Measurement System: A Systematic Review and Research Agenda," in International Journal of Production Economics, 183 (2017).
[16] K. Pandit, D. Buddhi, A. Averineni, M. S. Narayana, G. Jahnavi and K. Rohitha, "Detailed Investigation of Influence of Internet of Things and Big Data on Digital Transformation in Marketing," 2023 10th International Conference on Computing for Sustainable Global Development (INDIACom), New Delhi (2023).
[17] M. J. Land, M. Thürer, M. Stevenson, L. D. Fredendall and K. Scholten, "Inventory Diagnosis for Flow Improvement—A Design Science Approach," in Journal of Operations Management, 67 (2021).
[18] G. Cavone, M. Dotoli, N. Epicoco, D. Morelli and C. Seatzu, "Design of Modern Supply Chain Networks Using Fuzzy Bargaining Game and Data Envelopment Analysis," in IEEE Transactions on Automation Science and Engineering, 17, 3 (2020).
[19] M. Brandenburg, "Design and Implementation of a Measurement and Management System for Operational and Supply Chain Performance," in IEEE Engineering Management Review, 46, 3 (2018).
[20] J. L. García Alcaraz, R. D. Reza, E. J. Macías, R. P. I. Vidal, F. J. F. Montalvo and A. S.-T. Ledesma, "Effect of the Sustainable Supply Chain on Business Performance—The Maquiladora Experience," in IEEE Access, 10 (2022).
[21] R. Wang, C. Yu and J. Wang, "Construction of Supply Chain Financial Risk Management Mode Based on Internet of Things," in IEEE Access, 7 (2019).

[22] Z. Wu, Z. Meng and J. Gray, "IoT-Based Techniques for Online M2M-Interactive Itemized Data Registration and Offline Information Traceability in a Digital Manufacturing System," in IEEE Transactions on Industrial Informatics, 13, 5 (2017).
[23] H. Rahmani, D. Shetty, M. Wagih, Y, Ghasempour, V. Palazzi, N. Carvalho, R. Correia, A. Costanzo, D. Vital, F. Alimenti, J. Kettle, D. Masotti, P. Mezzanotte, L. Roselli and J. Grosinger, "Next-Generation IoT Devices: Sustainable Eco-Friendly Manufacturing, Energy Harvesting, and Wireless Connectivity," in IEEE Journal of Microwaves, 3, 1 (2023).
[24] C. Paniagua and J. Delsing, "Industrial Frameworks for Internet of Things: A Survey," in IEEE Systems Journal, 15, 1 (2021).
[25] N. Sahay, M. Ierapetritou and J. Wassick, "Synchronous and Asynchronous Decision Making Strategies in Supply Chains," in Computers & Chemical Engineering, 71 (2014).
[26] H. Schiele, P. Hoffmann and T. Koerber, "Synchronicity Management: Mitigating Supply Chain Risks by Systematically Taking Demand Changes as Starting Point—A Lesson from the COVID-19 Crisis," in IEEE Engineering Management Review, 49 (2021).
[27] Kristina, Supply Chain Agility: 6 Strategies to Improve Agility (2021), www.shipbob.com/blog/supply-chain-agility/.
[28] Y. Xiong, J. Wang, Y. Yang and L. Wei, "Modeling and Bullwhip Effects Control of Internet+ Supply Chain," 2019 Chinese Control and Decision Conference (CCDC), Nanchang (2019).
[29] P. M. Reyes, J. K. Visich and P. Jaska, "Managing the Dynamics of New Technologies in the Global Supply Chain," in IEEE Engineering Management Review, 48, 1 (2020).
[30] T. B. Sousa, J. S. Yamanari and F. M. Guerrini, "An Enterprise Model on Sensing, Smart, and Sustainable (S^3) Enterprises," in Gestão & Produção, 27 (2020).
[31] W. Zhou, Y. Jia, A. Peng, Y. Zhang and P. Liu, "The Effect of IoT New Features on Security and Privacy: New Threats, Existing Solutions, and Challenges Yet to Be Solved," in IEEE Internet of Things Journal, 6, 2 (2019).
[32] Y.-C. Tsao, Q. Zhang and Q. Zeng, "Supply Chain Network Design Considering RFID Adoption," in IEEE Transactions on Automation Science and Engineering, 14, 2 (2017).
[33] A. J. Dweekat, G. Hwang and J. Park, "A Supply Chain Performance Measurement Approach Using the Internet of Things: Toward More Practical SCPMS," in Industrial Management & Data Systems, 117 (2017).
[34] C. L. Garay-Rondero, J. L. Martinez-Flores, N. R. Smith, S. O. Caballero Morales and A. Aldrette-Malacara, " Digital Supply Chain Model in Industry 4.0," in Technology Management, 31, 5 (2020).
[35] A. Phase and N. Mhetre, "Using IoT in Supply Chain Management," in International Journal of Engineering and Techniques, 4, 2 (2018).
[36] A. Gupta, G. Prakash and J. Jadeja, "Supply Chain in the Public Procurement Environment: Some reflections from the Indian Railways," in Procedia—Social and Behavioral Sciences, 189 (2015).
[37] https://indianrailways.gov.in/railwayboard/uploads/directorate/stat_econ/2023/PDF%20Year%20Book%202021-22-English.pdf
[38] L. Ardito, A. Petruzzelli, U. Panniello, and A. Garavelli, "Towards Industry 4.0: Mapping Digital Technologies for Supply Chain Management-Marketing Integration," in Business Process Management Journal, 25, 2 (2018).

9 Supply Chain Management in the Digital Age for Industry 4.0

Nikunj R. Patel

9.1 INTRODUCTION: INDUSTRY 4.0 AND SUPPLY CHAIN MANAGEMENT

The arrival of Industry 4.0 creates a lively environment throughout the global industrial landscape, forever altering the trajectory of how we make, manage, and transport things. This digital transformation, dubbed the Fourth Industrial Revolution, pervades nearly every aspect of our modern society, from schools of tomorrow to hospitals of today, and reaches a climax in Supply Chain Management (SCM). This ubiquitous technological integration revolves around three important pillars: hyper-connectivity, automation, and data-driven decision-making. A symphony of innovative technologies, each playing a different yet harmonic role, is powering this change. Consider the Internet of Things (IoT) to be a massive orchestra—a network of interconnected physical objects ranging from sensor-laden robots tending tomato crops in MIT's Distributed Robot Garden to smart home appliances sitting snugly in our kitchens, all humming with real-time data collection [1, 2]. This flood of data is then channeled through Artificial Intelligence (AI), with sophisticated algorithms and machine learning techniques used by myself and my colleague GPT from OpenAI, analyzing, learning, and ultimately, making autonomous decisions that guide and optimize every cog in the industrial machine. Consider the content suggestion algorithms that curate your social media feeds effortlessly or the conversational fluency of massive language models like us—those are just miniature glimpses into the transformative power of AI [3, 4]. However, much like a silent score, raw data requires interpretation and extraction of significant information. This is where Big Data comes into play, with its huge and complex datasets drawn from a variety of sources such as IoT sensors, financial transactions, and educational records, offering fertile ground for specialized tools and methodologies [1, 5]. It is used by financial firms to forecast market trends, by educators to tailor learning experiences, and by governments to make data-driven policy decisions. Big Data presents an unmatched image of our linked world, illuminating trends and possible problems [5]. Finally, picture tying these technical threads together to create a tapestry of intelligent systems—this is the essence of Cyber-Physical Systems (CPS) [6]. Consider these to be

DOI: 10.1201/9781003473886-9

conductors that not only assess data and make judgments but also smoothly transfer those decisions into real actions. Consider the delicate interplay of intelligent traffic signals that adjust to real-time traffic flow or self-optimizing production lines in smart factories, where robots and machines choreograph an efficient industrial ballet [2]. These technologies combine to produce a strong force in Industry 4.0, converting old factories into dynamic ecosystems of smart equipment, intelligent logistics networks, and flexible supply chains. The end result? A quantum leap forward in efficiency, responsiveness, and resilience. The days of static, compartmentalized processes are over; today's world is one in which data flows freely, choices are made in real time, and flexibility reigns supreme. The future of industry does not rest in yesterday's rigidity but in the dynamic dance of interconnected technologies, and those who embrace its beat will be the ones who lead the way.

Traditional SCM, or supply chain management, focuses on optimizing internal operations, managing inventory levels, and coordinating logistics. There are some issues that come with sticking with traditional methods, including limited visibility, reactive decision making, and fragmented processes. Limited visibility means that the lack of real-time data on supply chain activities and asset locations often leads to inaccurate forecasting, inventory mismanagement, and operational inefficiencies. Reactive decision making refers to the reliance on historical data and manual analysis that leaves companies vulnerable to disruptions and unable to adapt to real-time changes in demand or supply. Fragmented processes means that siloed information and disconnected systems across different supply chain partners hinder collaboration and agility. Thankfully, integrating industry 4.0 into the supply chain brings solutions to these problems. For example, real-time data from IoT sensors enables precise tracking of materials, inventory levels, and product movement, improving visibility and responsiveness [1]. This solves our issue of limited visibility. Industry 4.0 also improves predictive analytics and optimization, since AI algorithms can analyze vast amounts of data to predict demand, optimize routes, and make informed decisions, leading to improved agility and resource allocation. Reduced fragmentation is also another benefit. For example, smart robots and automated systems can handle repetitive tasks, while connected platforms facilitate seamless collaboration and information sharing between supply chain partners [2].

Traditional Supply Chain Management (SCM), a sturdy fighter in the industrial landscape, has traditionally concentrated on fine-tuning internal processes, methodically maintaining inventory levels, and orchestrating the complicated logistical dance. However, like any seasoned warrior, it has the wounds of time, which are seen in a few residual limits. Traditional SCM's Achilles' heel is, frequently, its lack of visibility. Consider traversing a complex battlefield blindfolded, depending on out-of-date maps and gossip. Many firms are in this situation because they lack real-time data on essential supply chain operations and asset locations. The fog of war produces erroneous forecasts, mismanaged inventory, and operational inefficiencies that may destroy even the most formidable supply chain. Another weakness is the reliance on reactive decision-making. Consider contacting aged oracles and examining antique parchments to forecast enemy actions. Previously, judgments were made using historical data and painstaking manual analysis. This leaves corporations exposed to interruptions in the lightning-fast speed of modern industry, hampered by their incapacity to adjust to real-time fluctuations in demand or supply. Finally, conventional SCM is haunted by the phantom of fragmented processes. Consider fighting with dispersed forces, each acting in isolation, with communication links cut. Siloed information and fragmented systems among supply chain partners become an internal enemy, impeding cooperation and agility. This divided landscape exposes firms to internal conflict and wasted opportunity. But have no fear, for a brave new champion has entered the fray: Industry 4.0, armed with cutting-edge technologies that provide effective solutions to these age-old challenges. The introduction of real-time data from a huge network of IoT devices shifts the tide [1]. Imagine scouts returning with new information, pinpointing the position of every commodity, laser-tracking inventory levels, and recording the precise flow of things. This enhanced insight enables firms to manage the supply chain with unprecedented clarity, forecasting movements, anticipating demands, and optimizing operations. Furthermore, Industry 4.0 unlocks the power of predictive analytics and artificial intelligence. Consider a legion of powerful computers relentlessly examining large data sets to estimate demand with amazing accuracy. These quick thinkers optimize routes, manage resources with surgical precision, and shepherd enterprises through the market's ever-changing landscape. Agility and responsiveness are no longer just goals but the essential foundation of operations. Finally, the disjointed environment yields to the uniting power of linked platforms and intelligent automation. Robots, the digital age's relentless foot soldiers, take over repetitious jobs, freeing up human minds to strategize and communicate [2]. Cloud-based technologies provide seamless information exchange across partner ecosystems, fostering genuine unity and agility. The internal adversary is defeated, and a symphony of shared knowledge and synchronized action takes its place. The supply chain battleground of today needs more than traditional tactics. Industry 4.0 is the new norm, with its cutting-edge technology and disruptive solutions. So, will you drive your supply chain into the future, equipped with the power of digital intelligence, or will you stand back and watch as your competitors dominate the market with nimble precision? You have an option. Take up the mantle of Industry 4.0 and create a supply chain that not only survives but thrives in the digital age's tumultuous waters.

The seductive call of Industry 4.0 appeals to the realm of supply chain management, weaving its tune of efficiency and agility. However, like any seductive chorus, it has dissonant murmurs inside its harmony—possible snares that must be noticed on the path to digital transformation. A similar shadow descends over the human worker. Consider expert hands, polished by years of practice, suddenly confronted with the sharp contrast of automated systems, their language the cold logic of code. Industry 4.0 necessitates a skill transformation; a shift from the known to the unknown. Training and adaptation become buzzwords, a necessary expenditure to provide personnel with the capabilities to handle and maintain these digital steeds. Without it, the gap between man and machine becomes wider, threatening to strand millions in the sea of change. Another specter emerges from the invisible realm: the terrifying grasp of cybersecurity dangers. As the tendrils of connectedness spread farther, slithering across systems and databases, vulnerabilities appear like hidden reefs on a predetermined path. Cyberattacks, like digital krakens, lurk in the dark depths, ready to disrupt and damage the supply chain's meticulous choreography. In this digital battlefield, robust security measures and incident response mechanisms serve as shields and swords, vital for repelling invisible attackers and safeguarding the integrity of critical activities. Finally, ethical murmurs join the clamor of progress. Concerns about data privacy reverberate across the halls of digital decision-making. Who owns the massive amounts of data created by linked equipment and sensors? How is it used, secured, and safeguarded? Algorithmic bias, a specter of unfairness built into the fabric of artificial intelligence, also raises its flag. Can we be convinced that these digital oracles administer justice fairly, or do they hold hidden prejudices that might corrupt verdicts and penalize groups? And there is the looming shadow of job displacement, which plagues many people's minds: will the development of robots herald the end of human labor? These are not just murmurs but important questions that require meaningful responses before we can move on. This chapter goes deeper into the symphony of advantages and disadvantages, providing a detailed examination of the dance between progress and hazard. For the voyage to an Industry 4.0-powered supply chain is a careful waltz, with each step calculated and risk analyzed, as we aim to establish a future where technology empowers rather than disrupts and where the benefits of development reach all coastlines.

9.2 DIGITAL TRANSFORMATION OF THE SUPPLY CHAIN

As mentioned in the introduction, the traditional supply chain is characterized by linear workflows, siloed information, and limited visibility. Industry 4.0 offers solutions to these pitfalls and ushers in a new age of unprecedented efficiency, empowering organizations to navigate the complexities of the global marketplace with agility and resilience. There are three domains where Industry 4.0 will radically change the global supply chain. These include interconnected networks and real time visibility, smart automation and robotics, and data driven optimization and collaboration.

Industry 4.0 drives interconnected networks and real time visibility through the implementation of Internet of Things (IoT) devices [1]. Metaphorically, each device can be seen as a "neuron" in a "neural network" that makes up the overall supply chain. Sensors on devices, in factories, in shipping ports, and so many more places can continuously "whisper" data to other devices. This data can include information on the location, movement, and condition of products on a particular supply chain in real time. This instant transmission of supply chain data creates a situation where historical data is rendered obsolete when it comes to data-driven decision-making, since real time data about the entire supply chain is much more valuable than historical data. This can be taken a step further by integrating blockchain technology into the supply chain network. Blockchain technology is defined as a database that keeps an immutable list of information and digital transactions [7]. In this case, integrating blockchain technology into our network of Internet of Things devices would help boost transparency and traceability across the supply chain, leading to improved trust and accountability in complex supply chains [7]. Digital twins are also useful assets. A digital twin is a virtual representation of a real object or process. These allow companies to virtually test and optimize their supply chains before real world implementation, increasing efficiency, reducing risk, and fostering innovation [8]. Overall, integration of Internet of Things devices in the supply chain would foster interconnectedness within the supply chain that empowers a proactive and adaptive approach, allowing organizations to anticipate disruptions, optimize resource allocation, and respond in real-time to the dynamic pulse of the market.

Smart automation and robotics are also a significant aspect of the implementation of Industry 4.0 in supply chain management. Supply chains often have repetitive tasks that are traditionally done by human hands. In a world where Industry 4.0 is integrated, smart automation technologies and robots can carry out repetitive tasks that are vital to the supply chain with immense precision and tirelessness [2]. Autonomous vehicles guided by advanced computer algorithms can take care of typical warehouse tasks at all hours of the day without any human supervision. Self-driving trucks can also ship products long distances with minimal stoppage time and with less error's than humans can. When such technology is implemented, we will see increased speed and accuracy in these tasks while simultaneously reducing reliance on human labor, which fosters a safer and more productive supply chain. Artificial intelligence also plays a key role in automation of the supply chain. Advanced algorithms can be used by companies to generate predictive analytics and drive intelligent decision making. For example, they can analyze vast datasets from sales records, market trends, and sensor data, enabling precise demand forecasting, optimized inventory management, and proactive disruption mitigation. Cobots, or collaborative robots, are also another tool for enhancing the supply chain. In manufacturing, they work alongside human workers, enhancing productivity, quality control, and worker safety, kind of like a robot companion that assists humans to do their jobs better and faster [2].

The unsung heroes of data-driven optimization are big data analytics [5]. These advanced tools probe into large oceans of data collected from throughout the supply chain, discovering hidden patterns, identifying areas for improvement, and forecasting future occurrences with amazing accuracy. This enables firms to negotiate market uncertainty with newfound confidence by enabling educated decision-making for demand forecasting, inventory management, and risk mitigation. Cloud-based solutions enable secure and scalable data storage, sharing, and collaboration among many supply chain participants. This promotes unprecedented openness, increased communication, and coordinated optimization throughout the ecosystem, allowing for seamless synchronization and optimal resource allocation. Finally, machine learning algorithms automate decision-making processes and improve resource allocation in a variety of supply chain features, such as intelligent route planning for logistics networks. These algorithms continually learn and adapt, ensuring ongoing improvement and optimization within the modern supply chain's changing terrain [3, 4].

The digital transformation of the supply chain, facilitated by Industry 4.0 technology, represents a fundamental shift in how things travel from conception to consumption. Interconnected networks promote trust and informed decision-making by providing real-time visibility and transparency. Automation and robots improve efficiency and safety, while data-driven optimization increases agility and resilience [2]. This convergence of breakthroughs gives firms who embrace it a competitive edge, allowing them to traverse the complexity of the global marketplace with unparalleled agility, resilience, and a dedication to sustainable practices. The future of the supply chain lies in leveraging the revolutionary force of Industry 4.0, which will pave the way for a more efficient, responsive, and linked world in which the flow of commodities reflects the needs of a dynamic and ever-changing marketplace.

9.3 IMPACT ON KEY SUPPLY CHAIN FUNCTIONS

Under the influence of Industry 4.0 technologies, the supply chain, which was previously a rigid engine of industrial advancement, is undergoing a seismic change. This digital tsunami reshapes every aspect of the industry, from raw material procurement to consumer smiles. Let's take a closer look at how Industry 4.0 is transforming five critical functions: procurement and sourcing, production and manufacturing, logistics and transportation, inventory management, and customer service.

The days of monolithic supplier relationships and opaque sourcing procedures are over. Industry 4.0 heralds a new age of flexible sourcing tactics powered by digital markets and powerful analytics. Platforms such as Alibaba Cloud and Sourceful connect customers with a worldwide pool of pre-screened suppliers, promoting transparency and real-time pricing comparisons. SpendEdge and other AI-powered solutions evaluate massive amounts of supplier data, forecasting future performance and eliminating potential hazards. Consider Unilever's AI-driven supplier evaluation, which ensures ethical sourcing while minimizing environmental effects. Alternatively, Siemens' blockchain-powered provenance monitoring ensures transparency across the supply chain [7]. These developments enable firms to save expenses, acquire dependable partners, and establish long-term buying policies.

The manufacturing floor is undergoing a transformation from inflexible production lines to adaptive ecosystems. Flexible robotics, such as Kuka and ABB robots, undertake monotonous tasks, allowing human workers to focus on higher-value jobs such as design and quality control [2]. Ultimaker and HP 3D printing technologies provide quick prototyping and customization, adapting to specific tastes while minimizing waste. Predictive maintenance, which is powered by sensor data and algorithms such as those from GE Predix, predicts equipment faults before they occur, reducing downtime and assuring production continuity. Companies such as Rolls-Royce use this technology to remotely monitor and service aircraft engines, improving performance while lowering costs. This trend toward smart manufacturing promotes not just efficiency and agility but also a safer and more stimulating work environment.

Roads and skies are transitioning from simple infrastructure to sophisticated mobility platforms. Autonomous vehicles, both terrestrial and aerial, such as Waymo's self-driving cars and Zipline's drones, are ready to transform transportation. Drones effectively carry critical products to remote regions, while self-driving trucks navigate highways with unrivaled precision, lowering emissions and accidents. Faster and more cost-effective transportation is ensured by hyper-efficient delivery networks optimized by AI algorithms such as ONYX Robotics' route planning software, pleasing clients with reduced delivery times. Amazon's Prime Air drone delivery service exemplifies the technology's promise, while DHL's collaboration with Volkswagen to create self-driving delivery vehicles illustrates its rising acceptance. These innovations not only streamline delivery routes and save costs but they also open up prospects for last-mile delivery solutions, meeting the ever-increasing need for convenience and speed.

Warehouses that are overcrowded and projections that are wrong are becoming a thing of the past. Blue Yonder's dynamic inventory optimization, backed by big data analytics and machine learning, forecasts demand swings with surprising precision [3–5]. This enables firms to keep ideal inventory levels while reducing the risk of stockouts and overstocking. Real-time insight into inventory levels across the network, offered by Zebra Technologies RFID technology, improves efficiency and agility by enabling fast modifications to production and distribution plans. For example, Walmart's RFID-enabled supply chain enables real-time insight into inventory levels, resulting in a 15% decrease in out-of-stocks and a 10% boost in sales. Similarly, Nike uses artificial intelligence-powered demand forecasting to manage inventory levels for their famous footwear, reducing waste and increasing consumer happiness. This change to data-driven inventory management not only cuts costs but also assures product availability and market response.

The client journey is no longer a static, one-way street; it is, rather, a dynamic, interactive experience. Industry 4.0 enables businesses to customize client interactions by delivering personalized suggestions, real-time monitoring data, and predictive maintenance services. AI-powered chatbots and virtual assistants, such as LivePerson's platform, provide 24/7 customer care, addressing issues swiftly and effectively. Personalized suggestions based on unique consumer interests are used by companies such as Amazon and Netflix to improve the purchasing experience and drive client loyalty. Similarly, Tesla's predictive maintenance technology warns consumers about possible automobile problems before they arise, assuring safety and convenience. These innovations not only improve client happiness but also open new avenues for deeper connection and brand loyalty.

9.4 MANAGING THE CHANGE: CHALLENGES AND OPPORTUNITIES

The digital tsunami of Industry 4.0 is sweeping across enterprises, delivering enormous growth but necessitating cautious navigation. Embracing this shift brings with it a slew of obstacles, from technological stumbling blocks to ethical quandaries, but it also brings with it a slew of unparalleled opportunity. Let's look at the crucial intersections where problems and opportunities meet and how leaders may navigate the critical decisions that will define their Industry 4.0 journey.

While smart manufacturing and AI-powered logistics seem appealing, traversing the technological trenches needs foresight. Infrastructure investment is one of the primary challenges. The backbone of Industry 4.0 is robust internet connectivity, cloud computing capacity, and sensor networks, which necessitate considerable financial investments. Companies such as Ford are investing billions of dollars in constructing intelligent factories, underlining the scale of the task. Another significant obstacle is data security. The flood of data created by linked devices and systems needs strong cybersecurity safeguards. Breach can cause not just financial damage but also a loss of customer confidence. Regulations like GDPR (General Data Protection Regulation) and CCPA emphasize the importance of responsible data governance, mandating modern encryption and access control methods. Finally, integrating old systems into the new digital ecosystem might be a difficult jigsaw to solve. Many firms use a hodgepodge of outdated IT infrastructure, which causes interoperability concerns and data silos. Siemens is leading efforts such as MindSphere, an open cloud platform aimed to bridge the gap between old systems and Industry 4.0 technology, providing a possible path for effective integration.

Implementing Industry 4.0 is a major cultural transition, not just a technological one. As automation and AI change the labor environment, the question of worker reskilling looms big. Employees must learn new skills ranging from data analysis and programming to collaborating with intelligent robots [2]. Training programs like those provided by Amazon Web Services and SAP are significant tools for upskilling the workforce and ensuring that they do not fall behind in the face of technological innovation. Another critical issue is change management. To overcome opposition to new technology and adjust organizational structures to embrace data-driven decision-making, excellent communication and leadership are required. Unilever, for example, has effectively integrated "agile ways of working" frameworks, encouraging a culture of continuous learning and experimentation and proving the necessity of a flexible and responsive approach to change.

While Industry 4.0 offers efficiency and comfort, it also presents important ethical concerns. Transparency is essential, as algorithms influence everything from employment choices to product suggestions. Companies must guarantee that their AI models are fair and explainable while also avoiding prejudice and discrimination. Concerns about data privacy must also be addressed. Regulations such as GDPR give people more control over their data, demanding appropriate data gathering and usage procedures. The possibility of job displacement is perhaps the most fundamental ethical concern. Certain activities will undoubtedly be replaced by automation, raising concerns about reskilling efforts, social safety nets, and the future of labor. Open debate and collaborative initiatives including governments, industry, and academics are critical for navigating this difficult terrain and ensuring a fair transition for all.

Despite the limitations, Industry 4.0 offers a limitless scope for collaboration. Companies may combine resources and knowledge through open innovation, speeding development and overcoming individual limits. Initiatives like Siemens' MindSphere platform promote collaboration across different enterprises on the industrial IoT landscape, demonstrating the power of collective creativity [1]. Collaboration with partners becomes another effective method. Organizations may tap into varied viewpoints and skill sets by collaborating with startups, institutions, and even rivals, unlocking new business models and applications. The "Maker Movement," with its emphasis on cooperation and open-source innovation, offers a glimpse of collaborative creation in the Industry 4.0 age. The last frontier of potential is the creation of completely new company models. Data-driven platforms, subscription services, and on-demand manufacturing are putting traditional business models to the test. Companies like Rolls-Royce, which offer engine "power-by-the-hour" services rather than selling engines outright, highlight Industry 4.0's disruptive potential. Accepting the flexibility and adaptability that these new models provide is critical to prospering in the new paradigm.

While navigating the choppy seas of Industry 4.0 necessitates careful assessment of possible problems, the appeal of the opportunities that await is unmistakable.

Businesses can leverage the potential of this technology transformation to chart a road to a vibrant, affluent future by investing in digital infrastructure, upskilling their staff, adopting ethical issues, and forming creative collaborations. The time for hesitating has passed; the sails of advancement have been lifted, and the wind of opportunity is blowing.

9.5 THE FUTURE OF SCM IN INDUSTRY 4.0

Once a rigid network of manual procedures, the supply chain is fast evolving into a dynamic digital ecosystem. Industry 4.0 technologies are the brushstrokes that will shape a future dominated by smart networks, self-optimizing processes, and hyper-personalization. Let's look at the rising trends that will shape the future of Supply Chain Management (SCM) in this digital age.

Blockchain technology, the digital ledger that underpins cryptocurrencies, has enormous promise in SCM. It's safe, unchangeable nature has the potential to revolutionize payments by speeding up cross-border transactions and eliminating fraud. Platforms like TradeLens, created in collaboration with Maersk and IBM, use blockchain to enable real-time insight into shipments, encouraging confidence and transparency across the supply chain [7]. Furthermore, blockchain has the potential to play a critical role in encouraging sustainability. It helps customers to make educated decisions and incentivizes sustainable manufacturing processes by tracing the provenance of materials and assuring ethical sourcing methods. Initiatives like the "Provenance Blockchain for Food" initiative show how blockchain may be used to monitor and verify the origin and sustainability of food items, hence increasing confidence and transparency in the food business [7].

The emergence of immersive technologies such as virtual reality (VR) and augmented reality (AR) ushers in a new era of collaborative SCM. VR may immerse users in virtual representations of warehouses, factories, and other supply chain facilities, allowing for real-time remote inspections, training, and troubleshooting. In contrast, augmented reality (AR) may superimpose digital information on actual settings, helping workers through complicated tasks and offering real-time data visualization. Consider a situation in which a maintenance engineer in Tokyo may digitally check faulty equipment in a plant in Berlin, using augmented reality to diagnose and advise on-site personnel through the repair procedure. This cross-border collaboration increases productivity, decreases downtime, and improves problem-solving abilities.

Individual requirements and tastes will be catered to in the future of SCM. Advanced analytics and artificial intelligence (AI) algorithms can analyze massive volumes of client data to forecast demand variations, tailor product offerings, and optimize delivery routes. This hyper-personalization improves not just customer happiness but also waste reduction and resource allocation throughout the supply chain. Consider a world in which your online grocery order modifies automatically depending on your shifting tastes and dietary needs, and your delivery drone navigates the most efficient path based on your current position and schedule. This level of personalization promotes a dynamic and flexible supply chain that caters to the modern consumer's ever-changing wants.

These new developments pave the way for the future's "smart supply chain"—a self-optimizing, networked ecosystem that reacts to real-time changes and disturbances. Transportation will be handled by autonomous cars and drones, while AI-powered systems will monitor inventory levels, forecast maintenance needs, and optimize logistics in real time. Sensors integrated throughout the supply chain will provide continuous data streams for analysis and decision-making, blurring the borders between physical and digital. Consider a smart factory in which machines interact with one another, altering production parameters depending on real-time demand and automatically ordering new components as they reach the end of their useful lives. This seamless integration of technology and physical processes results in a self-learning, flexible supply chain that can withstand unexpected difficulties while continually improving efficiency.

The path to effective SCM transformation in the Industry 4.0 age necessitates a proactive strategy. Here are a few significant conclusions and suggestions:

- Accept continual innovation: Don't be hesitant to try new things or accept new technology. The rate of change is quick, and staying ahead of the curve is critical for maintaining a competitive edge.
- Invest in people and training: Create a staff capable of working with and managing complicated technology. Initiatives for upskilling and reskilling are critical for adjusting to a changing context.
- Encourage cooperation by collaborating with other organizations and technology vendors to exchange expertise and resources. To realize the full potential of Industry 4.0 technology, openness and cooperation are essential.
- Utilize data and analytics to acquire insights into your supply chain and make educated decisions. Invest in a solid data architecture and analytics tools to get the most out of your data.
- Prioritize sustainability by including it in your supply chain strategy. Blockchain technology and other advancements can assist you in tracking and optimizing your environmental effect [7].

Organizations can tap the great potential of Industry 4.0 to build a future of smart, resilient, and sustainable supply chains by accepting these guidelines and managing new trends. The digital frontier is calling, and those who rise to meet its challenges will influence business's future. So, are you ready to go on this life-changing journey? The moment has come to act. Take the first step now to tap into the potential of Industry 4.0 to redefine your supply chain and push your company into a brighter, more connected future.

9.6 CASE STUDIES

The conventional supply chain is at a crossroads as the industrial landscape evolves. While manual procedures and compartmentalized information were formerly dominant, a new vision is emerging, one propelled by the revolutionary force of Industry 4.0. This collection of case studies delves into the real-world applications of these revolutionary technologies, demonstrating how organizations such as DHL, Bosch, and Unilever are reshaping their supply chains into agile, responsive, and sustainable ecosystems by leveraging interconnected networks, smart automation, and data-driven optimization. Prepare to be impressed by these trailblazers' inventiveness and bravery as they pave the way for a future in which efficiency, transparency, and resilience reshape the global flow of products.

9.6.1 DHL and Fetch Robotics: A Warehouse Dance with Robots

For decades, the warehouse dance resonated with the staccato movements of human hands, constantly choosing, packaging, and stacking items in a continual, albeit slow, waltz. However, this old beat is being recast, with a new conductor: technology. Logistics giant DHL and the leader of robot automation, Fetch Robotics, have collaborated to create a breakthrough performance—a warehouse where tireless robots waltz along aisles, changing the once-arduous human dance into a symphony of efficiency. Who is the antagonist in this story? Manual procedures' incessant inefficiency. Errors lurked in the shadows, stifling the flow of commodities and driving up expenses. Human tiredness weighed heavily on production, and the physical demands of the job took their toll. DHL saw the need for a new act, one in which technology would take center stage. Fetch Robotics, the choreographer of robotic automation, steps in. They provided an attractive as well as practical answer with their fleet of autonomous mobile robots (AMRs)—agile devices resembling metallic butterflies. These tireless dancers glided along aisles, led by laser precision and sophisticated algorithms, doing the most mundane jobs with steady accuracy—choosing products, traversing shelves, and delivering goods. The end result? A remarkable turnaround deserving of a standing ovation. Productivity increased by an astonishing 50%, owing to the robots' unrelenting energy. Labor expenses, which were formerly a considerable burden, have dropped dramatically due to being liberated from human weariness. The greatest part is that inventory accuracy, the elusive prima ballerina of warehouse operations, has now taken center stage, basking in the limelight of robot-driven precision [2]. The brilliance of this partnership, however, goes beyond simple numbers. It's in the human dancers who have been liberated from the monotony of repeated chores and are now able to focus on higher-level roles such as managing the robotic orchestra and guaranteeing the seamless flow of commodities. It's in the consumers who get their orders faster and with fewer mistakes, their delight reverberating like thundering applause. The collaboration between DHL and Fetch Robotics provides a look into the future of warehouses, where humans and robots coexist in a harmonic ballet of efficiency and precision. It's a narrative about a shared vision, not simply technological skill—a vision in which innovation transforms the mundane into the spectacular, and the once-slow waltz of logistics gives way to a dynamic, robot-powered pirouette [2]. This is more than simply a case study; it heralds a new age in warehouse operations—one in which technology takes the lead and the flow of products becomes a flawless, exciting performance.

9.6.2 Bosch Automotive: Where Machines Waltz with Data and Efficiency Sings

Consider a manufacturing floor to be a lovely waltz, not a cacophony of clanging metal and whirling gears, with each machine a dancer perfectly attuned to the beat of data. This, my friends, is the ambition realized by Bosch Automotive, the engineering behemoth, through their groundbreaking embrace of big data [5]. Downtime has tormented their production lines for years. Complex production processes obscured by a fog of inadequate data resulted in inefficiencies and unanticipated failures, disturbing the flow and casting a shadow on product quality. Bosch saw the need for a new conductor—one who could not only hear but also interpret the music of their machines. Enter big data—the information master. Bosch built a complex platform—a virtual orchestra pit—to gather and analyze data from hundreds of sensors placed in their devices [5]. These sensors—the system's attentive eyes—murmur temperature, vibration, and performance secrets and present a real-time picture of each machine's health and efficiency. With this new ensemble at their disposal, Bosch began rewriting the score. Predictive maintenance, which was previously a pipe dream, has now become a reality. The diligent analysts and algorithms combed the data, detecting minor alterations and forecasting probable breakdowns before they could interrupt the routine. Machines were no longer quiet collaborators, but rather, loud collaborators who provided insights into their own operation. The end result? The manufacturing line was changed. Downtime—the unwanted visitor—was respectfully shown the door and was reduced by an amazing 20%. Machine efficiency, which had previously been slow, increased by 15%, ushering in a new era of production. Above all, in the light of data-driven optimization, product quality, the final measure of the dance, achieved new heights. However, music transcends the confines of the workplace. Big data enables Bosch to predict market fluctuations, optimize inventory levels, and tailor production plans with previously inconceivable agility [5]. Once a rigid line, their supply chain now pulsates with the reactivity of a living creature. The Bosch Automotive tale is more than simply a case study; it is a statement of intent. It's a monument to big data's revolutionary potential, as robots and people interact in a dance of efficiency and insight [5]. It's a vision of a future in which factories sing the pleasant song of data-driven optimization, and the goods that emerge are more than simply things but testaments to technology's ability to rewrite the very score of manufacture. So, while the orchestra of invention continues to play, let Bosch's narrative serve as an inspiration—a reminder that even the most complicated machines can dance to the rhythm of data, and when they do, the outcome is a symphony of efficiency and quality that reverberates across industry.

9.6.3 Unilever: Where Palm Oil Whispers Its Origins and Transparency Takes Center Stage

One concern has always lurked, cloaked in obscurity, amid the convoluted maze of global supply chains: the opaque origins of raw commodities. Unilever, a proponent of mindful consumerism, was frustrated by the route of palm oil—a key component in many of their cherished goods. Consumers want openness, assuring themselves that their selections reflected their ideals. However, the web of suppliers and distributors stretched across countries, hiding this adaptable oil's route. But, ever the pioneer, Unilever refused to dance to the melody of uncertainty. They envisioned a new rhythm in which each drop of palm oil might tell its own tale, showing its origins and travels with complete transparency. This audacious concept was guided by blockchain technology, the digital ledger of truth. Unilever collaborated with IBM to create a digital tapestry that traced the journey of palm oil from source to supermarket shelf. The attentive scribe, Blockchain, scrupulously chronicled every step—from sustainable planting to responsible processing and ethical distribution [7]. Each transaction, each touchpoint, was inscribed in its immutable code, providing consumers with a glimpse into the core of their supply chain. The end result? A powerful symphony of faith and empowerment. Armed with greater knowledge, consumers might make educated decisions, matching their purchases with their ideals. Supplier collaboration developed as data-driven insights supported shared accountability and sustainable practices. Unilever could confidently stand behind their goods again, their commitment to ethical sourcing ringing loud and clear. However, the influence goes beyond basic openness. Unilever used blockchain to streamline their supply chain, finding inefficiencies and bottlenecks with pinpoint accuracy. As the digital trail shed light on areas for improvement, directing businesses toward more responsible sourcing methods, sustainability efforts gained support. As deforestation rates fell and biodiversity thrived under the careful eye of technology, the earth, too, echoed the rhythm of change. The Unilever tale is more than simply a case study; it is a credo for a more transparent future. It demonstrates blockchain's ability to revolutionize the core fabric of trade; not only track and record. It's a symphony of data, trust, and sustainability, all working together to reimagine how we source, consume, and care for our world [7]. Let Unilever's narrative encourage us all to raise our voices for a future where every product whispers its journey and every purchase resonates with the principles we hold dear as the world listens to this new tune.

9.6.4 Nestlé: Where Supply Chains Dance with Digital Doppelgangers

Consider a planet, its veins pulsing with the movement of products, a complicated dance of creation, transit, and distribution. This, my friends, is the complex dance of the supply chain, and navigating its twists and turns was a continual struggle for Nestlé, the global culinary master. Disruptions lurked like unexpected pirouettes, causing timetables to fall apart and supplies to fall behind. The necessity for a fresh choreographer—a visionary capable of predicting every blunder and gracefully guiding the flow—was apparent. Enter the digital twin, a computer reflection of the actual world's intricacies [8]. Nestlé, ever forward-thinking, created virtual versions of its vast supply lines, sophisticated simulations throbbing with the same data as their physical counterparts. Nestlé was able to test several situations with these artificial doppelgangers—the silent dancers in the wings—mimicking everything from unexpected ingredient shortages to port closures. Nestlé began to rework the supply chain's choreography with this fresh foreknowledge. Once inflexible routines, production timetables became fluid and sensitive, reacting to market fluctuations and disturbances with the agility of a seasoned dancer. Logistics routes, which were previously predictable waltzes, have evolved into dynamic jigs, rerouting around obstacles and maximizing delivery schedules with pinpoint accuracy. The end result? A supply chain buzzing with newly discovered visibility and agility. Disruptions, unpleasant visitors, were elegantly avoided, and their impact was reduced. Market shifts, which were formerly a cause of contention, were transformed into chances for improvisation, which were welcomed with the assurance of a great performer.

However, the digital twin's charm stretches beyond mere efficiency. It enables Nestlé to interact with partners in unprecedented ways [8]. Suppliers may coordinate their productions with Nestlé's demands by using the same virtual stage, guaranteeing a continuous supply of components. Distributors that understand the dance steps may optimize their own motions, ensuring items arrive at their destinations on time. The whole supply chain, which was previously a lonely performance, has evolved into a big, collaborative ballet, with each member held together by the beat of shared data and foresight. The Nestlé narrative foreshadows the future of supply chains, in which computerized twins orchestrate the flow of commodities with extraordinary accuracy and reactivity [8]. It shows the capacity of simulation to anticipate and empower. It's a symphony of data, cooperation, and agility, all working together to rewrite the score of global business. So, as this new era begins, let Nestlé's story inspire us all to embrace the digital twin and dance to the beat of a more efficient, responsive, and collaborative future [8].

9.6.5 Maersk: Where Currents Whisper Secrets and AI Steers the Course

Container ships previously danced to a single beat across the wide expanse of the oceans: the steady thrum of engines led by the charts and instincts of human commanders. This antiquated waltz would not do for Maersk, the king of the seas. Fuel use threw a lengthy shadow, pollutants billowed like unwanted sails, and inefficiencies lurked beneath the waves. A new conductor was required—one who could hear the ocean's whispers, read its currents, and steer these commerce behemoths on a more sustainable, efficient path. Enter artificial intelligence, the ever-changing data master. Maersk, ever the innovator, wove a digital tapestry onto the bridge—a complex software package tailored to the melody of the ocean. The relentless navigators, algorithms, consumed real-time meteorological data, whispering about currents, storms, and best routes. The system's seasoned oceanographers created a dynamic picture of tides and swells, providing insights no human eye could fathom. Maersk began rewriting the ocean's score with this fresh orchestra at its disposal. Once a static tune, route planning has evolved into a dynamic improvisation, altering, and adapting with the ocean's every breath. The unwanted refrain, fuel consumption, was reduced by an astounding 3%, a tribute to the AI's accuracy. Emissions, once a discordant symphony, have softened, mirroring a commitment to a cleaner future. Above all, operational efficiency, the elusive key to a successful journey, achieved unparalleled heights, driving these massive warships onward with incredible elegance. However, the influence of AI goes beyond the bounds of the spacecraft. Maersk improves port scheduling, minimizes congestion, and simplifies communication throughout their enormous fleet using data-driven insights. As AI discovers prospects for cleaner energy and more environmentally friendly behaviors, sustainability efforts gain traction. The ocean, too, is breathing a sigh of relief as lower emissions and more responsible navigation minimize the human impact on its fragile environment. The Maersk narrative is more than simply a case study; it's a statement of intent. It demonstrates AI's transformational capacity, not just to guide but also to optimize and protect. It's a symphony of data, efficiency, and sustainability, all arranged in perfect harmony to reshape the very rhythm of maritime trade. Let Maersk's story encourage us all to listen to the ocean's whispers, embrace the power of AI, and chart a path toward a cleaner, more efficient future for our seas and the huge network of trade that sails upon them as the sails of innovation catch the wind of change.

9.6.6 Siemens: Where Data Dances Across Silos and Innovation Takes Center Stage

Information frequently languishes in lonely rooms in the expansive labyrinth of a multinational organization, segregated data echoing like lost whispers in the hallways of bureaucracy. This fragmentation posed a challenge to Siemens, the engineering juggernaut, by stifling innovation and impeding cooperation throughout their large supply chain. There was an evident need for a new conductor—a platform that could connect these disparate islands and manage the flow of knowledge. Enter MindSphere, the digital bridge that spans information gaps. Siemens, ever the trailblazer, envisioned a platform that would go beyond the constraints of a database, transforming into a lively environment where data flowed freely, stimulating cooperation and lighting the flame of invention. MindSphere, the untiring interpreter, shattered departmental barriers by connecting robots, factories, and partners in a fluid interchange of knowledge. Production data, which was earlier confined to separate computers, now waltzed between departments, alerting engineers about possible problems and streamlining procedures with pinpoint accuracy. Customer feedback became a critical thread sewn into the fabric of product development, directing designs and molding services to meet changing demands. Siemens began to rewrite the score of collaboration with this fresh data symphony. Cross-functional teams, equipped with shared insights, solved issues with unprecedented agility in the days of siloed data. Suppliers synced their operations with Siemens' demands, guaranteeing a seamless flow of supplies and reducing interruptions, thanks to access to production estimates and real-time data. The whole ecosystem, which was previously a cacophony of disjointed voices, was transformed into a harmonic orchestra, with each instrument contributing to the great composition of invention. MindSphere's influence, however, goes beyond ordinary cooperation. It enables Siemens to create new data-driven services and products, stretching the frontiers of what a digital behemoth is capable of. Predictive maintenance, once a sci-fi fantasy, became a reality as algorithms analyzed machine data, whispering warnings of possible problems before they disrupted the production rhythm. Once a marketing pipe dream, personalized customer experiences grew as data insights revealed unique requirements and preferences, directing the delivery of tailored solutions and services. The Siemens tale is more than a blueprint for an integrated data platform; it is a manifesto for a future in which information bridges gaps and stimulates creativity. It demonstrates the ability of data to alter and empower, not merely connect. It's a symphony of cooperation, visibility, and invention, all arranged in perfect harmony to reinvent the very essence of what it means to be a global innovator. As the Industry 4.0 baton is passed, Siemens' story inspires us all to break down information barriers, release the power of data, and dance to the beat of a connected, collaborative, and eternally inventive future.

These different case studies give a vivid picture of Industry 4.0 technology's revolutionary impact. Each example demonstrates how businesses are changing their supply chains into nimble, responsive, and sustainable ecosystems, from warehouse robots pirouetting with efficiency to AI captains navigating ocean currents. Whether it's via data-driven optimization, seamless communication, or new solutions, one thing is certain: Industry 4.0 is more than a term; it's the beat of a reinvented future, and the pioneers highlighted here are leading the way. Remember that the power rests not only in these precise examples but in the seemingly limitless possibilities they inspire. So, embrace innovation, unleash the power of data, and join the revolution that is altering the flow of commodities as well as the fundamental nature of industry itself.

9.7 CONCLUSION

The winds of Industry 4.0 are blowing over the Supply Chain Management (SCM) landscape, heralding a transformational future. Emerging technologies weave a compelling tapestry of possibilities, from blockchain-powered transparency to immersive collaboration tools. This revolution is set to change the fundamental nature of how we manage the flow of products, orchestrating a symphony of linked networks, real-time data insights, and self-optimizing processes.

However, capturing the benefits of this digital transition necessitates smart planning. Leaders must become cartographers, mapping a path across unknown territory of technological complexity and organizational transformations. Building a strong infrastructure, investing in cutting-edge software, and fostering a competent team are all critical brushstrokes in the painting of a future proof SCM.

Aside from thorough preparation, adaptability is the key to success in the industry 4.0 age. Organizations must embrace a culture of constant adaptation—a dance with disruption in which open minds and quick feet negotiate the ever-changing rhythms of innovation. Fostering a learning environment in which talent thrives on unlearning and relearning, as well as being ready to harness cutting-edge technology like autonomous cars and hyper-personalization algorithms, assures a constant pirouette toward perfection.

The message is clear: embrace the digital tango of Industry 4.0 SCM. Invest in strategic projects, develop a staff that is future-ready, and foster a culture of continual adaptability. As a result, your supply chain will be transformed from a rigid chain to a lively ecosystem throbbing with agility, efficiency, and sustainability. Hesitation puts you on the sidelines of this transforming ballet. Begin by joining the digital dance and choreographing a supply chain revolution that will reverberate across your sector. Those that dare to march to the beat of innovation will be rewarded in the future. Are you up for the challenge?

REFERENCES

[1] Asadullah, M. M., & Ullah, F. (2019). Industrial IoT for Industry 4.0: A comprehensive survey. IEEE Access, 7(1), 12727–12748. DOI: 10.1109/ACCESS.2019.2890519

[2] Sadrfaridpour, B., & Wang, Y. (2018). Collaborative assembly in hybrid manufacturing cells: An integrated framework for human–robot interaction. IEEE Transactions on Automation Science and Engineering, 15(3), 1178–1192. DOI: 10.1109/TASE.2017.2748386

[3] Alpaydin, E. (2014). Introduction to machine learning (3rd ed.). MIT Press.

[4] Ray, S., Shah, K., & Kumar, V. (2019). A review of contemporary machine learning methods in artificial intelligence research: Empirical investigations. Applied Artificial Intelligence, 33(4), 389–474. DOI: 10.1080/08839514.2018.1545743

[5] Chen, M., Mao, S., & Liu, Y. (2014). Big data: A survey. Mobile Networks and Applications, 19(2), 171–209. DOI: 10.1007/s11036-013-0220-0

[6] Lee, E. A., & Seshia, S. A. (2008). Introduction to embedded systems: A cyber-physical systems approach. Cambridge University Press.

[7] Tapscott, D., & Tapscott, A. (2016). Blockchain revolution: Hyperledger for business. McGraw Hill Professional.

[8] Qi, Q., Tao, F., Liu, A., & Wang, L. (2019). Digital twin and its implementation for food manufacturing enterprises. Journal of Manufacturing Science and Engineering, 141(2). DOI: 10.1016/j.jmsy.2019.10.001

10 Artificial Intelligence, Computer Vision and Robotics for Industry 5.0

Gouthaman P., Rajasegar Rajendhiran Shanthi, Vijayakumar Ponnusamy, Arivazhagan N., and Nallarasan V.

10.1 ROLE OF ARTIFICIAL INTELLIGENCE (AI), COMPUTER VISION AND ROBOTICS

AI, Computer Vision (CV) and Robotics—they all are, some way or other, linked with each other because of their critical role in different industries. To start with, Artificial Intelligence (AI) does support, in different ways, problem solving, machine learning, natural language processing and automation. Firstly, AI facilitates the development of algorithms as well as model which supports machines to execute tasks where human intelligence is necessary; to paraphrase, pattern-recognition and decision making. Secondly, AI influences machine learning algorithms to assists systems in learning through data then enhance their performance deprived of explicit programming. Thirdly, AI is being utilized for building systems which can gain knowledge, construe, then create human-like language; to paraphrase, this can be beneficial for chatbots, virtual assistants, then applications for translating languages. Finally, AI is used for automating monotonous and rule-based activities so as to enhance efficiency, thereby permitting humans to concentrate on better creative tasks.

Next, computer vision (CV) does have certain roles to play; namely, image and video analysis, identification of objects, facial recognition and medical imaging. To start with, CV encompasses the constructing of algorithms so as to facilitate machines to construe and become aware of the world's visual details; namely, videos and images. Secondly, CV assists in recognizing objects, where machines work towards identifying as well as categorizing objects amongst images and streaming videos. Thirdly, CV is implemented within facial recognition systems so as to authenticate and identify individuals to maintain security. Lastly, CV in healthcare assists in investigating medical images and enabling tasks; namely, diagnosis of diseases and to plan treatment accordingly [1].

Robotics contributions are towards different aspects; namely, manufacturing and automation, robots for medical surgery, independent vehicles and dependable robotics. To begin with, robotics has become an integral part to automate tasks in various manufacturing industries for certain repetitive processes as well as risky tasks. Secondly, robotics in the medical field assists to perform minimally invasive surgical

 DOI: 10.1201/9781003473886-10

treatments, resulting in precision as well as enabling surgeons with control. Thirdly, robotics does play an imperative role in the building of autonomous vehicles, which not only includes self-driven cars but also drones utilized for various reasons. Finally, robotics are recently used for the construction of assistive devices to improve the quality of life for individuals with special needs.

10.1.1 Objectives of the Work

The significant impact created by AI, CV and robotics is explained; it is imperative to understand how they might generate a difference when integrated. There are applications where robots utilize computer vision to recognize objects and AI algorithms can improve their decision-making abilities. Robotic systems with AI can acclimate and interpret the environment using computer vision, permitting them to navigate as well execute tasks in an effective manner. The integration of AI, CV and robotics in transforming sectors will enhance efficiency and support the building of creative solutions for complex issues. In addition, these advanced technologies continue to grow and identify new applications within several fields, starting with healthcare, engineering, transportation and entertainment.

The Figure 10.1 portrays the consequence of Industry 5.0, where it involves computerization, digitization and humanization to be present as the significant aspect

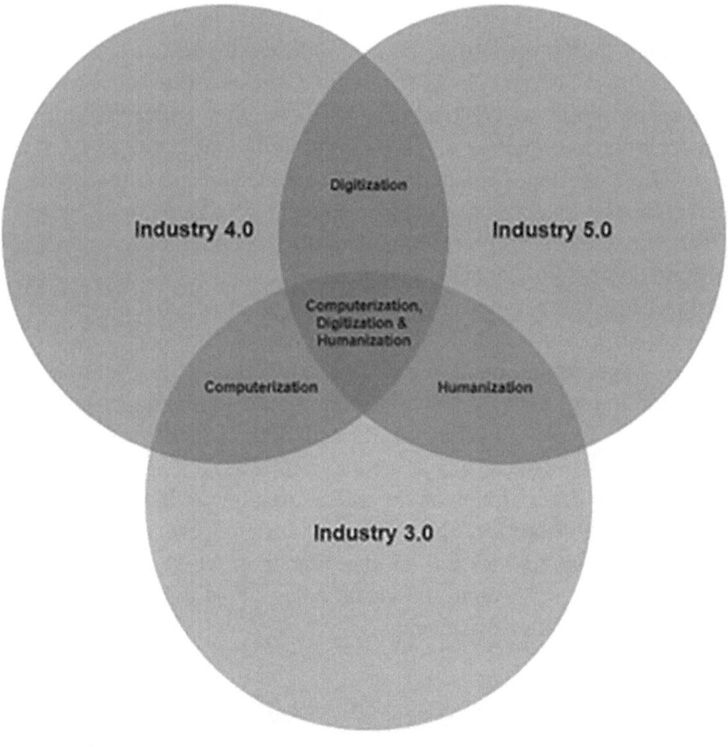

FIGURE 10.1 Unification of Industry 3.0, 4.0 and 5.0.

which are accomplished through every era of Industrial Revolution, which is expected to be the future of Industry 5.0.

10.2 ARTIFICIAL INTELLIGENCE (AI) IN INDUSTRY 5.0

Industry 5.0 is still in its developing stages, with 4.0 currently prevalent in the industries. On the contrary, Industry 5.0 has achieved enough attention, since it has been termed. Understanding it to be growth from Industry 4.0, it is significant to achieve insights regarding the basic trends from the integration of AI within industrial grounds. To name a few, collaborative robots is the amalgamation of AI in robotics, which has resulted in its creation and can assist humans to provide better productivity and safety. Furthermore, AI contributes to build the necessary products through unique customization along with flexibility in smart factories.

10.2.1 What is Artificial Intelligence?

AI means programming human intelligence within machines for executing tasks, which usually requires knowledge, perception, cognition, interpretation and recognition of speech. AI focuses on building systems which can execute activities in an intelligent way and mostly mirroring cognitive roles linked with human actions. There are two basic categories of AI—narrow and general. Narrow is built and trained for a specific task, since it performs particular tasks in a better way; however, they are deficient in comprehensive human abilities; for instance, virtual personal devices such as Alexa, Siri and Google assistant. General is the theoretical form of AI which contains the ability to interpret, become aware, then utilize that information to perform different tasks, which is way too similar to human intelligence, though it is still in its development stages and building it has some considerable scientific as well as ethical drawbacks. In addition, AI can be characterized as other categories as well through its approaches; namely, machine learning, deep learning, natural language processing and computer vision [2].

Machine learning can be referred to as a subset of AI which includes the building of algorithms that assists machines to obtain patterns through the available data. To paraphrase, supervised, unsupervised and reinforcement learning are of certain types. Deep learning is a form of machine learning which facilitates neural networks containing several layers—i.e., deep neural networks. This has been mostly appreciated for tasks like speech and image recognition. Natural language processing is a division of AI which concentrates on the interaction among humans and computers with natural language. Moreover, this helps machines to learn, construe and develop human language. Computer vision is another branch of AI which helps machines to learn and make decisions through the visual data. CV is immensely beneficial for facial recognition and analyzing images.

10.2.2 Key Components and Techniques in AI

AI involves a range of components and techniques, where each of them assists, for various purposes, towards developing intelligent systems. The components of AI are machine learning, deep learning, natural language processing and computer vision. Machine learning has different types and primarily supervised, unsupervised and reinforcement learning. Supervised learning is where models get their learnings through a labelled dataset in which the algorithm gains knowledge to map input data to appropriate output labels. On the other hand, unsupervised learning denotes a model to understand patterns and relationships present within data deprived of explicit labels, and these are mostly utilized for clustering as well as dimensionality reduction. Reinforcement learning is to permit agents to gain knowledge for making decisions through interaction with the environment then obtaining feedback in specific forms.

Deep learning is of many forms, and some significant ones are Neural, convolutional and recurrent neural networks. Neural networks are regarding algorithms simulated through the structure and the way a human brain functions. It involves multiple layers and is also referred to as deep neural networks. Convolutional neural networks are understood to be specialized neural networks utilized for recognizing images and to perform computer vision tasks. Recurrent neural networks are the ones built to manage sequential data; namely, natural language or time series. Natural language processing does have certain types; namely, speech recognition, text generation and sentiment analysis. Speech recognition is all about the machine's ability to interpret and decipher spoken language. Text generation refers to building coherent and contextually related contents that are usually viewed in chatbots or certain language models. Sentiment analysis means interpreting the emotions articulated either through written or spoken language.

Computer vision can be used in different forms, and some of them are image recognition, object detection and image generation. The recognition and classification of objects from available images or streaming videos are referred to image recognition. Object detection is where multiple objects get located and identified from an image or a video stream. Image generation is building realistic images through generative models such as generative adversarial networks [3].

The techniques of AI are data preprocessing, model training with optimization and ensemble learning. Data preprocessing is where feature engineering and data augmentation are performed. Feature engineering is regarding identification and transformation of related features within the dataset to enhance a model's performance. Next, data augmentation is where the range of training data is increased through transformations; namely, scaling or rotation. Model training and optimization is regarding gradient descent and hyperparameter tuning. The gradient descent is where an optimization algorithm is utilized to lower the error in models by tuning the metrics during training process. Next, hyperparameter tuning is where parameters are adjusted for optimizing model performance, since they are not learnt through training.

The history of revolution started from the year 1784, which was the First Industrial Revolution, which was mostly involved mechanization. The Second Industrial Revolution was more about mass production and electrification, which started from 1870. The Third Industrial Revolution started from 1969 and was the start of computers and automation. The current era is the Fourth Industrial Revolution, which incorporates digitalization and the Internet of Things. Figure 10.2 explains the different stages of industrial revolution.

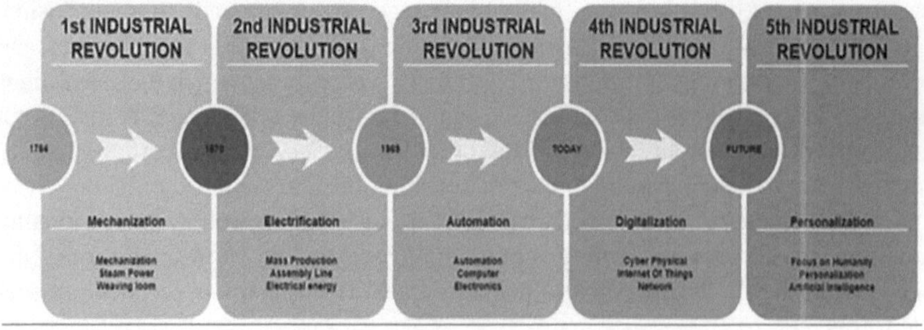

FIGURE 10.2 History of the Evolution of the Industrial Revolution.

Deep learning, being a part of machine learning, has been immensely important to theoretical as well as applied Artificial Intelligence. Deep learning algorithms ingrained with not only complex but also non-linear artificial neural systems perform better at obtaining comprehensive features from the available data. Deep learning has made complex tasks possible by human in real-world tasks; namely, medical diagnostics, in addition, has achieved results to existing complicated issues within different fields, such as robotics, computer vision and automation in industries [4].

10.2.3 Applications of AI in Industry 5.0

An autonomous vehicle created with stereo imaging systems and artificial intelligence applications can avoid obstacles. It necessitates a combination of a passive depth sensing system along with artificial intelligence-driven sematic segmentation in conjunction of a real-time obstacle avoidance process which applies the foregoing technologies united. Ultimately, the developed vehicle becomes facilitated to make semantic inferences regarding its surroundings, thereby effectively circumventing obstacles. It is to be understood that AI-driven augmented reality and virtual reality applications aid in maintenance and training processes through captivating and collaborating experiences. In order to enhance cybersecurity measures, AI is applied to protect critical networks, infrastructure and data from cyber-attacks.

10.2.3.1 Predictive Maintenance and Fault Detection

In industries such as healthcare, energy and manufacturing, predictive maintenance and fault detection are implemented as critical applications of AI. This is due to the fact that AI algorithms predict and avoid equipment malfunctions, lower downtime and improve the system's efficiency. To begin with, predictive maintenance encompasses data as well as analytics to forecast whether a piece of equipment might fail, and maintenance can be initiated during that period to avoid the failure. The advantages here are lowering downtime, cost-effectiveness and enhanced reliability of the equipment. Next, fault detection demands the discovery of deviations or issues in a piece of equipment so as to fix them in a timely manner. The benefits are early detection of faults and assuring safety standards.

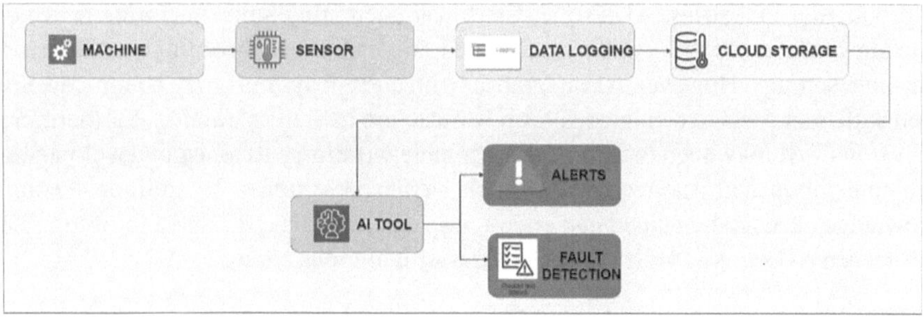

FIGURE 10.3 Process Workflow of Predictive Maintenance.

Having understood about predictive maintenance and fault detection as depicted in Figure 10.3, integrating them develops a complete technique to manage equipment wherein the issues that are known as well as unexpected are clearly addressed. The investigation of data obtained through predictive maintenance and fault detection for an organization lets them process continually then enhance their maintenance policies over time. Furthermore, this integration of predictive maintenance and fault detection reconstructs industry operations, thereby making them extremely reliable.

10.2.3.2 Intelligent Decision-Making and Optimization

In this digital era, algorithms with big data are acting as the base to achieve any smart approach, such as smart cities and any such environments. However, collecting real-time data for those smart applications from a specific location appears to be a hard-hitting undertaking in both technical and financial aspects. In order to solve this, open-source projects can be used, and hardware that is economically feasible can be used to develop a dashboard which supplies information obtained from various data streams. Next, AI systems assist in working with humongous data to gain decision-making processes. NLP aids AI systems to interpret and develop human language so as to let them communicate with users then supplying information to make decisions. To paraphrase, chatbots and virtual assistants available in various portals do communicate with users then help them take appropriate actions. They have rule-based systems in combination with knowledge bases which aid them with such recommendations through particular inputs [5].

10.2.4 Benefits and Challenges of AI in Industry 5.0

With the rising technological advancements familiarized by implementation of fully autonomous processes in the real-world situations, human intervention is still inevitable for various manufacturing industries. To paraphrase, small to medium enterprises are still finding it hard due to cost and application factors with many beneficial reasons. This is where artificial intelligence and computer vision can assist in achieving a fully autonomous manufacturing process. To begin with benefits, AI can streamline processes, automate usual activities, then improve the complete efficiency of production industries. AI also assists towards creating smart factories in which machines, processes and systems are linked and can work accordingly in different real-time settings. However, AI integration with current systems may be intricate and need sufficient costs to be incurred for infrastructure as well as training. Furthermore, inclusion of AI may need to upskill the available workforce to assist in development, implementation and maintenance of such advanced systems. In addition, enough knowledge needs to be gained and effective approaches to be followed for mitigating unforeseen risks linked with system failures or malicious attacks.

10.3 COMPUTER VISION IN INDUSTRY 5.0

With respect to industries, computer vision (CV) does provide a critical impact within several facets of manufacturing and production. To name a few under the Industry 5.0 setting, CV systems are utilized towards live control under the production department. Next, through analyzing the visual data obtained from machinery and equipment, CV algorithms can forecast regarding possibilities of machine failures. In addition, CV is vital during robotic systems guidance under the manufacturing line wherein robots built with cameras have the ability to work in complex situations, move around with objects accurately and assist the human workforce. Moreover, CV can assist in tracking and maintenance of inventories not only in the factory but also in the supply chain.

10.3.1 What is Computer Vision?

Computer Vision can be referred to as a multidisciplinary field which helps machines to learn and make appropriate decisions through the visual data within the environment. Also, it has the capacity to mimic human actions and performs various tasks by analyzing the available images and videos. Recently, deep learning algorithms—namely, convolutional neural networks—have contributed, in a powerful way, to progressing with tasks of computer vision, as they can learn hierarchical features through data automatically then enhance performance in image classification and object detection. CV has been utilized in various industries; namely, automotive through autonomous vehicles, healthcare for analyzing medical images, manufacturing for quality control and entertainment through augmented and virtual reality applications [6].

10.3.2 Core Concepts and Techniques in Computer Vision

Computer vision is comprised of diverse concepts and techniques which are crucial for learning and executing vision-based systems. To list a few, feature extraction includes recognizing and segregating necessary details or patterns from an image, filtering like convolution that assist in improving specific features within an image, which can aid in blurring or sharpening, object identification where recognition and classification of objects for an image is done and motion analysis which is regarding learning object movements from a set of images or streaming video. Furthermore, optical flow analysis and tracking techniques are critical for performing surveillance and creating robots, since this greatly benefits evaluating the system's accuracy and reliability.

10.3.3 Applications of Computer Vision in Industry 5.0

In this fast-paced lifestyle, eating healthy food items is extremely significant, and growing such foods—namely, mushrooms—with the implementation of computer vision alongside machine learning algorithms can be path breaking. There is quite a number of artificial intelligence and computer vision technologies that have been developed through research work in regards to edible fungi. However, the present techniques are unable to achieve the needs of digitization in the field of edible foods. Having said that, it is feasible to create digital phenotype determination, big data leading towards high-throughput breeding then delegating robots for mechanical harvesting.

Artificial Intelligence and other recent technologies are leaving no stone unturned, and in particular, the agricultural industry, where it is making a paradigm shift. To paraphrase, artificial intelligence, computer vision with deep learning assists several agricultural tasks being achieved automatically in a precise manner, resulting in smart agriculture. Computer vision, in tandem with good-quality image procurement through remote cameras, supports non-contact and well-organized technology-infused solutions for agriculture. It is significant to understand how latest advancements in computer vision—namely, vision transformers, generative adversarial networks and different known deep learning architectures—work to successfully accomplish this task. However, their success relies upon the way computer vision is used for developing a model with high-quality datasets, resulting in real-time solutions for agriculture.

10.3.3.1 Object Recognition and Tracking

The significance of locomotion tasks—that is, walking and climbing stairs—is when leg joints generate mechanical energy with activity-specific kinematic and moving patterns. Having said that, motion assistive devices must have such abilities and capable of adapting to task being performed; however, these tasks must be felt early enough for assuring smooth transitions with the device controller. This is where wearable vision sensors can play a significant role in predicting future tasks by identifying locomotion affordances within surroundings, which includes flat ground, stairs and inclines with a depth camera placed on user's upper body and process with a machine learning classifier [7].

10.3.3.2 Gesture and Facial Recognition

Significant attention has been gained through facial age estimation applications, but they are still deficient in identifying age consistently due to non-availability of enough training data comprised of precise age labels. To solve this issue, divergence-driven consistency training techniques can be utilized for improving efficiency as well as performance. With the hint of pseudo-labelling along with consistency regularization, pseudo labels can be allocated that were predicted through the teacher model towards unlabelled samples, thereby training the student model towards both labelled and unlabelled samples on the basis of consistency regularization.

10.3.3.3 Automated Inspection and Surveillance

Automated inspection and surveillance play significant roles in computer vision in an Industry 5.0 setting, since they greatly assist in enhancing efficiency, aiding quality control, providing safety and supporting smart decision-making. Automated inspection through CV facilitates the visual data usage and algorithms for analyzing and evaluating the quality of products in the absence of humans. A pictorial representation is given in Figure 10.4. Some of the benefits are uninterrupted monitoring of quality, flexibility with customization and predictive maintenance. Surveillance using CV utilizes cameras and image analysis algorithms for monitoring and investigating tasks under a specific environment.

The advantages are ensuring occupational safety, security with access control and efficiency in resource management. On the whole, automated inspection and surveillance with computer vision, implemented appropriately, can ensure the maximum outcome through these technologies for various industries.

10.3.4 Benefits and Challenges of Computer Vision in Industry 5.0

Smart garment manufacturing is where automated sewing systems are developed with machine vision abilities embedded within a customized sewing machine. The vision-system technology assists in gathering the image of the fabric pattern positioned across acrylic plates through a small opening and makes use of a deep-learning model for identification and segmenting this opening. Eventually, an expert algorithm identifies a narrow seam line amongst that image segmented then creates a stitching path in that seam line, thereby making sure a consistent distance is maintained. The sewing machine then precisely darns along that path, generated automatically. The mentioned innovation technique can play a crucial role in changing garment manufacturing into a smart industry.

Healthcare has been effectively facilitated with computer vision techniques with several applications to achieve greater benefits for patients. The imperative significance of computer vision is to enable surgical teams to enhance performance and to accomplish the safety norms of their patients. In other words, it is necessary to have operating room and endoscopic vigilance during surgery so as to ensure the individual's safety. It is extremely vital to leverage computer vision technology with

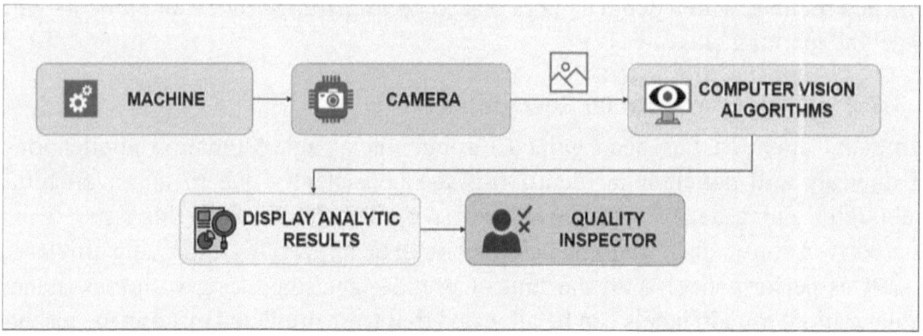

FIGURE 10.4 Workflow of Automated Inspection.

interdisciplinary alliance as well as latest techniques to gather data, model, interpret, then integrate, thereby leading towards creating an influential impact on different aspects such as safety, health and cost [8].

10.4 ROBOTICS IN INDUSTRY 5.0

Robotics in Industry 5.0 will clearly make waves through advancements from collaborative robotics (cobots), advanced automation, AI in robotics, sensors and perception then IoT integration.

10.4.1 What are Robotics and Industrial Automation?

It must be understood that robotics and industrial automation complement each other due to their relative usage of technology for performing tasks which are usually done with a human workforce in industries. Robotics can be referred to as a division of technology which involves the design, building, working and utilization of robots. In general, robots are machines which can be programmed to perform activities independently or with some human intervention. The benefits are assisting in the production line within manufacturing industries, healthcare, warehouse logistics and defence. Next, industrial automation means using control systems like computers or robots to perform various processes and using machinery substituting for the human workforce. The aim here is to enhance productivity, achieve efficiency and gain maximum operational performance.

10.4.2 Types of Robots in Industry 5.0

There are many varieties of robots, but common forms that can assist in industrial applications are listed here. Firstly, industrial robots such as articulated robots, which are robots with rotatory joints similar to human arm are versatile which assist in welding as well as assembly whereas delta robots are swift and precise with three arms connected to a central point. Next, collaborative robots, which are also referred to as cobots, are built to collaborate with humans and are built with sensors as well as safety aspects to avoid any dangerous events. Later, mobile robots are usually designed with mobility so as to move around independently within a specific location like warehouses and manufacturing settings. Next, humanoid robots are built to look like humans with body and similar movements and are mostly not used for industrial settings. Finally, there are autonomous vehicles and drones, which, in an industrial environment, can be used for transporting raw materials, maintenance of inventory and surveillance [9].

10.4.3 Applications of Robotics in Industry 5.0

Industry 5.0 may implement distributed production systems in which robotic objects and production units are to be interlinked. This distribution form aids in better flexibility and addressing of localized requirements. It is to be understood that innovative robotic systems can work with humans to perform complex activities; namely, accurate assembling, quality control and other complex production line tasks—portrayed

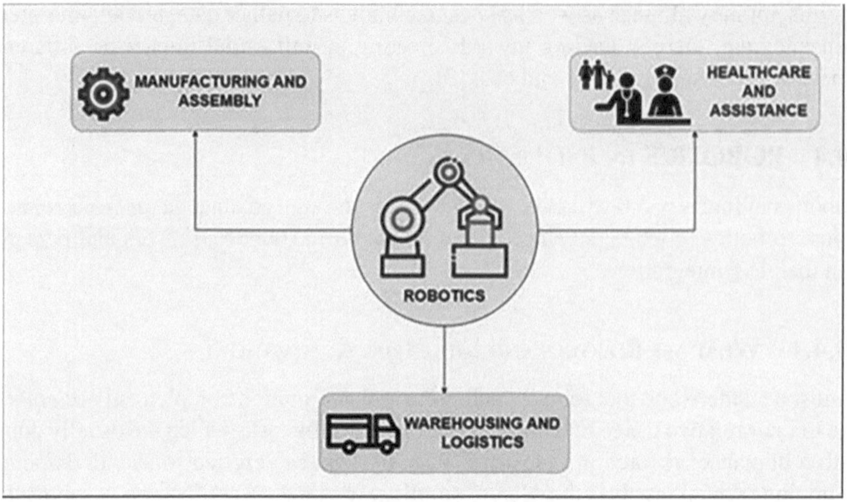

FIGURE 10.5 Applications of Robotics in Industry 5.0.

in Figure 10.5. Moreover, robots can be called upon for working on hazardous and risky activities, since it greatly reduces the risk to human life and achieves effective outcomes.

The visual recognition system can be achieved using computer vision along with machine learning algorithms, and this can assist in recognizing the right product amongst the defective ones. For instance, to lower the cost of labour and achieve effective tea production, a visual recognition system with a mechanical structure and a high-quality automatic tea-plucking robot with a motion control system can be implemented. In order to achieve this, algorithms—namely, particle swarm optimization supporting vector machines along with a deep convolutional neural model—can be settled upon to identify the tender tea shoots through global cameras then focus the plucking point as per the images shot with local cameras.

10.4.3.1 Manufacturing and Assembly

Possibly including robotics in production lines for manufacturing and assembly in the Industry 5.0 setting are agile manufacturing, digital twins and sustainability. Firstly, inclusion of highly versatile and agile manufacturing processes that can usually easily acclimatize to the changing demands and product specifications; hence, robotics here must have the possibility of easy modification and swift adaptation for different tasks; secondly, digital twins, where physical systems get virtualized and effectively used for manufacturing and assembling of robotics, since this virtual setting can provide immense benefits to achieve better outcomes during physical production; finally, sustainable manufacturing techniques during the manufacturing of robotic systems, since this may involve utilization of organic materials, low carbon manufacturing methodologies and reducing wastages during production. Industries have begun building robots which can ascend stairs to deliver packages. Figure 10.6 showcases one such robots that assists manufacturing industries [10].

FIGURE 10.6 Stair-Climbing Robot (Side & Top View).

10.4.3.2 Warehousing and Logistics

In a warehouse, utilization of autonomous mobile robots can be achieved in Industry 5.0, due to their capability of moving around without any human intervention for performing tasks; namely, picking and moving things around in an efficient way. In addition, these warehouses can be installed with IoT sensors and devices to build a smart environment where this communication aids in effective monitoring of inventory, products and the condition of the environment. Furthermore, logistics may need enhanced cooperation amongst humans and robots. To paraphrase, robots could work on loading-unloading materials while humans could concentrate on other creative work. Finally, Industry 5.0 may focus regarding sustainability in logistics with involvement of electric vehicles for moving goods and implementing environmentally friendly packaging processes.

10.4.3.3 Healthcare and Assistance

The digital era with state-of-the-art technologies have lead scene-understanding models with global and local relational reasoning to accomplish human-like scene investigation and understanding. To paraphrase, scene understanding assists improved semantic segmentation along with object-to-object interaction recognition. Within the medical industry, an effective surgical scene-understanding algorithm lets the automation with respect to surgical skill assessment, real-time monitoring of surgeon's action and evaluation post-surgery.

10.4.4 Benefits and Challenges of Robotics in Industry 5.0

The significance of robots in Industry 5.0 is how they work together with humans for accomplishing better productivity and a smart means to perform tasks. The collaboration of AI with robotics aids innovative automation, letting the systems learn and withstand changing environmental settings. In addition, robotics produce humongous data, and the real-time analytics offer actionable items for decision-making and accomplishing process efficiency. However, the downside is using advanced robotic systems needs a considerable amount of funds for learning technology, upskilling the workforce and building infrastructure. Another point is that the higher usage of robotics results in enormous amounts of data leading towards issues like data security and privacy. Furthermore, at the advent of automation and robotics, there are ethical concerns regarding job shift and impact on workers [11].

Robotic active vision has an extremely interesting trait of active interaction with environments wherein robots are permitted to proceed to support visual tasks. Currently, various artificial intelligence platforms have been made available to act as the synthetic instances in order to learn robotic active vision, reducing the necessity of real-world interaction. On the other hand, models trained in these platforms through synthetic data may face performance degradation in reality. To solve this, high-quality 3D detectors must generate 3D detection outcomes for a specific object from various viewpoints. This may extensively help the robot to learn it in an effective manner.

10.5 INTEGRATION OF AI, COMPUTER VISION AND ROBOTICS IN INDUSTRY 5.0

In this digital era, AI, ML and DL have greatly assisted the growth of advanced robotics. These technologies are complementing each other for making the robots perform complex tasks and act extremely intelligently. For instance, the applications of Artificial Intelligence and machine learning for advanced robotics are object recognition, self-governing navigation and predictive maintenance. In addition, this is being done in various industries; namely, transportation to accomplish passengers' safety, manufacturing to build robots and aviation to achieve better performance and gain customer satisfaction.

Safety is something that every individual looks forward to in this digital world, and with recent technological advancements of the Internet of Things, computer vision and artificial intelligence, surveillance systems can be developed with processing of real-time facial recognition. For instance, when a stranger is identified, alert notifications with an email of that image can be communicated to a room control through the Internet of Things.

10.5.1 Synergies and Interplay of Technologies

The following section details the way in which each of these technologies—AI, CV and robotics—are complementing each other. To begin with, AI and robotics—AI-driven robots have the ability to act upon intricate tasks with the capacity to adapt in changing conditions. The machine-learning algorithms support robots to get trained through experience and act appropriately so as to become better and better in different environments. Next, CV and robotics—CV lets robots to recognize and identify objects in real-time, since this feature is vital for warehouse situations where robots must identify and move objects across those settings. Finally, AI and CV—AI techniques, in combination with CV, does work with visual data to examine the quality of a product wherein products with concerns are taken into consideration and fixed in real-time, which lowers wastage, thereby enhancing the overall quality in production.

10.5.2 Intelligent Robotics and Automation Systems

Intelligent robotics and automation systems have the ability of the latest technological advancements to develop systems with a combination of AI, robotics and automation which perceive, learn and work in changing conditions. Such systems do contribute in a critical manner for several sectors, from production and logistics to healthcare and services. Firstly, sensor fusion processing data, gathered through various sensors—namely, cameras, radar and tactile sensors—supports developing better learning of the environment. Secondly, AI algorithms such as machine learning and deep learning assist robots to gain knowledge through data and experiences, and this makes them acclimatize to variable situations, perform efficiently and make smart decisions. Finally, intelligent robotics systems have the ability to independently plan and build their routes under dynamic environments, which is significant in applications such as drones and mobile robots. On the whole, development of intelligent robotics with automation systems facilitated towards diverse industries will pave the way for new possibilities and enhance productivity through the human-robot workforce.

10.5.3 Cognitive Vision and Perception

The computer vision community recently has been working on recognition of human action as well as predicting postures through the individuals' action and posture present in videos. This has gained immense attention from both industry and academia, as it is the prerequisite for intelligent interaction as well as human-computer interaction, thereby helping machines to distinguish their surroundings. It is believed that interactive cognition in self-driving cars can be taken as an instance to understand the significance of human action recognition and posture prediction [12].

10.5.4 Human-Robot Collaboration and Interaction

Human-robot collaboration and interaction provide powerful significance in the incorporation of robotics for different industries, so as to endorse safety and productivity. To paraphrase, this exceptional relationship requires humans and robots to cooperatively work to achieve better outcomes, complete activities and accomplish common objectives. The following are some instances of human-robot collaboration and interaction. One is physical collaboration, where humans and robots perform tasks in shared spaces in which it happens independently or together in particular tasks. Example—cobots. Another one is cooperative tasks in which humans and robots perform tasks together to reach a common goal, each of them portraying their special abilities. Example—a robot contributing to lifting a heavy weight for a human within a warehouse setting.

10.5.5 Case Studies: Successful Implementations in Industry 5.0

The following are case studies in Industry 5.0 under the manufacturing and healthcare industries. Initially, cobots are in assembly units in the manufacturing industry; namely, BMW and Audi, who have implemented cobots in their production lines. These robots contribute with humans through performing tasks, like heavy material lifting and accurate assembly. Next are intuitive surgical systems, utilized under a surgical platform which lets minimally invasive surgeries be performed under the supervision of surgeons who control robotic arms accurately, so as to improve their outcomes. The mentioned instances clearly outline the way in which AI, CV and robotics work efficiently in different industries, even though Industry 5.0 is still in its developing stages.

10.6 IMPACTS AND BENEFITS OF AI, COMPUTER VISION AND ROBOTICS IN INDUSTRY 5.0

AI, CV and robotics, integrated in Industry 5.0, will lead towards humongous benefits, changing industrial techniques and developing innovative solutions. The following are some of the ways in which it may aid the evolution of Industry 5.0. One point is efficiency in the production line will be clearly evident due to the automation of time-consuming tasks. This will, in tandem, lower the duration of manufacturing, accomplish accurate results and better utilization of resources. Another factor is AI-based analytics and robotics will eventually achieve effective production workflows and better allocation of resources. This will eventually lead to better yields and lower wastage and build quality products. It is imperative that organizations leverage these advanced technologies to achieve that edge and build a better environment [13].

10.6.1 Enhanced Productivity and Efficiency

The influence of technological advancements, such as expert systems and machine vision in agricultural domain, is recently flourishing since the health benefits of growing organic food in changing weather conditions has contributed towards achieving quantity as well as quality outcomes. In addition, research teams are heading towards farmers for understanding the requirements and facilitating them with artificial intelligence technologies for better seeds, safeguarding crops and fertilizers. The process of incorporating artificial intelligence in agriculture is to monitor crops and soil condition so that predictive analytics can be done, then initiating inclusion of robotics, sensors and soil sampling where necessary thereby able to collect data for building farm management systems to perform appropriate investigations. Eventually, this will help researchers to gain adequate knowledge on benefits and drawbacks regarding artificial intelligence applications and different methodologies implemented in smart farming; namely, image processing, expert systems and machine learning.

10.6.2 Improved Quality and Precision

The means through which integration of AI, CV and robotics contribute towards considerable level of improvements in quality and precision around different industrial settings are discussed here. One way is automated inspection and quality control through AI and CV, where CV systems are used to automatically investigate products for concerns and nonconformities from the specific quality metrics. Moreover, AI algorithms examine visual data to recognize issues so as to assure accurate manufacturing of products. Another way is predictive quality analytics, where not only historical data but also real-time data to understand and recognize patterns in relation to quality concerns. To reiterate, technologies contribute towards industries to let them produce effective, defect-free products and optimize processes in a continual style to achieve performance standards and customer satisfaction [14].

10.6.3 Safety and Risk Reduction

The process of maintaining safety and lowering risks in Industry 5.0 can be achieved through the integration of AI, CV and robotics. To name a few issues—workers safety, protecting materials and ensuring continuity of operations. There are some approaches to overcome these concerns. One strategic means is to perform risk evaluation and analyze hazards where a thorough risk evaluation is to be performed along with hazard analysis prior to the implementation of AI, CV and robotics systems. In addition, recognize possible risks linked with particular tasks within the setting. Another technological aspect is to build robotic systems with a range of sensors; namely, proximity, force and vision systems. With that, install safety-rated sensors which have the ability to identify obstacles and perform surveillance of the environment so as to assure safe operation. Furthermore, it is crucial to be constantly vigilant, comply to standards and maintain a proactive safety process.

10.6.4 Human Workforce Augmentation

The human workforce augmentation in regards to AI, CV and robotics in Industry 5.0 requires improving human possibilities under the integration of these mentioned technologies. It is significant to understand how these advanced technologies are complementing the human workforce in collaboration and effective use. An imperative aspect in this setting is AI-based decision making where AI algorithms analyze data to offer perceptions and aid human decision-making. In addition, humans utilize AI-assisted suggestions for additional support to make decisions. Organizations, when working on strategies, can lead them to accomplish better productivity, creativity and excellence.

10.6.5 Ethical and Social Issues

With all the benefits through this integration of AI, CV and robotics, it is critical to make a note on the ethical and social concerns that arise with these technologies. To begin with, when organizations utilize these advanced technologies, it may result in a job shift wherein there are routine and repeated tasks. Next, there is bias in AI and CV systems, where there is a possibility of inheriting and preserving biases regarding the data that is available to get trained [15]. Finally, AI and robotics being on the rise towards decision-making processes may lead towards building anxieties regarding human dignity and sovereignty. It is extremely vital that a proper collaboration is created within teams from industry, higher management and society to develop guidelines and frameworks which concentrate on ethical concerns.

10.7 CHALLENGES AND ISSUES IN ADOPTING AI, COMPUTER VISION AND ROBOTICS FOR INDUSTRY 5.0

The positive outcomes through the integration of AI, CV and robotics in the Industry 5.0 setting have been discussed, but there are drawbacks and concerns, as well, when adopted. Some of the key challenges are listed here. Firstly, the cost factor—the capital investment that is needed to develop AI, CV and robotics systems can be extensive. To paraphrase, most organizations—in particular, small and medium-sized entities—suffer financial issues when implementing these state-of-the-art technologies. Next, recruiting skilled workforce—to recruit and maintain such skilled professionals in these advanced technologies is a tough task, since organizations may find those individuals, but to make such talents stay for developing, installing and maintaining these latest technologies may not be feasible at all times. Having said that, organizations must build an effective approach to overcome these challenges with consultation from their stakeholders.

10.7.1 Technical Complexity and Integration

The combination of AI, CV and robotics for Industry 5.0 will eventually have certain complexities, due to the refinement that is involved behind these advanced technologies. One point is advanced algorithms, since their development and execution towards AI, CV and robotics needs proficiency in machine learning, deep learning and computer vision. In addition, creating algorithms which can manage intricate activities, interpret through data and make smart decisions leads to a level of technical complexity. Another aspect is gathering data through various sources in order to perform a comprehensive analysis so as to take appropriate decisions. It is significant for organizations to understand that talent as well as infrastructure is necessary to manage complexities that exists in implementation of such advanced technologies [16].

10.7.2 Safety and Security Concerns

Safety and security concerns need to be addressed effectively in order to properly implement advanced technologies in Industry 5.0 settings. Some of the safety and security concerns are listed as follows. Firstly, human-robot collaboration is something that needs to be managed diligently since there may be some physical interaction and it needs to be established properly. Next, autonomous navigation of robots may result in unexpected events due to some error when worked within dynamic conditions. In addition, vulnerabilities exploited through malicious attacks may result in data breaches and damaging systems. Furthermore, systems containing sensitive information gathered through AI and CV systems being accessed or misused lead to violation of privacy standards.

In order to mitigate these safety concerns, it is vital that appropriate safety features are being incorporated through robust path planning algorithms with fail-safe techniques. With respect to security concerns, robust encryption with consistent security audits are planned to manage systems. In addition, data protection laws are followed, along with appropriate access controls being maintained within the organization.

10.7.3 Workforce Adaptation and Skills Development

For industries to survive in a competitive environment with advanced technologies like AI, CV and robotics, it is vital to have an equipped workforce that can succeed in those situations. The way in which this can be achieved is discussed here. Firstly, continual learning processes can be accomplished by developing a training calendar in order to let the workforce be up-to-date on advanced technologies. Secondly, interdisciplinary skills can be adopted by offering training which integrates AI, CV and robotics, since it greatly benefits employees to get enough knowledge on the way in which these advanced technologies work. Finally, building skills which help employees to work in a collaborative way with AI, CV and robotics systems. When organizations take such measures, their workforce appears to be well-equipped to survive in a technologically advanced setting of Industry 5.0, and such process will clearly achieve better productivity for an organization [17].

10.8 FUTURE TRENDS AND OPPORTUNITIES

In this digital era, extensive implementation of the next generation technologies—namely, Artificial Intelligence, Internet of Things and blockchain technology—is contributing significantly towards this technological and industrial fast-paced environment. Furthermore, Artificial Intelligence has gained enough interest from government, universities and industries. This has led to the necessity of Artificial Intelligence-bound research and vital insights for researchers around the world to build a better future.

10.8.1 Advancements in AI, Computer Vision and Robotics

To begin with, reinforcement learning techniques are applied to robots to act upon tasks like gathering and administration, since this makes them gain knowledge with enough training by their experience and move appropriately in changing situations. Secondly, AI-based chatbots and virtual assistants are being experienced in various portals, providing better interactions, and these can be effective for serving customers in e-commerce sites and healthcare. Thirdly, CV and AI play a vital role in the manufacturing of self-driven vehicles, where gathering data through cameras, radar and various sensors aids in perception as well as decision-making abilities. Finally, the incorporation of AI techniques within edge devices working around the data source lowers latency and achieves real-time processing, which is extremely significant among IoT and robotics.

10.8.2 Human-Centric Robotics and Assistive Technologies

For enhancing quality of life by addressing the needs of differently abled people, not only human centric robotics but also assistive technologies provide humongous opportunities. To begin with, innovative prosthetic limbs and exoskeletons apply robotics to offer mobility for individuals who have limb loss or weakness in their muscles. Next, socially assistive robots are built to communicate with humans on a social as well as empathetic note. In addition, robotics and IoT technologies are integrated into smart homes in order to build accessible setting for individuals. Finally, smart canes with sensors and computer vision in-built guided robots may aid individuals with visual impairments to move around their environments without any support [18].

10.8.3 Explainable AI and Ethical Frameworks

Explainable AI (XAI) denotes the AI systems' ability to offer clear detailing with respect to their decisions, actions and outcomes. In addition, it focuses to make the decision-making process of AI models transparent, helping humans to understand. XAI assists in building trust with AI systems, thereby making developers and organizations accountable with respect to the effect of their AI applications. In addition, regulatory bodies direct explainability to follow ethical and fair usage of AI in industries. Ethical frameworks in AI clearly offer guidelines and processes for the creation, implementation and application of AI technologies, and these frameworks resolve concerns regarding bias, privacy and societal impacts. To paraphrase, AI systems are to be ensured so as to work with all individuals in a fair way and understanding how an individuals' personal data is to be responsibly handled. The integration of these elements aids in developing AI systems being used to comply with ethical standards and meet societal values [19].

10.8.4 Collaborative Robotics and Co-robotics

Humans and robots working in a collaborative manner at shared locations is referred to as collaborative robotics. The aim is to build a symbiotic relationship where the outcome is towards improving efficiency, productivity and safety. When compared with traditional robots, these robots are trained to stay alongside humans without the necessity of physical separation. Collaborative robotics have some benefits, and they are discussed here: to begin with, achieving better productivity by working alongside human workforce, since they automate repetitive or physically demanding activities so as to let humans to focus on better creative tasks. Next, safety aspects being embedded lowers the risk of accidents and injuries within the workplace, and this is in the manufacturing industry setting, since they both share the same workplace. Finally, these robots are cost-effective; in particular, small and medium-sized enterprises, since they may have the facility to go for large-scale automation [20].

10.9 CONCLUSION

In conclusion, the forecast of Industry 5.0 clearly portrays a transformative paradigm which creates a structure with the digital foundations provided by Industry 4.0. This forecast regarding industrial systems envisages an efficient collaboration amongst humans and machines in order to build them with human abilities. Having said that, many key themes arise. To start with, human-centric values get prioritized in Industry 5.0, focusing on the significance of a synergetic relationship between humans and latest technologies. The synergy of AI, CV and robotics intends to develop work environments which are efficient as well as provide significance for well-being and equipping the human workforce.

Next, Explainable AI ends up as a foundation for Industry 5.0, as it conveys the power of transparency and learning about the decision-making procedures regarding complex AI systems. The transparency mentioned here improves the trust in technology as well as makes sure of accountability, implementing it within applications in an ethical way. Furthermore, on connectivity, 5G technology is on the rise to provide better communication and exchange data, aiding swift and more reliable connectivity amongst devices and systems. This clearly assists in ensuring continuous operation of Industry 5.0 as well as unveiling innovative possibilities for augmented reality and the Internet of Things.

To sum up, the future of Industry 5.0 is clearly appearing to be in the comprehensive and human-focused strategy of industrial growth. It envisions building a trailblazing environment where creativity, symbiosis and collaboration congregate to reevaluate how the Fourth Industrial Revolution worked. It is extremely significant to build an efficient collaboration between human and cutting-edge technologies to perform exceptional innovations to build a better future which is sustainable as well.

REFERENCES

[1] C. Zhang and Y. Lu, "Study on Artificial Intelligence: The State of the Art and Future Prospects," J Ind Inf Integr, vol. 23, p. 100224, Sep. 2021, doi: 10.1016/j.jii.2021.100224.

[2] E. Elbasi et al., "Artificial Intelligence Technology in the Agricultural Sector: A Systematic Literature Review," IEEE Access, vol. 11, pp. 171–202, Jan. 2023. doi: 10.1109/ACCESS.2022.3232485.

[3] S. O. Bezrucav, N. Mandischer, and B. Corves, "Artificial Intelligence Task Planning of Cooperating Low-Cost Mobile Manipulators: A Case Study on a Fully Autonomous Manufacturing Application," Procedia Comput Sci, vol. 217, pp. 306–315, 2022, doi: 10.1016/j.procs.2022.12.226.

[4] J. M. Górriz et al., "Computational Approaches to Explainable Artificial Intelligence: Advances in Theory, Applications and Trends," Inf Fusion, vol. 100, p. 101945, Dec. 2023, doi: 10.1016/j.inffus.2023.101945.

[5] M. Alencastre-Miranda, R. M. Johnson, and H. I. Krebs, "Convolutional Neural Networks and Transfer Learning for Quality Inspection of Different Sugarcane Varieties," IEEE Trans Industr Inform, vol. 17, no. 2, pp. 787–794, Feb. 2021, doi: 10.1109/TII.2020.2992229.

[6] L. R. Kennedy-Metz et al., "Computer Vision in the Operating Room: Opportunities and Caveats," IEEE Trans Med Robot Bionics, vol. 3, no. 1, pp. 2–10, Feb. 2021, doi: 10.1109/TMRB.2020.3040002.

[7] Z. Bao, Z. Tan, J. Wan, X. Ma, G. Guo, and Z. Lei, "Divergence-Driven Consistency Training for Semi-Supervised Facial Age Estimation," IEEE Trans Inf Forensics Secur, vol. 18, pp. 221–232, 2023, doi: 10.1109/TIFS.2022.3218431.

[8] N. Ma et al., "A Survey of Human Action Recognition and Posture Prediction," Tsinghua Science and Technology, vol. 27, no. 6, pp. 973-1001, Dec. 2022, doi: 10.26599/TST.2021.9010068
[9] M. Soori, B. Arezoo, and R. Dastres, "Artificial Intelligence, Machine Learning and Deep Learning in Advanced Robotics, a Review," Cognitive Robotics, vol. 3, pp. 54–70, Jan. 2023. doi: 10.1016/j.cogr.2023.04.001.
[10] H. Hwang, C. Jang, G. Park, J. Cho, and I. J. Kim, "ElderSim: A Synthetic Data Generation Platform for Human Action Recognition in Eldercare Applications," IEEE Access, vol. 11, pp. 9279–9294, Jan. 2023, doi: 10.1109/ACCESS.2021.3051842.
[11] U. Ulusoy, O. Eren, and A. Demirhan, "Development of an Obstacle Avoiding Autonomous Vehicle by Using Stereo Depth Estimation and Artificial Intelligence Based Semantic Segmentation," Eng Appl Artif Intell, vol. 126, p. 106808, Nov. 2023, doi: 10.1016/j.engappai.2023.106808.
[12] A. Shams, A. Schekelmann, and W. Mülder, "A Proof of Concept for Providing Traffic Data by AI Based Computer Vision as a Basis for Smarter Industrial Areas," Procedia Comput Sci, vol. 201, pp. 239–246, Mar. 2022, doi: 10.1016/j.procs.2022.03.033.
[13] H. Yin, W. Yi, and D. Hu, "Computer Vision and Machine Learning Applied in the Mushroom Industry: A Critical Review," Comput Electron Agric, vol. 198, Jul. 2022. doi: 10.1016/j.compag.2022.107015.
[14] V. G. Dhanya et al., "Deep Learning Based Computer Vision Approaches for Smart Agricultural Applications," Artif Intell Agric, vol. 6, pp. 211–229, Jan. 2022. doi: 10.1016/j.aiia.2022.09.007.
[15] A. H. A. Al-Dabbagh, and R. Ronsse, "Depth Vision-Based Terrain Detection Algorithm During Human Locomotion," IEEE Trans Med Robot Bionics, vol. 4, no. 4, pp. 1010–1021, Nov. 2022, doi: 10.1109/TMRB.2022.3206602.
[16] Q. Zhao, L. Zhang, L. Wu, H. Qiao, and Z. Liu, "A Real 3D Embodied Dataset for Robotic Active Visual Learning," IEEE Robot Autom Lett, vol. 7, no. 3, pp. 6646–6652, Jul. 2022, doi: 10.1109/LRA.2022.3157028.
[17] H. Meddeb, Z. Abdellaoui, and F. Houaidi, "Development of Surveillance Robot Based on Face Recognition Using Raspberry-PI and IOT," Microprocess Microsyst, vol. 96, no. 8, p. 104728, Feb. 2023, doi: 10.1016/j.micpro.2022.104728.
[18] L. Seenivasan, S. Mitheran, M. Islam, and H. Ren, "Global-Reasoned Multi-Task Learning Model for Surgical Scene Understanding," IEEE Robot Autom Lett, vol. 7, no. 2, pp. 3858–3865, Apr. 2022, doi: 10.1109/LRA.2022.3146544.
[19] S. Ku, H. W. Choi, H. Y. Kim, and Y. L. Park, "Automated Sewing System Enabled by Machine Vision for Smart Garment Manufacturing," IEEE Robot Autom Lett, vol. 8, no. 9, pp. 5680–5687, Sep. 2023, doi: 10.1109/LRA.2023.3300284.
[20] H. Yang et al., "Computer Vision-Based High-Quality Tea Automatic Plucking Robot Using Delta Parallel Manipulator," Comput Electron Agric, vol. 181, Feb. 2021, doi: 10.1016/j.compag.2020.105946.

Self-assessment problems

1. On the basis of AI in relation to workforce, what approaches may be utilized to address the gap with respect to Industry 5.0?
2. Assess the ethical and privacy issues when implementing computer vision within Industry 5.0 and what steps can be taken to manage them.
3. Demonstrate the drawbacks linked with developing effective alliance between humans and robots in the evolution of Industry 5.0 within the manufacturing industry.

11 Data Analytics and Decision-Making in Industry 4.0

Amol P. Bhagat, Shrikant M. Harle, Chetan R. Ingole, and Vishalsinh V. Bais

11.1 INTRODUCTION: BACKGROUND

Data analytics in Industry 4.0 plays a pivotal role in transforming raw data into actionable insights. By analyzing historical data and real-time information from sensors, companies can predict when equipment might fail, allowing for proactive maintenance instead of reactive fixes. This not only minimizes downtime but also reduces overall maintenance costs. Analytics helps in monitoring and analyzing the production process, identifying patterns or anomalies that could signal potential defects. This ensures a higher level of product quality. From demand forecasting to inventory management, data analytics enables companies to optimize their supply chains. This results in cost savings, reduced lead times, and improved overall efficiency.

Smart analytics can be applied to monitor and optimize energy consumption in manufacturing processes. This not only contributes to sustainability goals but also reduces operational costs. Analytics allows companies to understand customer preferences and market trends, facilitating the customization of products to meet specific demands. This level of personalization is a hallmark of Industry 4.0. Continuous analysis of data generated during production processes helps in identifying areas for improvement. This leads to optimized workflows, increased productivity, and enhanced overall efficiency.

Instead of relying on gut feelings, decision-makers can now base their choices on concrete data. This not only reduces the margin of error but also ensures that decisions align with strategic goals. IoT and connected devices, data analytics can provide real-time monitoring of equipment and processes. This immediate feedback loop allows for quick responses to any deviations from the norm. There is literature available in the field of Industry 4.0 categorized by its applicability, shown in Table 11.1.

A brief review of the role of data analytics in Industry 4.0, along with some key references for further exploration, are presented in Table 11.1.

11.2 DATA ANALYTICS IN INDUSTRY 4.0

The approaches to data analytics in Industry 4.0 are diverse and dynamic. Descriptive analytics is the basic level, dealing with historical data to understand what happened

DOI: 10.1201/9781003473886-11

TABLE 11.1
Literature Available in the Field of Industry 4.0

Reference	Industry 4.0 Aspects Covered	Description
[Schwab, 2016]	Introduction to Industry 4.0	A comprehensive introduction to the Fourth Industrial Revolution
[Lu et al., 2017]	Technological Components	An exploration of the interplay between data analytics and the Internet of Things
[Davenport and Harris, 2007]	Data Analytics in Industry 4.0	A review of how data analytics is applied for predictive maintenance in Industry 4.0
[Lee et al., 2015]	Cyber-Physical Systems (CPS)	A review of how data analytics is applied for predictive maintenance in Industry 4.0
[Ashton, 2009; Zhang et al., 1996]	Security and Privacy Concerns	An examination of the challenges and solutions related to data security in the context of Industry 4.0
[Schuh et al., 2017]	Human-Machine Interaction	A study on how data analytics facilitates collaboration between humans and machines

in the past. It involves summarizing, aggregating, and presenting data to provide insights into trends and patterns. Moving a step further, diagnostic analytics aims to answer the question of why something happened. It involves digging deeper into data to uncover the root causes of certain outcomes or issues. Predictive analytics is where the crystal ball comes into play. Predictive analytics uses statistical algorithms and machine learning techniques to identify the likelihood of future outcomes based on historical data. Think of it as forecasting.

Beyond predicting, prescriptive analytics suggests actions to optimize a given outcome. It not only tells you what might happen but also provides recommendations on what actions to take to achieve a desired result. In Industry 4.0, real-time processing is crucial. Edge analytics involves analyzing data at the source (where it's generated), rather than sending everything to a centralized system. This reduces latency and allows for quicker decision-making. Cognitive analytics involves leveraging advanced technologies like artificial intelligence and machine learning to simulate human thought processes. It can understand, reason, and learn, providing a more intelligent and adaptive approach to analytics. With the increasing concern about data privacy, an approach that focuses on anonymizing and protecting sensitive information is crucial. This involves implementing encryption, access controls, and other security measures.

Combining the strengths of both humans and machines, human-machine collaboration approach involves using analytics to augment human decision-making rather than replacing it. It recognizes the unique capabilities of each and finds a harmonious balance. The effectiveness of these approaches often depends on the specific goals, challenges, and context of the Industry 4.0 implementation.

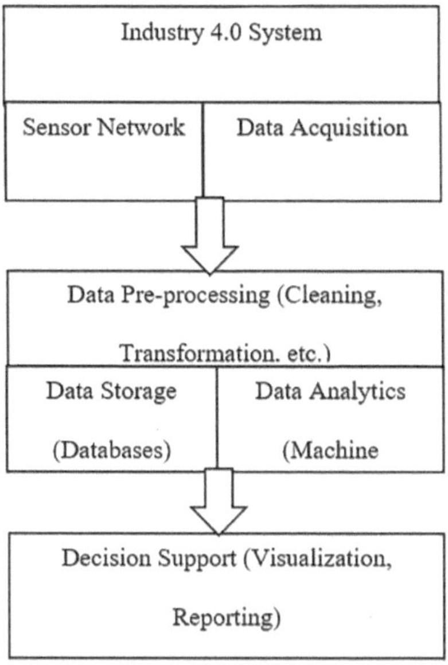

FIGURE 11.1 Industry 4.0 Data Analytics' Key Components.

TABLE 11.2
Algorithms Used in Data Analytics for Industry 4.0

Algorithm Category	Algorithms	Purpose
Machine Learning Algorithms	Linear Regression, Decision Trees, Random Forests	Used for predictive maintenance, quality control, and demand forecasting
	Support Vector Machines (SVM), Neural Networks	Applied in complex pattern recognition tasks, such as image and speech recognition in manufacturing processes
Clustering Algorithms	K-Means, Hierarchical Clustering	Employed for grouping similar data points, useful in identifying patterns and anomalies within large datasets
Time Series Analysis Algorithms	ARIMA (Auto Regressive Integrated Moving Average), Exponential Smoothing	Ideal for analyzing time-dependent data; often used in predicting equipment failures and production trends
Anomaly Detection Algorithms	Isolation Forest, One-Class SVM	Identify unusual patterns or outliers in data; crucial for detecting defects or irregularities in manufacturing processes
Association Rule Mining	Apriori Algorithm	Used for discovering interesting relationships and patterns in large datasets; applicable in supply chain optimization and market basket analysis

(Continued)

TABLE 11.1
(Continued)

Algorithm Category	Algorithms	Purpose
Natural Language Processing (NLP)	Word Embeddings, Text Classification	Applied to analyze unstructured data, such as maintenance logs, customer feedback, and documentation, to extract valuable insights
Reinforcement Learning	Q-Learning, Deep Reinforcement Learning	Applied in autonomous systems and robotics for optimizing processes and making real-time decisions
Optimization Algorithms	Genetic Algorithms, Simulated Annealing	Employed for optimizing production schedules, resource allocation, and supply chain logistics
Deep Learning Algorithms	Convolutional Neural Networks (CNN), Recurrent Neural Networks (RNN)	Used for image analysis, sequence prediction, and processing large amounts of sensor data
Fuzzy Logic	Fuzzy Inference Systems	Applied for systems with uncertainty and imprecision, such as quality control and decision-making in uncertain environments

Figure 11.1 shows the block diagram comprising the key components of Industry 4.0 data analytics. The process begins with a network of sensors and smart devices collecting data from various sources within the manufacturing environment. Raw data goes through pre-processing steps such as cleaning, transformation, and normalization to ensure its quality and readiness for analysis. Processed data is stored in databases or other storage systems to facilitate easy retrieval and accessibility.

Data analytics is the heart of the system where advanced analytics techniques such as machine learning, statistical analysis, and predictive modeling are applied to derive meaningful insights from the data. The results of the analytics phase are presented through visualization tools and reports, providing decision-makers with actionable insights. The entire system is part of the broader Industry 4.0 framework, contributing to the intelligent, connected, and data-driven manufacturing processes.

In Industry 4.0, various algorithms are employed for data analytics to extract valuable insights and support decision-making processes. The choice of algorithm depends on the specific use case and the nature of the data. Some common types of algorithms used in data analytics for Industry 4.0 are shown in Table 11.2. The choice of algorithm often involves a trade-off between computational complexity, interpretability, and the specific requirements of the Industry 4.0 application. The field is evolving, and newer algorithms and techniques continue to emerge to address the unique challenges posed by the integration of advanced analytics in manufacturing processes.

A simple algorithm for data analytics in an Industry 4.0 context is specified as follows. The library for data manipulation and for basic visualization can be utilized. This algorithm assumes a CSV file with sensor data is available for data analysis.

Algorithm: Data analytics in an Industry 4.0 context

1. Load dataset//Assuming dataset has columns like 'Sensorl', 'Sensor2', etc.
2. Extract relevant features for clustering.
3. Standardize the data.
4. Applying clustering algorithm.
5. Visualizing the clusters.

11.3 DECISION-MAKING IN INDUSTRY 4.0

Decision-making in Industry 4.0 is a critical aspect, often facilitated by advanced analytics and technologies. Decision-making in the context of Industry 4.0 is marked by the integration of real-time data, predictive analytics, and automation to enable agile and informed choices. It empowers organizations to respond swiftly to dynamic production environments. Data analytics plays a pivotal role in decision-making by providing actionable insights from vast amounts of data. Predictive maintenance, quality control, and supply chain optimization are areas where analytics contributes significantly [Davenport and Harris, 2007].

Integration of Artificial Intelligence (AI) in decision support systems, including machine learning algorithms, enhances decision-making by automating complex analyses and learning from historical data. Decision Support Systems (DSS) leverage AI to provide real-time recommendations [Turban et al., 2017]. Industry 4.0 emphasizes the collaboration between humans and machines. Decision-making benefits from human intuition and creativity, coupled with machine efficiency and data-driven insights. Algorithms for optimization, like genetic algorithms and simulated annealing, contribute to decision-making in resource allocation, production scheduling, and logistics optimization [Goldberg, 1989].

The integration of CPS in Industry 4.0 leads to automated decision-making based on real-time feedback from physical processes, enhancing efficiency and reducing response times [Lee et al., 2015]. Decision-making in Industry 4.0 brings challenges related to data security, privacy, and ethical considerations. Striking a balance between automation and human oversight is crucial [Ienca and Vayena, 2018]. The landscape of Industry 4.0 is dynamic, and ongoing research contributes to evolving practices in decision-making methodologies. Figure 11.2 shows the block diagram comprising the key components of Industry 4.0 decision making.

In Industry 4.0, data is continuously collected from sensors and devices in real-time, providing a stream of information about the manufacturing processes. Advanced analytics techniques, including predictive modeling and machine learning, are applied to historical and real-time data to forecast future events or outcomes. Decision support systems, often powered by AI and machine learning, assist human decision-makers by providing insights, recommendations, and predictions. Decision-making involves a combination of human expertise and automated processes. Human decision-makers bring domain knowledge, intuition, and creativity, while automated systems handle algorithmic decisions based on data.

The decisions made are implemented in the manufacturing processes. This could involve automated actions executed by machines, adjustments to production parameters, or human interventions based on the decisions. This block diagram illustrates

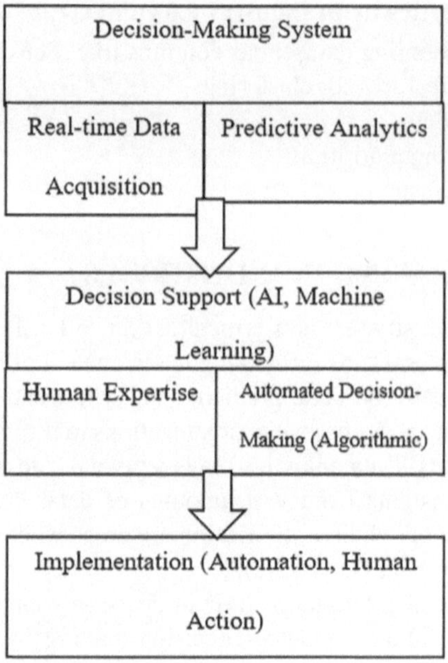

FIGURE 11.2 Industry 4.0 Decision-Making Key Components.

the flow from real-time data acquisition to decision support and the ultimate implementation of decisions. It emphasizes the collaboration between human expertise and automated decision-making systems in the Industry 4.0 context. The specifics of this diagram may vary based on the complexity and requirements of a particular Industry 4.0 system. Implementing a comprehensive decision-making system in Industry 4.0 involves various components, including data processing, analytics, and potentially machine learning. A simplified algorithm for data manipulation and a hypothetical decision-making scenario is specified below.

Algorithm: Decision making in an Industry 4.0 context

1. Load dataset//Assuming dataset has columns like 'Sensorl', 'Sensor2', etc.
2. Features (X) are sensor readings, and the target variable (y) is DecisionLabel.
3. Extract features and target variable.
4. Split the data into training and testing sets.
5. Initialize a classifier.
6. Train the model.
7. Make predictions on the test set.
8. Evaluate the accuracy.
9. Use this trained model for decision-making based on new sensor data.

In a real Industry 4.0 scenario, likely need more sophisticated pre-processing, feature engineering, and model tuning. Additionally, the decision-making process could involve more advanced techniques like reinforcement learning or optimization algorithms based on the specific nature of problem. Decision-making in Industry 4.0 is a multifaceted process that integrates advanced technologies, data analytics, and human expertise to optimize manufacturing operations. The key elements of decision-making in industry 4.0 are shown in Table 11.3.

Decision-making in Industry 4.0 is a dynamic and evolving field, with ongoing research contributing to the development of more sophisticated and efficient systems.

TABLE 11.3
Key Elements of Decision-Making in Industry 4.0

Element	Description	Objective	Technology
Real-Time Data Acquisition	Industry 4.0 relies on a network of sensors and IoT devices to collect real-time data from various stages of the manufacturing process	Obtain accurate and timely information about production, quality, and equipment status	Sensors, IoT devices, PLCs
Data Processing and Analytics	Raw data is processed, cleaned, and analyzed to extract meaningful insights using techniques such as machine learning and statistical analysis	Identify patterns, trends, and anomalies in the data to support informed decision-making	Data analytics tools, machine learning algorithms
Decision Support Systems (DSS)	AI-powered decision support systems provide real-time recommendations and predictions based on the analyzed data	Assist human decision-makers in making informed choices by presenting relevant insights	Machine learning models, AI algorithms
Human Expertise and Collaboration	Human operators and decision-makers bring domain knowledge, intuition, and creativity to the decision-making process	Combine human expertise with data-driven insights for well-rounded decisions	Collaborative platforms, augmented reality
Automated Decision-Making	Certain decisions can be automated based on predefined rules or machine learning models	Improve efficiency by automating routine and repetitive decisions	Rule-based systems, machine learning algorithms
Implementation and Action	Decisions are implemented through actions taken in the manufacturing environment, such as adjusting production parameters or initiating maintenance	Translate decisions into operational changes to optimize processes	Automation systems, robotics
Monitoring and Feedback	Continuous monitoring of implemented decisions with feedback loops for performance evaluation	Ensure the effectiveness of decisions and make adjustments as necessary	Monitoring systems, KPI dashboards

(*Continued*)

TABLE 11.1
(Continued)

Element	Description	Objective	Technology
Adaptability and Continuous Improvement	Industry 4.0 decision-making is dynamic, adapting to changing conditions and continuously improving based on feedback	Foster agility and responsiveness in the manufacturing processes	Adaptive algorithms, continuous improvement frameworks
Security and Ethical Considerations	Decision-making systems must address cybersecurity concerns and ethical considerations related to data privacy and algorithmic bias	Ensure the integrity, confidentiality, and ethical use of data in decision-making	Secure communication protocols, ethical AI frameworks
Collaboration between Humans and Machines	The synergy between human decision-makers and automated systems enhances overall decision-making capabilities	Leverage the strengths of both humans and machines for optimal outcomes	Human-machine interfaces, collaborative technologies

11.4 CONCLUSION AND FUTURE DIRECTIONS

In conclusion, data analytics in Industry 4.0 is the backbone of a transformative revolution in manufacturing. The integration of advanced analytics, machine learning, and real-time data processing reshapes how industries operate, optimize processes, and make informed decisions. In essence, data analytics in Industry 4.0 is not just a tool for processing information; it's a catalyst for innovation, efficiency, and strategic decision-making. As industries continue to embrace and refine these practices, the transformative impact on manufacturing processes will only intensify, shaping the future of industrial ecosystems.

Decision-making in Industry 4.0 represents a paradigm shift in manufacturing, integrating cutting-edge technologies, data analytics, and human expertise. This transformative approach enhances operational efficiency, agility, and adaptability, ultimately shaping the future of industrial processes. Decision-making in Industry 4.0 marks a significant shift toward intelligent, adaptive, and collaborative manufacturing ecosystems. As technologies continue to evolve, so, too, will the sophistication and impact of decision-making processes in the industrial landscape.

The future scope of data analytics in Industry 4.0 is incredibly promising, with ongoing advancements poised to revolutionize manufacturing processes and decision-making. The key areas that highlight the evolving landscape and potential developments are: Edge Analytics Advancements, AI and Machine Learning Integration, Exponential Growth in Data Sources, Explainable AI for Trust and Transparency, Integration of Blockchain for Data Security, Human Augmentation and Cognitive Analytics, Digital Twins for Simulation and Optimization, Quantum Computing

Impact, Ethics and Responsible AI Practices, Robust Cybersecurity Measures, Interoperability Standards, and Skill Development and Workforce Training.

The future scope of decision-making in Industry 4.0 is dynamic and expansive, driven by advancements in technology, data analytics, and the evolving needs of manufacturing ecosystems. The key areas that outline the potential developments and trends in the future of decision-making are: Exponential Growth in Data Integration, Advanced Predictive Analytics, Autonomous Decision-Making Systems, Explainable AI for Trust and Accountability, Human-Centric Decision Support, Dynamic Adaptive Systems, Interconnected Decision Ecosystems, Ethical and Responsible Decision-Making, Human Augmentation Technologies, Robust Cyber-Physical Security Measures, Continuous Learning and Adaptability, and Quantum Computing Impact.

REFERENCES

Ashton, K. (2009). That 'Internet of Things' thing. RFID Journal. http://www.rfidjournal.com/articles/view?4986

Davenport, T. H., & Harris, J. (2007). Competing on analytics. Analytics and Data Science, January 2006.

Goldberg, D. E. (1989). Genetic algorithms in search, optimization, and machine learning. Boston, MA: Addison-Wesley Longman Publishing Co., Inc.

Ienca, M., & Vayena, E. (2018, April). On the responsible use of digital data to tackle the COVID-19 pandemic. Nat Med;26(4):463–464. doi: 10.1038/s41591-020-0832-5.

Lee, J., Bagheri, B., & Kao, H. A. (2015). A cyber-physical systems architecture for Industry 4.0-based manufacturing systems. Manuf Lett; 3:18–23.

Schwab, K. (2016). The fourth industrial revolution. https://www.weforum.org/about/the-fourth-industrial-revolution-by-klaus-schwab/

Turban, E., Sharda, R., Delen, D., & Aronson, J. E. (2017). Decision support and business intelligence systems. https://archive.org/details/Decision-Support-And-Business-Intelligence-Systems_201808

Zhang, K., Sandhu, R., & Ahn, G. J. (1996). Role-based access control (RBAC) in cyber-physical systems (CPS): A case study. IEEE Comp;29(2):38–47.

12 Evolving Landscape of Industrial Engineering in the Modern Era

Ida Hector, Rukmani P., and Benson Edwin Raj

12.1 INTRODUCTION

The Fourth Industrial Revolution, often referred to as Industry 4.0, is an era characterized by the merging of technologies that bring together the physical, digital, and biological worlds. It defines a tremendous improvement in employment, lifestyle, and interaction with our environment. Artificial intelligence (AI), the Internet of Things (IoT), nanotechnology, automation, biotechnology, along with other cutting-edge sciences are the driving forces behind this transformation. In this chapter, we'll examine Industry 4.0's salient characteristics and talk about its significant social implications.

12.1.1 Definition of the Fourth Industrial Revolution

Originally, the "Industrial Revolution" refers to dramatic changes in manufacturing and production techniques. The Fourth Industrial Revolution builds upon the digital advancements of the previous era and goes a step further by integrating technologies like AI, machine learning (ML), robotics, IoT, nanotechnology, biotechnology, and others [1]. Various sectors are being transformed by the Fourth Industrial Revolution, which is characterized by connectivity, automation, AI, a data-driven decision making, smart systems, and human-machine interaction. The increasing use of high-speed internet and smart gadgets has made data exchange and communication between people, machines, and things much easier. The sharing and use of information has been revolutionized by this interconnection.

AI and ML-based automation technologies have given robots the ability to complete difficult jobs on their own. These technological developments have rendered unnecessary the need for considerable human involvement in several operations, from autonomous automobiles to intelligent systems. Data-driven decision making has also been made possible by the capacity to analyze massive volumes of data using sophisticated algorithms, informing, and improving decision-making processes across sectors. The IoT is used by the Fourth Industrial Revolution to link and coordinate physical systems, equipment, and things. Due to this interconnection, operations can be monitored, controlled, and optimized in real-time, which improves resource allocation, efficiency, and user experiences.

 DOI: 10.1201/9781003473886-12

12.1.2 Historical Background

The Fourth Industrial Revolution marks a new stage in technical development and builds on the achievements of the preceding three. Let's quickly review the previous industrial revolutions to comprehend its historical context.

First Industrial Revolution (Late 18th—Early 19th century): Agricultural and skilled labour-based economies gave way to mechanization and mass manufacturing during the First Industrial Revolution. It all started in Britain with the adoption of water and steam-powered technology in the textile industry, which resulted in higher production and economic expansion.

Second Industrial Revolution (Late 19th—Early 20th century): Electrical innovation, railway network growth, and the advent of mass manufacturing were the main features of the Second Industrial Revolution. The rise of globalization was sparked by inventions like the postal service, telephone, and the factory floor, which revolutionized production and communication methods.

Third Industrial Revolution (Mid-20th century): This Industrial Revolution (also known as the Digital Revolution) witnessed the growing deployment of digital devices, electronic devices, computers, and robotics. With the rise of desktop computing devices, the discovery of the incorporated circuit, and the creation of the internet were significant developments. This shift resulted in important developments in connectivity, computer technology, and the digitization of several sectors.

The Emergence of the Fourth Industrial Revolution: The Fourth Industrial Revolution, also referred as I4.0, has acquired popularity over the past decade as an outcome of developments in digital technology, the internet, and the fusion of numerous industries. Its increase is a result of several important reasons. Complex statistical analysis, ML, and AI are now possible due to the rapid rise in computational power, the creation of strong systems, and advances in data storage. International exchange of information, exchange of data, and cooperation have been made significantly easier due to the widespread use of the internet and the growing accessibility of high-speed connectivity. The internet has developed into a critical piece of infrastructure for facilitating the digital transformation of several businesses. The gathering of enormous volumes of data produced by computers, sensors, and linked devices has opened doors for sophisticated analytics and decision-driven insights. Big data analytics makes it possible to extract useful data and patterns that weren't previously available. The I4.0 incorporation of sophisticated robotics and AI is another distinguishing feature. The robotics industry, AI, and ML have all contributed to the creation of fully autonomous systems, cognitive devices, and sophisticated algorithms that can carry out challenging tasks [2]. Furthermore, by linking physical items and technologies to the internet, the IoT is essential to I4.0. Due to the real-time data collecting, remote monitoring, and control made possible by this interconnectivity, processes are more efficient and chances for innovation are increased. Industry 4.0 is the result of the fusion of several technological areas, including AI, robotics, nanotechnology, biotechnology, 3D printing, and renewable energy. The revolution is further advanced by the synergies and amplified revolutionary potential brought forth by this confluence. Together, these elements have formed I4.0, revolutionizing industries and opening the path for higher productivity, economic development, and sector-changing breakthroughs. Figure 12.1 visualizes the different eras of industrial revolution.

12.1.3 Importance of the Fourth Industrial Revolution

The IoT, robots, automation, and other digital technologies are being integrated into different facets of society and the economy as part of the Fourth Industrial Revolution (4IR). It signifies a substantial change in the way we operate, engage, and live. Numerous important characteristics help us comprehend the significance of the Fourth Industrialization.

Economic Impact: I4.0 offers the prospective to significantly boost both productivity and economic growth. In addition to altering current industries and business models, it may also generate new ones. AI, automation, and IoT technologies help businesses to automate tasks, increase productivity, and create new goods and services. Improved profitability and employment growth may follow from this.

Societal Transformation: Through improvements in healthcare, transportation, and communication, I4.0 raises the standard of living. For instance, digitized medical and telemedicine systems can improve accessibility to medical assistance, particularly in rural regions. Smart towns and cities may increase sustainability, resource efficiency, and urban experience in

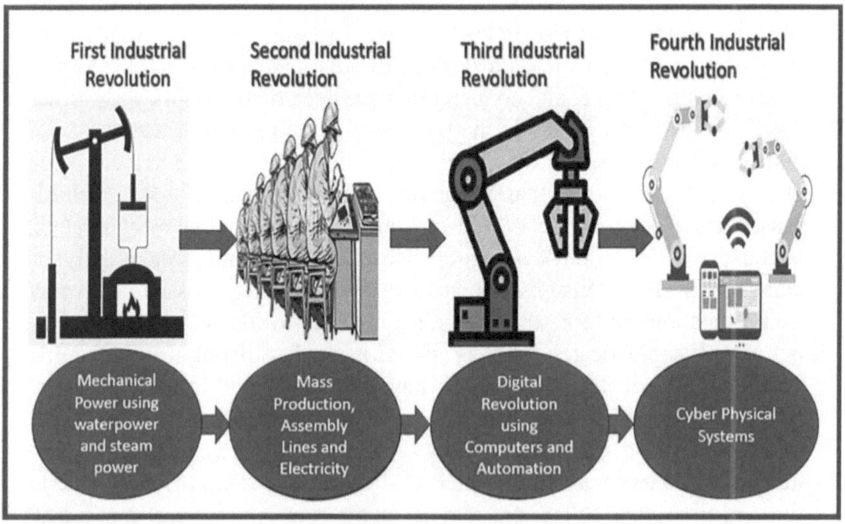

FIGURE 12.1 Industrial Revolution.

general [3]. However, it is essential to make sure that these changes are comprehensive and beneficial to society.

Technological Advancements: Technological development is advancing rapidly during the Fourth Industrial Revolution. As AI and ML algorithms advance, computers will be able to absorb and analyze enormous volumes of data, predict, and gain insight from their mistakes. Manufacturing is being transformed by robotics and automation, which is also changing sectors like logistics and agriculture. These scientific and technical developments can progress many sectors and tackle difficult challenges [4].

Workforce Transformation: The nature of employment and the skills needed to thrive in the labour market are changing because of I4.0. Routine and repetitive work are rapidly being mechanised as ML and AI technologies improve [5]. Employees must adjust to this and learn new technologies-friendly abilities, such as innovation, analytical problem-solving, and digital literacy. Initiatives for workforce transformation and upskilling are now necessary to make sure people are ready for occupations of the future.

12.2 KEY TECHNOLOGIES OF THE FOURTH INDUSTRIAL REVOLUTION

The Fourth Industrial Revolution (4IR) encompasses a range of groundbreaking technologies that are revolutionizing industries. These interconnected technologies, such as AI, IoT, robotics and automation, Big Data and analytics, blockchain, and 3D printing, are driving rapid advancements, and are pictured in Figure 12.2. AI enables machines to perform tasks that require human intellect, while IoT facilitates the exchange of data between interconnected objects. Robotics and automation are transforming industries through the integration of AI and advanced sensors [6]. Big Data and analytics enable organizations to draw insights from vast amounts of data, while blockchain ensures secure and decentralized transactions. Lastly, 3D printing allows for efficient and customizable production of physical objects.

These technologies are converging and integrating, leading to profound consequences for industries, societies, and economies. Nanotechnology, biotechnology, quantum computing, renewable energy, and augmented reality/virtual reality are other significant technologies contributing to the Fourth Industrial Revolution. As these technologies continue to evolve and interact, they shape the future of various sectors, enabling innovative solutions, improving efficiency, and driving economic growth.

12.2.1 Artificial Intelligence

Industry 4.0 is characterized by the incorporation of technological advances and digitization into production and industrial operations. AI permits machines and systems to imitate cognitive abilities, analyze huge quantities of statistics, make intelligent choices, and streamline tasks.

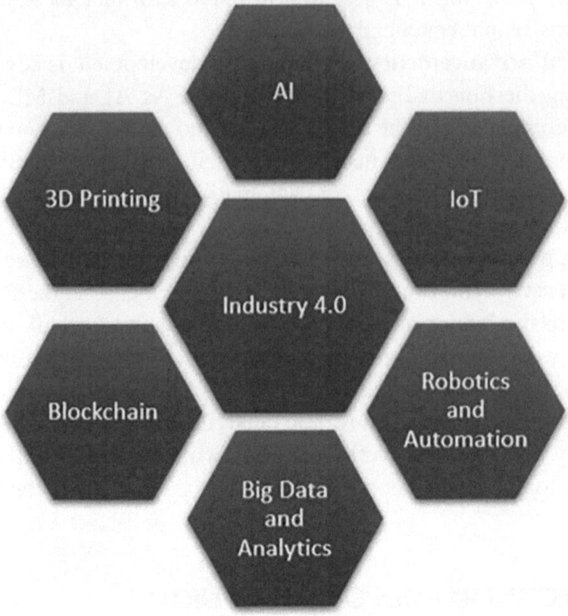

FIGURE 12.2 Key Technologies of Industry 4.0.

Predictive Maintenance [7]: Sensor data and ML algorithms are used by AI-powered predictive maintenance systems to anticipate machinery faults. AI can foresee maintenance requirements, plan service calls, and reduce downtime by examining previous data and spotting patterns. This strategy improves operating efficiency overall, lowers costs, and optimizes maintenance routines.

Quality Control [8]: By spotting flaws and irregularities in real-time, AI systems can improve quality control procedures. Visual examination of items and components is made possible by computer vision algorithms, guaranteeing that quality requirements are met. To enhance product quality and decrease waste, AI-powered systems may analyze sensor data, spot trends, and send out warnings when deviations or irregularities are found.

Autonomous Robots: Robots that are autonomous and capable of carrying out jobs that were previously performed by people can be created thanks to AI [9]. These robots employ AI algorithms for vision, navigation, and decision-making, enabling them to operate independently or in collaboration with people in challenging production situations. Robots that operate autonomously improve efficiency, productivity, and safety by automating dangerous or repetitive jobs.

Supply Chain Optimization: To improve supply chain operations, AI systems can analyze enormous volumes of data from several sources. Demand forecasting, inventory optimisation, logistics route optimization, and supply chain efficiency may all be improved by AI [10]. AI assists in lowering costs, enhancing delivery times, and improving overall supply chain efficiency by seeing patterns and trends in data.

Smart Manufacturing and Process Optimization [11]: AI makes it possible to create linked, productive manufacturing systems that integrate and optimize systems, processes, and equipment. AI systems may continually monitor and analyze data from sensors, equipment, and manufacturing lines to increase performance overall, streamline operations, and detect bottlenecks. As a result, there is less waste, better resource utilization, and more production.

Natural Language Processing and Chatbots: Machines can comprehend and communicate with people using spoken or written language thanks to artificial AI-powered natural language processing (NLP) [12]. Through the usage of chatbots, NLP is used in customer care and assistance. Chatbots may aid with basic problem-solving, comprehend consumer inquiries, and deliver pertinent information, improving the customer experience while lightening the strain on human agents.

12.2.2 Internet of Things

Industry 4.0, which refers to the incorporation of digital technology into manufacturing and industrial processes, heavily relies on the IoT. The IoT is a network of physically linked sensors, systems, and objects that gather and share data [13]. IoT facilitates the development of intelligent, networked systems that support automation, data-driven decision-making, and optimization in the framework of Industry 4.0.

Smart Monitoring and Control: Real-time monitoring and control of industrial processes and equipment are made possible by the IoT. Sensors integrated into equipment and devices gather data on variables like temperature, pressure, humidity, and energy consumption. This data is transmitted over the IoT network, allowing operators and managers to remotely monitor the status and performance of equipment. Real-time monitoring aids in spotting anomalies, predicting maintenance requirements, and enhancing operations.

Predictive Maintenance [7]: To anticipate equipment breakdowns, IoT-enabled predictive maintenance frameworks use data gathered from sensing and other components. IoT and ML algorithms can predict maintenance needs and arrange repairs in advance by examining past data and spotting patterns. This strategy helps to minimize unexpected service interruptions, cut expenses, and optimize maintenance routines.

Supply Chain Optimization: Total transparency as well as supply chain optimization are made possible by IoT. Real-time statistics on the state of inventory, position monitoring, and working conditions may be gathered by integrating IoT devices and sensors at many locations in the supply chain, including warehouses, logistics, and transportation [10]. Better demand forecasting, inventory control, and logistics planning are made possible as a result, which lowers costs, increases productivity, and improves customer satisfaction.

Energy Management and Sustainability [14]: Intelligent energy management and sustainable business practices are made possible by IoT. Businesses may spot energy inefficiencies and adopt energy-saving measures by installing IoT devices and sensors to monitor energy use trends. IoT can automatically change settings to save waste and optimize energy use based on demand. As a result, energy costs decrease sustainability is increased, and environmental effect is decreased.

Product Tracking and Tracing [15]: IoT makes is possible to track and trace end-to-end product in the supply chain. Businesses may track the location, state, and movement of items from production through delivery by employing RFID or IoT-enabled tags. This offers transparency, allowing real-time inventory management. Aids in finding any manufacturing or logistical process bottlenecks or inefficiencies.

Data Analytics and Optimization: Large volumes of data are produced by IoT from linked devices and sensors. Companies may gain useful insights from this data by utilizing sophisticated analytics, ML, and AI [16]. Making data-driven judgements and streamlining procedures are both possible with these revelations. Data analytics, for instance, may optimize energy usage, production schedules, and failure trends in equipment.

12.2.3 Big Data and Analytics

Industry 4.0 refers to the incorporation of digital technology and automation into manufacturing and industrial processes, and big data and analytics play a crucial part in this movement. Analytics entails drawing conclusions and patterns from this data to guide decision-making and spur innovation. Big Data is the enormous volume of organized and unstructured data produced from diverse sources.

Data gathering and fusion: Industry 4.0 depends on the gathering and fusing of data from a variety of sources, including as sensors, devices, production lines, supply chain systems, and consumer interactions [17]. These many datasets may be stored, processed, and integrated using big data technology, producing a thorough and unified perspective of operations.

Real-time Monitoring and Predictive Analytics: Real-time monitoring and predictive analytics are made possible by big data analytics in industrial operations. Patterns and abnormalities may be found by continually analysing streaming data from sensors and equipment, which enables the early detection of possible failures or problems. Algorithms for predictive analytics can foresee equipment breakdowns, optimize maintenance schedules, and boost operational effectiveness by using historical data and ML approaches [18].

Process Optimization: Through the analysis of vast amounts of data to pinpoint bottlenecks, inefficiencies, and opportunities for development, Big Data analytics allows process optimization. Analytics may offer insights into simplifying supply chain operations, eliminating waste, enhancing quality control, and optimizing production schedules by seeing patterns and correlations in the data. This results in increased productivity, lower expenses, and better overall results.

Quality Control and Defect Detection [8]: Dataset from sensors, equipment, and manufacturing lines are analyzed using Big Data analytics to improve quality control procedures. Sophisticated statistical techniques can find trends and abnormalities in the data, allowing for early defect or quality standard deviation identification. To ensure constant product quality and minimize waste, real-time warnings and notifications may be issued.

Demand Forecasting and Inventory Management: Big Data analytics aids to increase the precision of demand forecasts and enhance inventory control. Analytics algorithms can forecast demand patterns and variations by examining past sales data, market trends, consumer behaviour, and other pertinent aspects [19]. In turn, this helps businesses to decrease costs and enhance customer satisfaction by optimizing production planning, adjusting inventory levels, and avoiding stockouts or overstocking.

Customer Insights and Personalization [20]: Big Data analytics helps businesses to learn important things about the behaviour, interests, and demands of their customers. Companies may comprehend client mood, spot trends, and personalize products and services by analyzing data from customer interactions, social media, and other sources. This aids in developing targeted marketing campaigns, creating better products, and enhancing the customer experience, all of which promote customer loyalty and boost market competition.

12.2.4 Robotics

The Fourth Industrial Revolution, referred to as Industry 4.0, is heavily reliant on automation and robotics. The use of digital technology in manufacturing and industrial processes is what defines this revolution. Automation and robotics are essential for changing industries and promoting production, efficiency, and innovation.

Enhanced Productivity and Efficiency [21]: Robotics and automation technologies make it possible to automate manual, repetitive processes that were previously done by people. Robots are capable of continuous, arduous labour, which boosts productivity and efficiency in manufacturing and industrial operations. They can complete jobs quickly, accurately, and with great precision, which leads to better quality and faster output.

Improved Safety [21]: Robots and automation technologies help to increase safety in potentially dangerous or physically demanding work conditions. Robots lower the risk of workplace accidents and injuries by substituting for people in hazardous activities like operating large machinery or working in harsh environments. By doing this, possible health and safety issues are reduced and workers' working environments are made safer.

Flexible and Scalable Production [22]: Robotics and automation make it possible to produce goods in a flexible and scalable manner. Robotics allows manufacturing lines to be readily adjusted and modified to meet shifting customer or market needs. Automation systems can adapt to changes in production levels, enabling effective scaling up or down without a lot of manual work. This adaptability improves industrial operations' agility and reactivity.

Collaborative Robots (cobots): Collaborative robots, also known as cobots, are introduced by Industry 4.0 and can collaborate with people in a shared workspace. Cobots are automated assistants that work alongside people to help them complete tasks and increase productivity [23]. Cobots provide new opportunities for automation in a variety of sectors since they are outfitted with cutting-edge sensors and security measures to ensure a secure and productive human-robot cooperation.

Enhanced Quality Control [8]: Automation and robotics technology help industrial operations have better quality control. Automated systems can carry out accurate and reliable inspections, guaranteeing compliance with quality requirements and minimizing flaws. Real-time monitoring, anomaly detection, and alarm generation are made possible by sensors and machine vision systems. These systems may also identify deviations from anticipated quality. Customer satisfaction and product quality are therefore improved.

Human-Machine Collaboration and Skill Development [24]: Collaboration between machines and the development of skills Industry 4.0 places a strong emphasis on human-machine cooperation. Employees may concentrate on more complicated and value-added jobs that call for cognitive abilities, problem-solving, and creativity as automation replaces routine work. This change in jobs opens chances for the workforce to reskill and upskill, allowing people to collaborate with robots and contribute to higher-value activities.

12.2.5 Blockchain Technology

Industry 4.0 may use the disruptive innovation of blockchain technology, which has attracted a lot of interest. Blockchain is a distributed and decentralized digital ledger that keeps track of transactions across several computers in an open, safe, and unchangeable way.

Supply Chain Transparency and Traceability [25]: Blockchain improves supply chain management transparency and traceability. Stakeholders can trace and confirm the items' provenance, legitimacy, and journey by documenting every transaction and movement of commodities on the blockchain. This aids in lowering counterfeiting, assuring moral sourcing, strengthening customer confidence, and improving quality control.

Decentralization and Peer-to-Peer Networks [26]: Because of its decentralized structure, blockchain transactions do not require centralized authority or middlemen. Direct peer-to-peer contacts are made possible, reliance on middlemen is decreased, and participant confidence is increased. Blockchain can enable direct communication and teamwork between machines in Industry 4.0, facilitating self-sufficient decision-making, data exchange, and safe interactions in decentralized networks.

Data Security and Privacy [25]: Blockchain uses cryptographic methods to offer strong data security and privacy. Blockchain is extremely hard to attack and tamper with because of its decentralized and unchangeable nature. Industries can maintain the integrity and security of vital information by storing sensitive data on the blockchain. This is especially useful in fields that deal with sensitive data, including banking, healthcare, and intellectual property.

Supply Chain Efficiency and Automation: By automating procedures, simplifying documentation, and minimizing paperwork, blockchain technology improves supply chain efficiency. Businesses may streamline and speed up procedures like customs clearance, logistics, and inventory management by digitizing and safely storing papers and information on the blockchain. As a result, the supply chain is more efficient overall, transactions are completed more quickly, and administrative expenses are decreased.

Data Management and Monetization: With the use of blockchain, people may own and own their data. Users may choose which individuals or organizations have access to their data, and they can get payment for doing so. As a result, people are given more autonomy over their personal data and more prospects for data monetization. Blockchain provides fair and transparent data transactions, encourages safe data exchange, and rewards data collaboration in Industry 4.0.

Interoperability and Standards [26]: Interoperability and standardization between many platforms and systems are encouraged by blockchain technology. Industries can improve seamless data interchange, interoperability between various technologies, and integration of diverse systems by using common protocols and standards. This makes it possible for stakeholders in Industry 4.0 ecosystems to collaborate, share data, and operate together effectively.

It's critical to recognize that blockchain technology is still in its early stages and that there are still issues to be solved, such as those related to scalability, energy usage, legal frameworks, and adoption hurdles. Ongoing research and development initiatives are, however, concentrated on overcoming these issues and maximizing the potential of blockchain in Industry 4.0.

12.3 IMPACTS OF THE FOURTH INDUSTRIAL REVOLUTION ON SOCIETY

The Fourth Industrial Revolution has brought about significant impacts on society, with both advantages and concerns. It has reshaped the economy by transforming industries, employment positions, and business models, leading to productivity gains, economic growth, and the emergence of new businesses and job opportunities. However, it also presents challenges in terms of job displacement, skill requirements, and potential inequalities. The revolution necessitates a highly competent and flexible workforce, driving the need for lifelong learning and continuous skill development. It also highlights the digital divide and the importance of ensuring equal access to technology and digital literacy to prevent further inequalities. The societal transformation brought by the revolution changes how people interact and communicate, offering new connectivity opportunities but also raising concerns about security, privacy, and misinformation.

Ethical and legal considerations arise in areas such as algorithmic bias, data protection, privacy, and cybersecurity, calling for comprehensive frameworks and regulations. The healthcare and well-being industries benefit from digital technologies but face challenges related to data security and privacy. Environmental sustainability opportunities exist through technologies like smart grids and renewable energy, but energy consumption and electronic waste need to be addressed responsibly. Governance and policy implications require adaptations to regulatory frameworks and collaborative efforts among governments, businesses, academia, and civil society. In conclusion, the Fourth Industrial Revolution has had a profound impact on society, encompassing economic, societal, ethical, environmental, and governance aspects. Embracing its potential while addressing challenges is crucial to ensure inclusive and responsible development. Proactive measures, collaboration, and a holistic approach are essential to maximize the benefits and navigate the complexities of this revolution. The impacts of industry 4.0 are depicted in the below Figure 12.3.

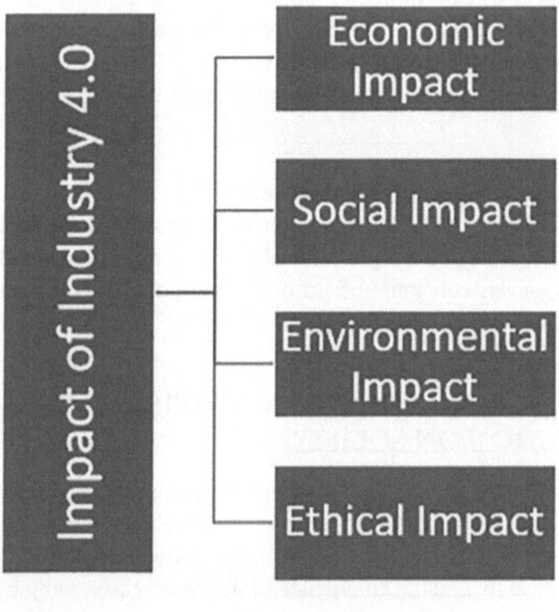

FIGURE 12.3 Impact of Industry 4.0.

12.3.1 Economic Impacts

Industry 4.0 has a big impact on different facets of economic development since it integrates cutting-edge technologies and digitalization into the economy.

Increased Efficiency and Productivity [27]: Automation, robotics, and AI (AI) are examples of Industry 4.0 technologies that improve productivity and efficiency in industrial and manufacturing operations. Automation lowers human error rates, speeds up production, and allows for continuous operations, increasing output levels while using fewer resources. The increased productivity supports economic expansion and competitiveness.

Job Transformation and Creation: Despite worries about job loss brought on by automation, Industry 4.0 also causes change and the emergence of new work opportunities. New employments are created in fields like data analysis, cybersecurity, robotics programming, and system maintenance even though repetitive and routine labour may be mechanized. The use of Industry 4.0 technology also promotes the expansion of associated sectors and generates new job possibilities.

Innovation and New Business Models [27]: Industry 4.0 fosters the emergence of new business models. Through digital technologies, data analytics, and networking, businesses can gather and analyze massive amounts of data, producing insights that foster innovation. This makes it simpler to develop novel products, services, and business models to meet shifting consumer preferences and needs.

Customization and Personalization: Mass customization and personalization are made possible by Industry 4.0. Advanced analytics and 3D printing are two examples of digital technologies that enable the affordable manufacture of highly customized products. The transition from mass production to personalized manufacturing satisfies the growing consumer demand for individualized items while also improving customer experiences.

Entrepreneurship and Small Business Empowerment: Industry 4.0 lowers entry barriers and gives small enterprises and entrepreneurs more influence. Startups and small businesses may now access global markets, compete with established players, and use digital tools for growth, thanks to advanced technologies, cloud computing, and e-commerce platforms. This encourages creativity, diversity in the economy, and job growth.

Economic Resilience and Adaptability [27]: Because technology enables quick adjustment to shifting market conditions, Industry 4.0 increases economic resilience. Digital technologies enable adaptable manufacturing procedures, speedy production line reconfiguration, and real-time customer response. In a dynamic corporate world, this adaptability aids organizations in remaining competitive and nimble.

Sustainable Development: By enhancing resource efficiency and minimizing environmental impact, Industry 4.0 encourages sustainable development. Businesses may optimize energy use, cut waste, and use predictive maintenance by integrating smart sensors, IoT devices, and data analytics. This supports the development of an economy that is more environmentally friendly and sustainable.

While there are many economic advantages to Industry 4.0, it's critical to address issues including worker migrations, privacy worries, cybersecurity dangers, and the digital divide. To maximize the beneficial economic impact of Industry 4.0, it is crucial to ensure a favourable policy climate, invest in education and skill development, and promote cooperation between the public and private sectors.

12.3.2 Social Impacts

With the digitalization and integration of cutting-edge technologies into various facets of society, Industry 4.0 has a significant societal impact. While it offers many advantages and opportunities, it also poses problems and prompts crucial questions.

Skills Development and Education: Industry 4.0 demands a change in how people are educated and improve their talents. Digital literacy, critical thinking, problem-solving, and adaptability are skills that are in greater demand. Institutions of higher learning must modify their curricula to provide students with the knowledge and abilities needed to participate in the digital economy. Initiatives for lifelong learning and upskilling programmes are essential for closing the skills gap.

Social Inclusion and Digital Divide [28]: Concerns concerning social inclusion and the digital divide are brought up by Industry 4.0. The availability of digital technologies, including cell phones, computers, and high-speed internet, varies across geographical and social boundaries. To avoid escalating already-existing disparities and excluding people from the advantages of Industry 4.0, it is crucial to guarantee that everyone has equitable access to technology and a working knowledge of the internet.

Ethical Considerations: Industry 4.0 raises important ethical considerations. Privacy, data protection, and security become crucial in a digitalized society. The collection, use, and analysis of vast amounts of personal data require robust ethical frameworks and regulatory measures to ensure individuals' rights are protected. Transparent and responsible use of emerging technologies, such as AI, is necessary to address concerns regarding bias, fairness, and accountability.

Social Relationships and Well-being [29]: Industry 4.0 reshapes social relationships and well-being. Digital technologies offer new connectivity, enabling people connect across distances and facilitating social interactions. However, concerns arise regarding the impact of excessive screen time, digital addiction, and the potential erosion of face-to-face social interactions. Balancing digital engagement with personal well-being becomes essential.

Access to Healthcare and Services: Access to services and healthcare is improved through Industry 4.0. By telemedicine, remote patient monitoring, and digital health technology, people can now obtain healthcare services without ever leaving the comfort of their own homes. Additionally, access to a range of services, such as e-commerce, entertainment, and education, is made possible via digital platforms, increasing ease and choice.

By fostering diversity, addressing inequality, ensuring privacy and data protection, increasing digital literacy, and embracing ethical practices, it is crucial to control the societal implications of Industry 4.0. To fully utilize Industry 4.0 for the benefit of all societal members, cooperation is required between the government, business, civil society, and educational institutions.

12.3.3 Environmental Impacts

With its blending of cutting-edge technologies and digitization, Industry 4.0 has both positive and bad environmental effects.

Resource Efficiency: Resources can be used more effectively in manufacturing and production processes thanks to Industry 4.0 technologies. Companies may optimize resource use, cut waste, and lower energy use using real-time monitoring, sensors, and data analytics. Lower carbon emissions, less resource depletion, and increased environmental sustainability result from this.

Sustainable Supply Chains [30]: Sustainability in supply chain management is encouraged by Industry 4.0. Businesses may improve their connectivity, data sharing, and transparency to streamline logistics, shorten transportation routes, and cut down on packaging waste. Real-time tracking of commodities is made possible by smart sensors and IoT devices, increasing efficiency and lowering environmental impact.

Renewable Energy Integration: Renewable energy sources can more easily be incorporated into industrial operations thanks to Industry 4.0. Companies can optimize energy use, track consumption trends, and include the creation of renewable energy, thanks to smart grids, energy management systems, and sophisticated analytics. This lessens dependency on fossil fuels, lowers greenhouse gas emissions, and promotes the development of an energy system that is more environmentally friendly.

Predictive Maintenance [31]: By sensors, data analytics, and ML algorithms, Industry 4.0 offers predictive maintenance. Companies can reduce downtime and maximize energy efficiency by proactively identifying and addressing maintenance issues by continuously monitoring the condition of machinery and equipment. This lessens the need for unneeded repairs, increases the equipment's lifespan, and has a positive effect on the environment.

Environmental Monitoring and Compliance: Industry 4.0 technology makes environmental compliance and monitoring easier. Real-time data on environmental factors including air quality, water quality, and noise levels can be gathered via IoT devices and sensors. Improved environmental management and protection result from the use of this data by businesses and regulators to monitor and enforce compliance with environmental standards.

Electronic Waste [32]: Electronic waste may rise because of Industry 4.0's extensive adoption of digital technologies. It is important to safely manage and dispose of electronic trash as more equipment and devices are used. To reduce the environmental impact of electronic waste and to advance the circular economy, proper recycling and disposal procedures are needed.

Life Cycle Assessment [33]: The use of life cycle assessment (LCA) methods to measure the environmental impact of products and processes is made possible by Industry 4.0. Companies can find chances for environmental improvement and make knowledgeable choices about material selection, production techniques, and waste management by considering the full life cycle, from raw material extraction to end-of-life disposal.

While there are chances for environmental sustainability provided by Industry 4.0, it's critical to address any potential problems and unforeseen repercussions. These can include how much energy digital technologies use, how electronic trash is disposed of, and how the extraction of raw materials for cutting-edge technologies affects the environment. To reduce Industry 4.0's environmental impact, it is essential to take a comprehensive approach that integrates sustainable practises, circular economy principles, and ethical manufacturing practises throughout the value chain.

12.3.4 Ethical Considerations

Important ethical and legal questions are raised by the fourth industrial revolution, which is characterized by the fusion of cutting-edge technologies and digitalization, and it is critical to address the following ethical and legal consequences as society depends on these technologies.

Privacy and Data Protection: Massive volumes of personal data are produced by the Fourth Industrial Revolution through a variety of sources, including IoT devices, social media, and digital transactions. Privacy protection for persons and secure data management are top priorities. To protect people's rights and ensure responsible data handling, strict privacy rules and regulations, such the General Data Protection Regulation (GDPR), are in place.

Data Ownership and Consent: The Fourth Industrial Revolution's use of data collection and analysis raises concerns regarding data ownership and consent. People should be able to manage their personal data and be aware of how it will be used. Clear data ownership frameworks, informed consent methods, and transparent data practices all contribute to building confidence and guaranteeing that people's rights are upheld.

Algorithmic Bias and Fairness: Concerns about algorithmic bias and fairness develop as AI and ML algorithms play a prominent role in decision-making processes. Biased algorithms have the potential to enhance already-existing inequities and discriminatory practises. The creation and application of algorithms should be guided by ethical principles, which include recurrent audits, transparency, and the use of a range of diverse and representative datasets.

Cybersecurity and Data Breaches: The risk of cybersecurity risks and data breaches grows as connectivity and digitalization expand. It's crucial to defend sensitive data, key infrastructure, and individual information against hackers. To stop unauthorized access, data breaches, and potential harm to people and organizations, strong cybersecurity safeguards, encryption methods, and ongoing monitoring are required.

Ethical Use of Technology: The importance of ethical considerations in the creation and application of technology is highlighted by the Fourth Industrial Revolution. Organizations and individuals can be guided by ethical principles, conduct codes, and frameworks for responsible innovation to make sure that technology is created and applied in a way that respects human rights, social values, and environmental sustainability.

Access and Digital Divide: Concerns concerning the digital gap and technological access are brought up by the Fourth Industrial Revolution. To stop exclusion and growing inequities, it is crucial to guarantee that everyone has access to digital technologies, connectivity, and digital literacy. Promoting inclusivity and equal opportunity by bridging the digital divide through programmes like infrastructural development, educational activities, and accessible technology is important.

Governments, regulatory organizations, industry stakeholders, and civil society must work together to address these ethical and legal issues. To ensure that the advantages of the Fourth Industrial Revolution are reaped while minimising risks and harm to people and society, it is essential to establish clear ethical standards, update regulatory frameworks, and promote responsible and inclusive innovation.

12.4 OPPORTUNITIES OF THE FOURTH INDUSTRIAL REVOLUTION

Industry 4.0 (the Fourth Industrial Revolution) offers a variety of prospects in several industries. Innovative goods and business models are made possible, like driverless vehicles and smart houses. It demands upskilling in fields like data analysis and cybersecurity and produces new work positions. It also encourages sustainability by reducing waste and optimizing energy use. Challenges include a lack of necessary expertise and ethical issues. However, there are a lot of potential advantages that might change communities, economies, and businesses.

12.4.1 Innovation and Growth Opportunities

Significant innovation and expansion prospects are provided by Industry 4.0 in several industries. Complex and customized items may be produced more quickly and cheaply, thanks to advanced manufacturing processes like 3D printing, robots, and automation. Connectivity and the IoT enable real-time data collecting and analysis, process optimization, and the development of new goods and services. Operational effectiveness and personalized consumer experiences are driven by data analytics and AI. Innovations in data privacy and cybersecurity maintain trust and dependability in the digital economy. In addition to transforming user experiences, augmented reality (AR) and virtual reality (VR) networks and platforms link stakeholders. Environmental issues are addressed via sustainable solutions. These prospects may be unlocked by embracing digital transformation and encouraging innovation, which will promote economic development and provide value in the changing industrial landscape.

In Industry 4.0, advanced manufacturing technologies revolutionize production, enabling the creation of complex and customized products with improved efficiency and cost-effectiveness. The IoT and connectivity facilitate real-time data collection and analysis, optimizing processes and enabling new product and service offerings. Data analytics and AI provide valuable insights for informed decision-making and drive operational efficiency. Cybersecurity and data privacy innovations ensure trust and reliability in the digital economy, fostering growth. Digital platforms and ecosystems connect stakeholders, while augmented reality (AR) and virtual reality (VR) transform user experiences and open new possibilities. Sustainable solutions addressing environmental concerns are also a crucial part of Industry 4.0, driving growth in green industries. Businesses may take advantage of these chances, promote economic growth, and add value in the changing industrial landscape by embracing digital transformation and promoting innovation. By embracing digital transformation and fostering innovation, businesses can seize these opportunities, spur economic growth, and create value in the evolving industrial landscape.

12.4.2 Increased Efficiency and Productivity

The main benefits of Industry 4.0, made possible by the integration of digital technology and automation, are increased production and efficiency. First, by assigning repetitive and manual duties to machines, automation and robots play a crucial role in simplifying operations. These speeds up operations and reduces human mistake. Continuous operations are also made possible through automation, which lowers downtime and boosts overall productivity. Robots accomplish tasks with accuracy, consistency, and efficiency whether they are operated alone or in conjunction with humans.

Second, Industry 4.0 makes use of data-driven decision-making by producing a lot of data from diverse sources, including sensors and machines. This information is gathered, examined, and used to help in decision-making. Businesses may get useful insights, spot trends, and streamline operations thanks to advanced analytics techniques like ML and predictive analytics. Organizations may increase productivity and create efficiency gains by utilizing real-time data analysis.

Industry 4.0 uses the IoT to enable real-time monitoring and control of manufacturing processes and equipment. Data on equipment performance, ambient factors, and quality characteristics is collected by sensors and linked devices. Real-time monitoring makes it possible for preventative maintenance, early anomaly identification, and quick solutions. Unplanned downtime is decreased; therefore, equipment utilization is enhanced and productivity is significantly increased. Furthermore, Industry 4.0 offers additional advantages such as improved supply chain management through enhanced visibility and connectivity. Connected systems and IoT-enabled sensors provide real-time insights into inventory levels, demand patterns, and logistical processes, allowing for better inventory management, demand forecasting, and optimized production planning. Collaboration is fostered through digital platforms and systems, facilitating seamless communication, data sharing, and coordination among different departments, suppliers, and customers. By embracing continuous improvement and optimization through feedback loops and data analysis, organizations can identify bottlenecks, address inefficiencies, and refine processes, leading to sustained productivity gains.

Finally, Industry 4.0 leverages automation, data-driven decision-making, in-the-moment monitoring, enhanced supply chain management, collaboration, and continuous improvement to boost production and efficiency. In the constantly changing industrial scene, leveraging digital technology and automation enables firms to reach better levels of operational efficiency, cost reduction, and productivity improvements.

12.4.3 Improved Quality of Life

Industry 4.0 brings about improvements in the quality of life through various advancements and transformations. Here's an explanation of how Industry 4.0 contributes to a better quality of life:

Enhanced Accessibility and Convenience: Industry 4.0 innovations increase usability and accessibility in many facets of daily life. Smart homes are made possible by connected gadgets and the IoT, where automation and remote control make it possible to handle lighting, temperature, security, and entertainment systems with ease. This makes life more commodious by improving comfort, energy effectiveness, and security. Additionally, developments in e-commerce and digital platforms make it simple for individuals to purchase online, saving them time and effort while providing access to a variety of goods and services from the comfort of their homes.

Personalized Experiences and Healthcare: In several sectors, Industry 4.0 allows personalized experiences. For instance, customized entertainment, news, and advertising recommendations and content improve customer pleasure. Personalized treatment plans, remote patient monitoring, and telemedicine services are made possible in the healthcare industry through digitalization and linked technologies. This enhances health outcomes and quality of life by enabling people to access timely and convenient healthcare services, regardless of where they are physically located. Additionally, wearable technology and health tracking tools enable people to take control of their wellbeing by maintaining their fitness and health levels.

Individuals gain from enhanced accessibility, convenience, and personalized experiences in their daily lives by utilising Industry 4.0 technology. Better health outcomes are made possible by medical innovations that allow for personalised therapy and remote monitoring. The potential for improving the quality of life through Industry 4.0 technologies will increase as the industrial environment continues to change.

12.5 FUTURE OUTLOOK

Technology developments, interconnected ecosystems, intelligent systems, human-machine collaboration, moral behaviour, and workforce change will power Industry 4.0's bright future. By embracing these trends, industry will be able to seize new chances, stimulate innovation, and improve society. Industry 4.0 will revolutionize sectors, economies, and how we live and work as the industrial environment continues to change. The following significant factors will influence Industry 4.0's future:

12.5.1 Predictions and Projections for the Future of the Fourth Industrial Revolution

Increased Connectivity and IoT Expansion: With more devices and items connecting to the IoT, the connectivity landscape will continue to grow. As a result, the amount of data created from diverse sources will increase exponentially, enabling even more sophisticated data analytics and decision-making abilities. Smart infrastructure, smart homes, and smart cities will proliferate, improving productivity, sustainability, and quality of life.

Integration of Cyber-Physical Systems: The distinction between the physical and digital worlds will become less distinct as cyber-physical systems are increasingly seamlessly integrated. As a result, intelligent factories and completely interconnected supply chains will be developed, enabling real-time collaboration between machines, systems, and people. Production procedures will be improved, waste will be reduced, and quick responses to shifting market demands will be possible thanks to real-time data sharing and analytics.

Advancements in 3D Printing and Additive Manufacturing: Technologies like 3D printing and additive manufacturing will keep developing, making it easier to produce intricate and unique things. Decentralized and on-demand production will be possible because of this upending conventional manufacturing processes and supply chains. Widespread use of 3D printing will result in lower prices, quicker prototyping, and greater sustainability due to less material waste.

Focus on Sustainability and Circular Economy: A larger emphasis on sustainability and the circular economy will result from Industry 4.0. Reduced environmental effect will result by energy-efficient technologies, renewable energy sources, and resource optimization. The ability to reuse, recycle, and repurpose materials will be made possible by closed-loop technologies and product lifecycle management, fostering a more sustainable method of production and consumption.

Social Impacts and Workforce Transition: The implementation of Industry 4.0 technology will cause social changes and necessitate a change in the workforce. While automation and AI may result in some job losses, they will also open new positions and opportunities. To prosper in a digital and automated environment, the workforce will need to adapt and upskill. Initiatives for lifelong learning and reskilling will be essential to close the skills gap and guarantee a smooth transition.

12.5.2 Possible Policy Interventions and Regulations

Data Privacy and Security: The creation of comprehensive data protection and cybersecurity rules needs to be given top priority by policymakers.

Regulations should protect against cyberthreats and breaches while ensuring the responsible collection, storage, and use of personal data. To keep the public's faith in the digital ecosystem, clear rules regarding data privacy, encryption requirements, and risk management procedures are necessary.

Ethical AI and Automation: As automation and AI expand, legislators should set up ethical frameworks and standards for its creation and application. Regulations should cover topics like algorithmic bias, openness in decision-making, and responsibility for AI-enabled systems. Policies can also encourage the ethical use of automation technology to strike a balance between the expansion of the economy, the creation of jobs, and human well-being.

Workforce Transition and Reskilling: Policymakers should concentrate on measures that encourage lifelong learning and workforce transition. To assist people in adjusting to the shifting employment market, this may entail programmes like reskilling efforts, upskilling subsidies, and job placement services. It is essential for governments, educational institutions, and businesses to work together to guarantee that the workforce has the skills needed for occupations of the future.

12.6 CONCLUSION

Every element of our lives will be impacted by the transformational changes brought about by the Fourth Industrial Revolution (Industry 4.0). Policymakers, corporate leaders, and people must take proactive measures to adapt to and prepare for this transformation. Funds and resources must be allocated to encourage lifelong learning and give people the knowledge and abilities they need to prosper in the digital era. Pertinent training programmes and initiatives should be developed by working with companies and educational institutions. Opportunities should be identified to leverage Industry 4.0 technologies to streamline operations, enhance customer experiences, and drive business growth. Resources must be allocated for research and development and collaborate with technology partners to stay at the forefront of innovation. It is necessary to recognize the importance of acquiring new skills and knowledge to stay relevant in the changing job market. Therefore, take advantage of online courses, vocational training, and other educational resources to upskill and reskill. By taking collective action and embracing the opportunities presented by the Fourth Industrial Revolution, we can shape a future that is inclusive, sustainable, and beneficial for all. Policymakers, business leaders, and individuals must work together to navigate the challenges, seize the possibilities, and create a positive impact in the era of Industry 4.0.

REFERENCES

[1] Tsaramirsis, G., Kantaros, A., Al-Darraji, I., Piromalis, D., Apostolopoulos, C., Pavlopoulou, A., Alrammal, M., Ismail, Z., Buhari, S.M., Stojmenovic, M. and Tamimi, H., 2022. A modern approach towards an Industry 4.0 model: From driving technologies to management. *Journal of Sensors*, *2022*(2022), 1–18.

[2] Thumfart, J., 2023. The democratic offset: Contestation, deliberation, and participation regarding military applications of AI. *AI and Ethics*, pp. 1–16.

[3] Lee, J., Babcock, J., Pham, T.S., Bui, T.H. and Kang, M., 2023. Smart city as a social transition towards inclusive development through technology: A tale of four smart cities. *International Journal of Urban Sciences*, *27*(supl), pp. 75–100.

[4] Hayat, A., Shahare, V., Sharma, A.K. and Arora, N., 2023. Introduction to Industry 4.0. In *Blockchain and Its Applications in Industry 4.0* (pp. 29–59). Singapore: Springer Nature.

[5] Kamaruzaman, F.M., Hamid, R., Mutalib, A.A., Rasul, M.S., Omar, M., and Zaid, M.F.A.M., 2023. Exploration and verification of fourth industrial revolution generic skills attributes for entry-level civil engineers. *International Journal of Evaluation and Research in Education*, *12*(1), pp. 121–130.

[6] Chaka, C., 2023. Fourth industrial revolution—a review of applications, prospects, and challenges for artificial intelligence, robotics and blockchain in higher education. *Research and Practice in Technology Enhanced Learning*, *18*(2), pp. 1–9.

[7] Gawde, S., Patil, S., Kumar, S. and Kotecha, K., 2023. A scoping review on multi-fault diagnosis of industrial rotating machines using multi-sensor data fusion. *Artificial Intelligence Review*, *56*(5), pp. 4711–4764.

[8] Kaswan, K.S., Dhatterwal, J.S., Kumar, N., Awasthi, S. and Chauhan, S.S., 2023. Real-time decision-making techniques using Artificial Intelligence and cloud computing. In *2023 International Conference on Disruptive Technologies (ICDT)* (pp. 355–358). Greater Noida, India: IEEE.

[9] Soori, M., Arezoo, B. and Dastres, R., 2023. Artificial intelligence, machine learning and deep learning in advanced robotics, a review. *Cognitive Robotics*, *3*, pp. 54–70.

[10] Guo, W., 2023. Exploring the value of AI technology in optimizing and implementing supply chain data for pharmaceutical companies. *Innovation in Science and Technology*, *2*(3), pp. 1–6.

[11] Shadravan, A. and Parsaei, H.R., 2023. Impacts of Industry 4.0 on smart manufacturing. In *Proceedings of the 13th International Conference on Industrial Engineering and Operations Management (IEOM Society)* (pp. 7–9). Manila, Philippines: IEOM Society.

[12] Dohale, V., Verma, P., Gunasekaran, A. and Akarte, M., 2023. Manufacturing strategy 4.0: A framework to usher towards Industry 4.0 implementation for digital transformation. *Industrial Management & Data Systems*, *123*(1), pp. 10–40.

[13] Ahmadi, M., Pahlavani, M., Karimi, A., Moradi, M. and Lawrence, J., 2023. the impact of the fourth industrial revolution on the transitory stage of the automotive industry. In *Sustainable Manufacturing in Industry 4.0: Pathways and Practices* (pp. 79–96). Singapore: Springer Nature.

[14] Shenkoya, T., 2023. Sustainable Urban development: An evaluation of the impact of IoT on sustainable development and energy management in SMART cities. DOI: https://doi.org/10.21203/rs.3.rs-1962624/v1

[15] Zhou, X., Zhu, Q. and Xu, Z., 2023. The role of contractual and relational governance for the success of digital traceability: Evidence from Chinese food producers. *International Journal of Production Economics*, *255*, p. 108659.

[16] Bag, S., Wood, L.C., Xu, L., Dhamija, P. and Kayikci, Y., 2020. Big data analytics as an operational excellence approach to enhance sustainable supply chain performance. *Resources, Conservation and Recycling*, *153*, p. 104559.

[17] Xu, K., Li, Y., Liu, C., Liu, X., Hao, X., Gao, J. and Maropoulos, P.G., 2020. Advanced data collection and analysis in data-driven manufacturing process. *Chinese Journal of Mechanical Engineering*, *33*(1), pp. 1–21.

[18] Koh, L., Orzes, G. and Jia, F.J., 2019. The fourth industrial revolution (Industry 4.0): Technologies disruption on operations and supply chain management. *International Journal of Operations & Production Management, 39*(6–8), pp. 817–828.

[19] Fu, W. and Chien, C.F., 2019. UNISON data-driven intermittent demand forecast framework to empower supply chain resilience and an empirical study in electronics distribution. *Computers & Industrial Engineering, 135*, pp. 940–949.

[20] Mihardjo, L., Sasmoko, S., Alamsjah, F. and Elidjen, E., 2019. Digital leadership role in developing business model innovation and customer experience orientation in Industry 4.0. *Management Science Letters, 9*(11), pp. 1749–1762.

[21] Sherwani, F., Asad, M.M., and Ibrahim, B.S.K.K., 2020, March. Collaborative robots and Industrial Revolution 4.0 (ir 4.0). In *2020 International Conference on Emerging Trends in Smart Technologies (ICETST)* (pp. 1–5). Karachi, Pakistan: IEEE.

[22] Arrais, R., Veiga, G., Ribeiro, T.T., Oliveira, D., Fernandes, R., Conceição, A.G.S. and Farias, P.C.M.A., 2019. Application of the open scalable production system to machine tending of additive manufacturing operations by a mobile manipulator. In *Progress in Artificial Intelligence: 19th EPIA Conference on Artificial Intelligence, EPIA 2019, Vila Real, Portugal, September 3–6, 2019, Proceedings, Part II* (pp. 345–356). Cham: Springer International Publishing.

[23] Knudsen, M. and Kaivo-Oja, J., 2020. Collaborative robots: Frontiers of current literature. *Journal of Intelligent Systems: Theory and Applications, 3*(2), pp. 13–20.

[24] Rotatori, D., Lee, E.J. and Sleeva, S., 2021. The evolution of the workforce during the fourth industrial revolution. *Human Resource Development International, 24*(1), pp. 92–103.

[25] Bai, C., Quayson, M. and Sarkis, J., 2022. Analysis of blockchain's enablers for improving sustainable supply chain transparency in Africa cocoa industry. *Journal of Cleaner Production, 358*, p. 131896.

[26] Kassen, M., 2021. Understanding decentralized civic engagement: Focus on peer-to-peer and blockchain-driven perspectives on e-participation. *Technology in Society, 66*, p. 101650.

[27] Su, C.W., Qin, M., Tao, R. and Umar, M., 2020. Financial implications of fourth industrial revolution: Can bitcoin improve prospects of energy investment? *Technological Forecasting and Social Change, 158*, p. 120178.

[28] Cheng, Y., Awan, U., Ahmad, S. and Tan, Z., 2021. How do technological innovation and fiscal decentralization affect the environment? A story of the fourth industrial revolution and sustainable growth. *Technological Forecasting and Social Change, 162*, p. 120398.

[29] Halbrook, Y.J., O'Donnell, A.T. and Msetfi, R.M., 2019. When and how video games can be good: A review of the positive effects of video games on well-being. *Perspectives on Psychological Science, 14*(6), pp. 1096–1104.

[30] Muthu, S.S., 2020. Assessing the Environmental Impact of Textiles and the Clothing Supply Chain. Sawston, Cambridge: Woodhead Publishing.

[31] Pech, M., Vrchota, J. and Bednář, J., 2021. Predictive maintenance and intelligent sensors in smart factory. *Sensors, 21*(4), p. 1470.

[32] Patil, R.A. and Ramakrishna, S., 2020. A comprehensive analysis of e-waste legislation worldwide. *Environmental Science and Pollution Research, 27*, pp. 14412–14431.

[33] Nwodo, M.N. and Anumba, C.J., 2019. A review of life cycle assessment of buildings using a systematic approach. *Building and Environment, 162*, p. 106290.

13 Artificial Intelligence (AI)-Enabled Digital Twin Technology in Smart Manufacturing

Vijayakumar Ponnusamy, Dilliraj Ekambaram, and Nemanja Zdravkovic

13.1 INTRODUCTION

Smart manufacturing is bringing about a paradigm shift in the age of Industry 4.0 by incorporating state-of-the-art technologies to improve the efficiency [1, 2], effective global scheduling process [3], sustainability [4], and productivity of conventional industrial processes [5]. The use of Digital Twins made possible by AI is a driving force behind this transformation. Virtual twins are representations of physical systems that are designed to imitate real-world scenarios to increase system performance, eliminate physical losses, and ensure user safety [6]. The complexity and variability of production settings have increased the demands placed on production management, particularly in terms of resource allocation. A self-regulating and adaptive scheduling method is necessary for the optimal distribution of scarce resources [7]. Multiple input features and high-resolution computed variables make real-time fine evolution difficult. Virtual testing and evaluation of alternative designs and characteristics would be considerably simplified with the help of the digital twin (DT). With the use of DT solutions, managers can create better goods, find physical problems earlier, and make more accurate predictions by creating a 3D digital model of the physical object [8]. This cutting-edge method gives producers extraordinary knowledge and command over the whole production lifecycle by enabling them to build virtual versions of real assets, processes, and systems [9–11].

Several sectors have benefited from developing the digital twin idea, which bridges the gap between the digital and physical realms to boost knowledge, efficiency, and creativity. While digital twins have been around since the dawn of CAD systems, it wasn't until recent developments in connection, data analytics, and AI that their full potential was realized. As computer-aided design (CAD) systems evolved in the middle of the 20th century, the idea of a DT began to take shape. Initially, some tools let designers and engineers make digital models of real-world items. This was useful for visualizing and testing ideas before building physical prototypes. The foundation was created by these early CAD systems, but modern understanding of a digital twin means going beyond static representations. The rapid development of digital twins

DOI: 10.1201/9781003473886-13

is crucial for composites life cycle analysis because it allows for the scalable and extremely informative interaction of real objects with virtual twins [2]. The timeline of DT development is shown in Figure 13.1.

During the early 2000s, the phrase "digital twin" became popular, and NASA was instrumental in popularizing it. To better monitor and analyze spacecraft and systems, NASA made use of digital twin technology [9] to build virtual replicas of them. We can now understand complicated systems on a deeper level, thanks to this method's ability to facilitate in-depth simulations, real-time monitoring, and troubleshooting. Digital twins advanced rapidly with the IoT in the early 21st century. With sensors and connected devices, physical objects and systems can generate real-time data. AI and IoT technologies exhibit great promise in terms of operational efficiency and accuracy as well as their usefulness in real-world applications [12].

For smart manufacturing to work, people and robots must be able to precisely record their three-dimensional movements [13]. This will allow for safe and flexible collaboration between the two. Digital twins are crucial for better human-robot collaboration because they enable physical-virtual connection. Dynamic assessments reflecting physical states are generated by digital twins through real-time interaction [14]. To deal with heterogeneity caused by disparate protocols and application management solutions as well as by various manufacturing tools and enterprise services, the digital twin (DT) approach is becoming more popular as the Internet of Things (IoT) is implemented in industrial settings [15]. An efficient learning algorithm is developed to model, monitor, and enhance the entire manufacturing process in the DT system. This algorithm utilizes features integrated and fused from both shallow and deep layers to recognize multiple small items [16]. The benefits of DTs include their capacity to change in tandem with the physical object, which helps with decision-making, which in turn streamlines production and improves product service quality [17, 18]. The manufacturing sector's increasing reliance on cyber-physical

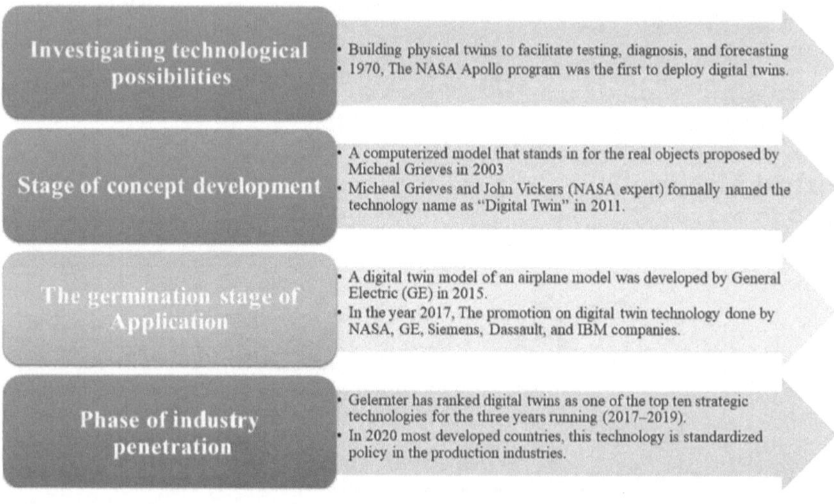

FIGURE 13.1 Evolution of Digital Twin Technology.

systems (CPS), the Internet of Things (IoT), big data analytics, machine learning (ML), and cloud computing has opened the door to the systematic and cost-effective application of DT [19, 20]. Digital twins are anticipated by product development leaders to expedite procedures, enhance outcomes, and decrease costs. Because the digital-twin technologies business is predicted to grow at a CAGR of 60% from 2019 to 2027, reaching $73.5 billion, they are preparing to invest swiftly in the notion [21].

13.1.1 How Do Digital Twins Work?

An object or system's "digital twin" is an interactive, online duplicate of the real thing. When it comes to smart manufacturing, these digital copies are more than just 3D models; they include predictive analytics and real-time data feeds. With this all-encompassing method, producers may track, evaluate, and enhance their processes with unmatched accuracy. There are three key parts to the digital twin concept model: a) real products in Real Space; b) virtual products in Virtual Space; and c) the data and information links between the real and virtual products [22]. Figure 13.2 shows the general structure of digital twin.

Digital twins (DT) are a novel combination of artificial intelligence (AI) and the idea of making exact digital copies of real stuff. Central to it is a painstaking data integration procedure that collects data from various sources, including sensors, IoT devices, and other data streams. This combination lays the groundwork for building real-time, all-encompassing digital models that mimic the physics, geometry, and behavior of their corresponding physical counterparts. An enhanced virtual reproduction that goes beyond a static representation is what a digital twin is all about.

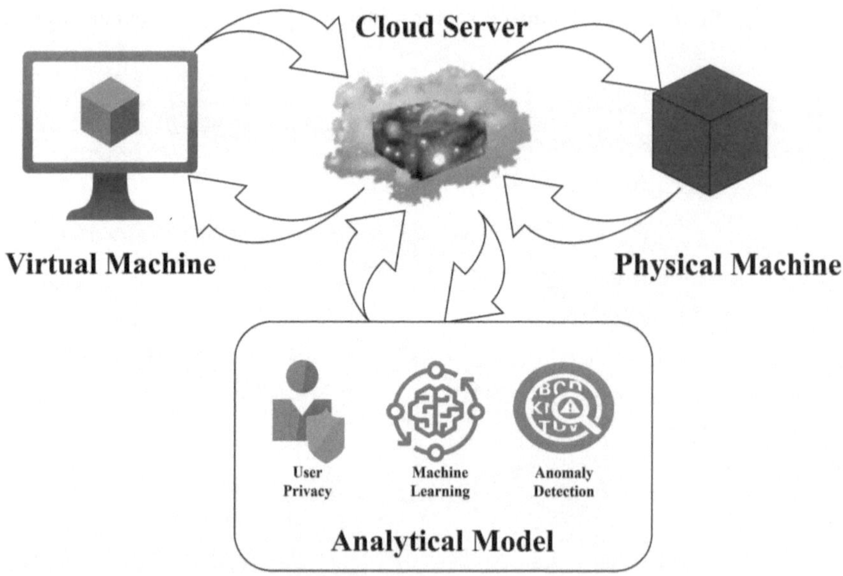

FIGURE 13.2 The General Structure of the Digital Twin.

Using machine learning algorithms and predictive analytics, both real-time and historical data may be continuously analyzed. The analytical skills of digital twins enable them to make predictions, identify patterns, and dynamically enhance operations, allowing them to build a responsive and adaptive framework.

13.1.2 The Confluence of Artificial Intelligence and Digital Twins

There is a revolutionary synergy developing between digital twins and artificial intelligence (AI), and it will impact creativity and efficiency across numerous industries. Their integration represents a paradigm shift in how we design, monitor, and optimize complex systems; it goes beyond basic collaboration as we delve into the subtle relationship between these two technologies. Industry 4.0 is centered around the interconnection of smart cyber-physical systems (CPS) that allow for mass customization in strong and adaptable Smart Factories. This movement is propelled by the digital revolution and the exponential growth of computing power, data processing capabilities, smart sensors, and AI [4]. By combining digital twins with AI, their powers are greatly enhanced. Artificial intelligence algorithms, fueled by data analytics and machine learning, allow these digital copies to adapt and gain knowledge from actual situations [5]. Ultimately, operational efficiency, predictive maintenance [6], and decision-making all benefit from this mutually beneficial partnership.

The data analytics, machine learning, and predictive modeling capabilities of AI give digital twins a degree of intelligence and autonomy that goes beyond their static beginnings. Aligning digital twins more closely with their physical counterparts, this integration boosts their potential to adapt, learn, and make educated decisions in real-time [7]. The digital twin and AI algorithms rely on this constant flow of real-time data captured by these gadgets from the physical environment [8]. In response, the AI algorithms examine this deluge of data for trends, patterns, and outliers that help fill in the gaps in our knowledge of the system under replication. The intersection of artificial intelligence and digital twins has revolutionary and far-reaching potential. By analyzing past data and sensor inputs, AI is transforming predictive maintenance; for example, by allowing equipment failures to be predicted in advance. Manufacturing processes are optimized to run at peak efficiency through real-time modifications based on AI insights. Digital twins' sophisticated data analysis allows supply chain management to obtain end-to-end visibility and agility.

13.1.3 Human-Focused Digital Twin

Human-focused DTs are redefining the relationship between people and technology, and their advent signifies a radical break from conventional wisdom in the field of digital innovation. A human-centric digital twin is a dynamic, data-driven depiction of an individual that goes beyond physical entity reproduction and into the very fabric of our personal and professional lives. A basic definition of a human-focused DT is a computer simulation that mimics an actual person's online actions, tastes, and habits. Personalized and immersive digital twins of people are the focus of the human-focused DT, in contrast to its industrial counterparts that mainly aim to replicate systems and technology. Everything from physiological and biological data to individual tastes, work patterns, and mental and emotional states is encompassed in this portrayal. Incorporating state-of-the-art technologies like sensors, wearables, and IoT forms the backbone of HRDs. Various parts of a person's life are monitored by these devices in real time, creating a steady flow of data that forms the basis of the digital copy. With the use of biometric sensors that record vital signs and smart devices that record motion, interactions, and ambient variables, a comprehensive dataset is created, enriching the human-focused DT. The structure of human-focused digital twins in shown in Figure 13.3.

13.1.4 Collaboration between Human-Robot

Accurately recording human and robot three-dimensional motions is critical for smart manufacturing's safe and adaptable human-robot collaboration [13]. To improve human-robot collaboration, digital twins are essential because they allow for physical-virtual interaction to take place. Digital twins generate dynamic judgments based on real-time interaction, reflecting physical states [14]. Human-robot collaborative systems enable humans and robots to operate in close quarters, sharing workspaces and doing shared activities. Collaborative systems rely on robots for their power, precision, and repeatability, while humans offer the agility to solve challenges with their cognitive competence [23]. Figure 13.4. Shows the simple process flow of human-robot collaboration digital twins.

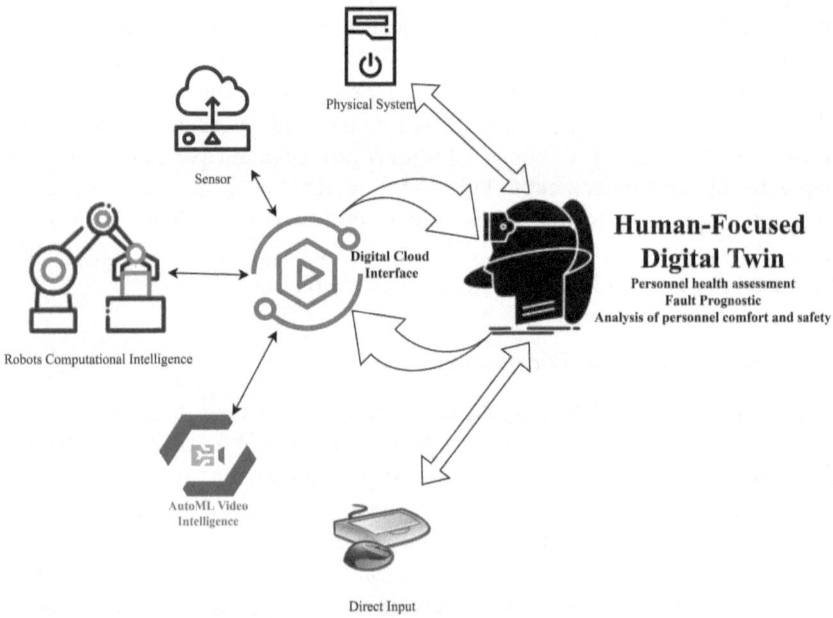

FIGURE 13.3 The Structure of Human-Focused Digital Twins.

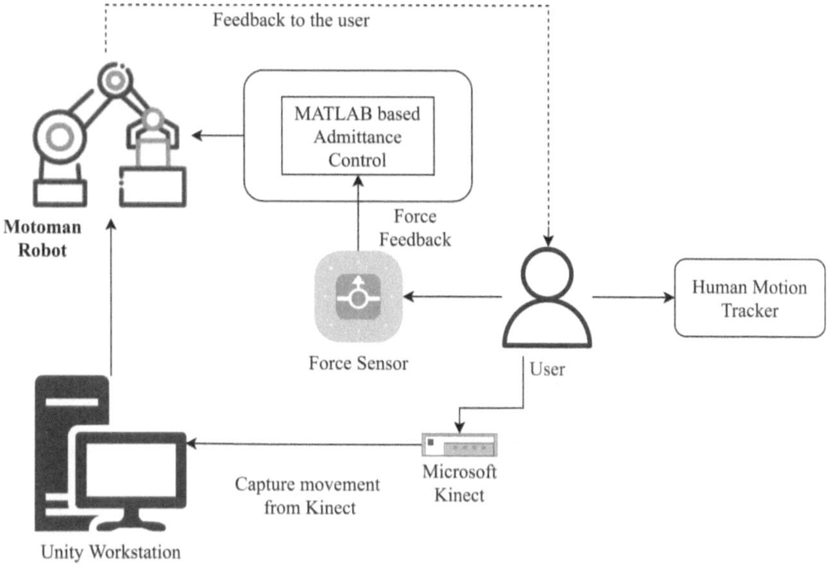

FIGURE 13.4 The Simple Process Flow of Human-Robot Collaboration Digital Twins.

13.1.5 Digital Twins and Cyber-Physical Systems

A DT of a real machine can be built using computer simulations using multi-physics modeling; this allows data to flow in both directions between the virtual and physical systems. These DTs can be used for a cyber-physical production system (CPS) flexible manufacturing quick testing, optimization, and deployment [4]. Recent advances in artificial intelligence (AI), larger data bandwidth and processing capacity, and widespread use of sensors have made it possible to construct virtual versions of CPS that are completely autonomous and capable of making complex decisions. The term "digital twin" describes these completely self-sufficient digital copies. Computational collaboration systems that combine DT and cyber-physical systems could help individuals comprehend the production process more effectively and provide resilience to evidence-based decision-making [16]. A cyber-physical manufacturing system can receive operational status and situational data from DTs. This data can improve the analytical evaluation, predictive diagnosis, and performance optimization capabilities of a manufacturing system. DTs are indeed a key component of the smart manufacturing paradigm [24]. A schematic of cyber-physical smart manufacturing is represented in Figure 13.5.

The key points and definitions of DT discussed in the introduction are presented in this chapter. Section 2 provides a summary of relevant literature on human-focused, human-robot collaborative, and cyber-physical systems that make use of AI-enabled DTs in smart manufacturing. In Section 3, we discuss the key components of an AI-powered digital twin. The applications of smart manufacturing are reviewed in Section 4. Section 5 discusses the pros and cons of DTs powered by AI. We wrap up

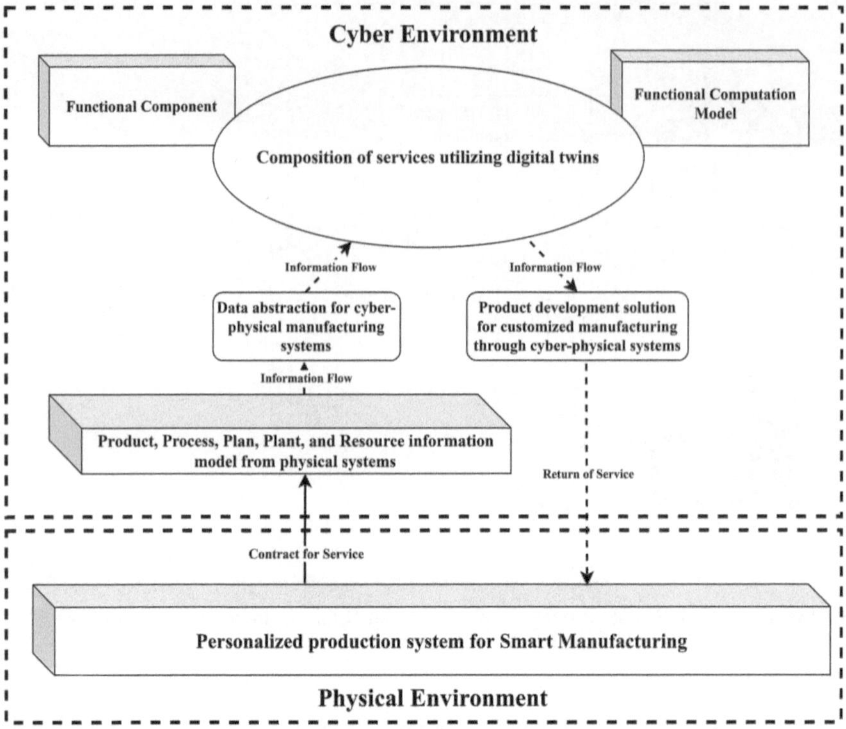

FIGURE 13.5 A Schematic of Cyber-Physical Smart Manufacturing.

the chapter by outlining several key areas for further study in AIDT-enabled smart manufacturing.

13.2 RELATED WORK

Digital twins have been the subject of extensive research into their theoretical foundations and practical applications in a wide range of fields. These works of literature improve smart manufacturing decision-making, predictive maintenance, and efficiency by providing a thorough overview of the history and concepts of digital twin technologies [1–6]. Researchers have studied many potential industrial uses of digital twins. These uses highlight the adaptability and efficacy of digital twin deployments, which extend from product creation and design to production optimization. There is a substantial amount of literature on digital twins as an environment for applying machine learning algorithms; this literature explores the creation and improvement of algorithms for smart manufacturing system data analytics, predictive modeling, and anomaly detection [7–11]. Investigation of AI-powered methods for smart manufacturing settings' like adaptive control strategies, feedback loops, and continuous monitoring are the important constraints for DT [12–16]. Some kinds of literature and its works are explored in this section.

With the use of semi-supervised deep learning and digital twins, S. Wang et al. (2023) [1] created a framework for human-robot cooperation that can recognize objects, monitor collaboration, and validate and evaluate safety assurance using DT in different lighting conditions. The results of this research include a semi-automated annotation tool [25] and publicly available datasets produced by our digital twin of a Universal Robot 10 (UR10) robot [26], which is available online. This framework measures the effectiveness of a convolutional neural network (CNN) trained on various datasets and in various lighting situations by calculating three different average precisions (APs): mean AP (mAP), AP at the intersection of the union (IoU) over 50% (AP50), and 70% (AP75) [27]. The evaluation measures are done with the following mathematical expressions.

The AP is the main evaluation metric for this framework [28]. It is evaluated as,

$$AP = \int_0^1 p(r)dr \tag{13.1}$$

Where, $p-$ *Precision* calculated as, $p = \dfrac{True\ Positive}{True\ Positive + False\ Positive}$ [29],

$r-$ *recall* calculated as, $r = \dfrac{True\ Positive}{True\ Positive + False\ Negative}$ [29]

When recall and precision are both high, the average precision is large; when one of these metrics is small, the average precision is small. The mean average precision (mAP) is calculated by averaging the average precision for all classes that are considered, while the average precision (AP) is found for each class independently.

The IoU [30] is evaluated by the following mathematical expression,

$$IoU = \frac{P \cap G}{P \cup G} \tag{13.2}$$

Where P-Anticipated objects bounding box; G-Actual bounding box of an object.

This work utilizes a laptop equipped with an Nvidia RTX 2070 GPU to execute the detection algorithm. With a detection pace of approximately 20 frames per second (fps), the physical system is capable of monitoring both the robot and the operator in real-time, the mAP outcomes for a human, and a UR10 robot working in varying illumination conditions. In contrast to the network trained on actual data under optimal lighting conditions (full light and semi-light), the faster R-CNN trained with the synthetic and semi-supervised models also outperforms the latter. Regardless of the illumination conditions (full and semi-lighting), the faster R-CNN achieves impressive mAP_{UR10} when trained exclusively with synthetic data. In fluctuating illumination conditions, the semi-supervised model exhibits more resilient performance. In addition, the semi-supervised algorithm achieves the highest score in terms of mAP_{human} in comparison to models that are exclusively trained with real or synthetic data. The mAP_{human} is above 77% under full lighting and also achieves 61% under the dark circumstance.

An AI-driven approach for digital twinning of intricate composite structures is illustrated by X. Xu et al. [2] Initially, three distinct categories of deep neural networks are constructed, employing autoencoders as representative models if desired. The architecture and data processing techniques of these networks are modeled after the rule-of-mixture for composites. The efficiency and precision of the forecasts are evaluated numerically as well as qualitatively. With sensing data like pressure, temperature, and loading displacement, it is feasible to immediately predict stress fields and 3D displacements. Ultimately, the evolving digital twin is determined to be the most suitable architecture. In conclusion, the digital duplicate is informed of the real-time interactive experiments, which showcase its capability to evolve online and engage in interactions with physical objects precisely. The Monte Carlo cross-validation method, which uses a sampling rate α ranging from 50% to 80%, is one of the selections used to evaluate the average predictive performance of the H-DNN architecture. This signifies that the dataset is arbitrarily divided into two sections, D_{train} and $D_{validation}$, which are utilized to train and validate the model, respectively. A conventional desktop computer equipped with an NVIDIA GeForce RTX 3090 GPU and an Intel I9–9900K 3.60 GHz 16-core processor is utilized for validation and testing.

The mathematical expression of MAE, RMSE, NMAE and NAE-Peak is,

$$MAE = \frac{1}{N}\sum\nolimits_{n=1}^{N}\left\| D_i^{(n)} - \hat{D}_i^{(n)} \right\| \tag{13.3}$$

$$NMAE = \frac{MAE}{\max\{D\} - \min\{D\}} \times 100\% \tag{13.4}$$

$$RMSE = \sqrt{\frac{1}{N}\sum\nolimits_{n=1}^{N}\left\| D_i^{(n)} - \widehat{D}_i^{(n)} \right\|^2} \tag{13.5}$$

$$NAE - Peak = \frac{\max\left(\left\| D_i^{(n)} - \hat{D}_i^{(n)} \right\|\right)}{\max\{D\} - \min\{D\}} \times 100\% \tag{13.6}$$

Where, n-index of training data; N-size of the training data; i-displacement and stress field component; $\left\| D_i^{(n)} - \widehat{D}_i^{(n)} \right\|$ verifies the magnitude of the displacement in absolute terms, while the von Mises stress is calculated using deep neural networks and finite element models.

Jialin Chen et al. [6] primary goal is to develop and test human biomechanics models of persons with disabilities who use assistive devices. We will then use machine learning to identify cases of reduced mobility. To assure the safety of human-assistive robot interaction, we will simulate edge scenarios before deploying the robots. This study adds to the growing body of knowledge on digital twins, which can improve the security of assistive robots in everyday use. An explicitly exploratory and asynchronously updated proximal policy optimization algorithm (E2APPO) was suggested by Gan X M. et al. [7] as an adaptable scheduling approach that utilizes DT and is built upon the RL algorithm for improved proximal policy optimization. First, a DT-enabled scheduling system architecture was built to facilitate communication among online and offline job shops, which, in turn, strengthened the scheduling model's ability to self-regulate. Two more things: first, we proposed an asynchronous updating approach to improve optimization, and second, we improved the scheduling model's ability to learn on its own. Finally, extensive testing was conducted on the proposed scheduling model, in comparison to other scheduling methods that use reinforcement learning, heuristic and meta-heuristic algorithms (such as genetic algorithms), and popular scheduling rules. It is shown in Table 13.1.

Scheduling with DT allows for on-demand model regulation through virtual-real testing and verification. With an average reduction of 8.9%, E2APPO's makespan is far better than MDQN for all instances. Since E2APPO's performance is normally distributed and does not exhibit any outlying extremes, it is evident that it is the most stable option. It verifies that the suggested method of asynchronous updates works.

TABLE 13.1

Input	π_θ actor-network with theta as a trainable parameter; N—Batch Size; v_w critic-network with 'w' as a trainable parameter; μ-clipping ratio; K—numerous updates Comparison of critic and actor frequency; gamma is a discounting factor; c_p, c_v, c_e -policy, value, and entropy loss coefficients.
Step1	Design of environmental states (S_t), actions (a_t), and rewards (r_t) using Markov modeling
Step2	Initialize $\pi_\theta, \pi_{\theta old}, and\ v_w$
Step3	$for\ i = 0,1,2,\ldots N\ do$
Step4	$for\ j = 0,1,2,\ldots J\ do$
Step5	Observe $S_{i,j}$, select action $a_{i,j}$ based on innovation strategy $\pi_{innor}(a_{i,j} \mid S_{i,j})$
Step6	Receive reward $r_{i,j}$ and next state $S_{i,j+1}$
Step7	Estimate advantages: $\widehat{A_{i,j}} = \sum_0^j \gamma^j r_{i,j} - V_\varphi(S_{i,j}), r_{i,j}(\theta) = \frac{\pi_\theta(a_{i,j} \mid S_{i,j})}{\pi_{\theta old}(a_{i,j} \mid S_{i,j})}$
Step8	If $S_{i,j+1}$ is terminal then,
Step9	break;
Step10	Collecting $\{S_{i,j}, r_{i,j}, a_{i,j}\}$
Step11	End
Step12	$L_i^p(\theta) = \sum_0^j \min(r_{i,j}(\theta)\widehat{A_{i,j}}, clip(r_{i,j}(\theta), 1-\mu, 1+\mu)\widehat{A_{i,j}})$
Step13	$L_i^v(w) = \sum_0^j (v_w(S_{i,j}) - \widehat{A_{i,j}})^2$
Step14	$L_i^s(\theta) = \sum_0^j S(\pi_\theta(a_{i,j} \# S_{i,j}))$
Step15	Aggregate losses: $L_i(\theta, w) = c_p L_i^p(\theta) - c_v L_i^v(w) + c_e L_i^s(\theta)$
Step16	Update critic $w \leftarrow \sum_{min} (y - Q_{\theta_i}(s,a))^2$
Step17	If $i\%k = 0$
Step18	Update actor θ by a gradient method:
Step19	$\theta = argmax\left(\sum_i^N L_i(\theta, w)\right)$
Step20	$\pi_{\theta old} \leftarrow \pi_\theta$
Step21	End
Step22	End

N. Jyeniskhan et al. [10] uses Unity, OctoPrint, and Raspberry Pi for real-time control and monitoring and suggest a digital twin system framework for additive manufacturing that incorporates machine learning models. Specifically, the system achieves a 92% average precision (AP) score when detecting defects using machine learning models, a 91% specific performance metric for defective objects, and a 94% specific performance meter for non-defected objects, indicating great efficiency. For simple additive manufacturing process monitoring, the Unity client user interface is also built for control and visualization. Alshathri et al. [11] provide a system that optimizes fault diagnosis utilizing performance metrics including recall, accuracy, precision, and F-measure while using a genetic algorithm (GA) to enhance classification accuracy.

The evaluation of performance metrics is expressed in mathematical form as follows,

$$Accuracy = \frac{TP + FP}{TP + TN + FP + FN} \tag{13.7}$$

$$F-measure = \frac{2\times\ Precision \times\ Recall}{Precision + Recall} \tag{13.8}$$

Where, TP-True Positive, TN-True Negative, FP-False Positive, FN-False Negative, Precision and Recall expressions discussed already in equation 13.1. The triplex pump fault diagnosis is used to thoroughly investigate this structure. Based on the experimental results, it was shown that the hybrid GA-ML technique outperforms other ML methods such as Logistic Regression (LR), Naïve Bayes (NB), and Support Vector Machine (SVM). With the hybrid GA-SVM in use, the proposed framework attains a maximum accuracy of 95%.

According to Shuming Yi et al. [13], digital twins can be used as a starting point for human and robot collaboration in assembly. A person-recognition system utilizing depth camera is being investigated inside this framework, and it is based on a model that includes deep learning. This method is designed to anticipate important areas for the human skeleton model and offer precise human localization in a setting where humans and robots work together. To accomplish physical-virtual mapping that is both dynamic and consistent, X Ma et al. [14] suggest a certain approach. To begin, we provide the DTM_{HRC} architecture, which takes into account critical components of a collaborative setting and collaboration relationships, to illustrate an HRC scenario. Next, DTM_{HRC} is used to suggest a consistency technique that incorporates model validation, evolution, and consistency checks. Model parameter evolution and model structure evolution are the two main categories of consistency method implementations as determined by model deviation analysis.

Two goals are outlined by Bellavista et al. [15] application-driven digital twin networking middleware: 1) simplifying the interaction among heterogeneous devices and enhancing the expressiveness of packet content via data enrichment via well-defined standards. DTs can now take advantage of IP-based protocols instead of specialized industrial ones; 2) using software-defined networking to dynamically manage network resources in industrial edge environments, taking advantage of the communication mechanisms best suited to application needs, whether that's native IP or more articulated dependent on packet content. Zhou et al. [16] concentrate on a compact DT object detection model to achieve real-time synchronization between a real-world production system and its digital twin. Three primary environmental parameters—equipment, product, and operator—are used to model and forecast the evolving characteristics and real-time changes in building a general DT system for a smart manufacturing workshop. A hybrid deep neural network model is constructed, utilizing MobileNetv2, YOLOv4, and Openpose to ascertain the present state of both the actual and virtual production environments.

A learning method is developed to accomplish efficient multitype tiny item recognition by integrating and fusing data from both shallow and deep layers. This approach will aid in the modeling, monitoring, and optimization of the entire manufacturing process in the DT system. A higher detection accuracy for DT in smart manufacturing might be achieved by using the suggested strategy, according to evaluations and experiments conducted in three distinct use scenarios. The five services provided by Park et al. [31] as part of their architectural framework operate on data derived from the P4R information model and provide solutions to the performance barrier associated with personalized manufacturing. Using object orientation and the "type and instance" notion, this information model presents manufacturing abstraction for individualized production at a detailed level. In contrast to earlier research on the concept of digital twins, which concentrated on individual facilities and application development, this study examines CPPS architecture and operation from a system-of-systems viewpoint, with a focus on the digital twin as a central technological component of the whole system. A state-of-the-art method for the customized manufacture of different goods is offered by the integration of the digital twin-based CPPS into a micro smart factory (MSF). By utilizing the CPPS services that were put into place in the MSF, an average improvement in makespan of about 26.87% was made.

In the smart manufacturing industry, these review papers stress the importance of AI-enabled digital twins (AIDT). The literature review is the basis for the next sections, which identify and discuss important enabling technologies and application domains for AIDTs.

13.3 ESSENTIAL ELEMENTS OF AI-ENABLED DIGITAL TWIN (DT)

A revolutionary new way to comprehend, analyze, and optimize physical entities or systems is available with AI-enabled digital twins, which are a high-level combination of digital twins and artificial intelligence (AI). Skillful management of several data streams, including data collected from sensors and Internet of Things devices, is central to this breakthrough. To build digital models that are as accurate representations of their real-world equivalents as possible, it is necessary to have strong data integration. As a foundation for digital twins enabled by AI, these high-fidelity models capture the geometry, physics, and behavior of the physical system. Machine learning algorithms and predictive analytics [1,2,4] also play a crucial role by carefully examining both past and present data. With their impressive analytical capabilities, digital twins can anticipate future events, spot trends, and adapt to changing situations by optimizing operations in real time.

In a highly competitive market, industries must continuously adapt to new technologies, think outside the box, and innovate to stay ahead of the curve. Industrialization in literature is typically categorized under periods called "Industrial Revolutions," which are marked by significant changes in technology and paradigm [32]. One thing that sets these digital twins apart is that they use closed-loop control systems. With its ability to spot outliers and identify problems, this closed-loop paradigm greatly aids predictive maintenance while also improving efficiency. Most importantly, digital twins may adapt and get better with time because of adaptive learning techniques [7]. With the constant flow of new data, these systems learn and improve, making them more efficient at optimizing processes in complicated settings. Key to human-machine collaboration [13, 14] is the user interface, which allows for interaction and understanding of the digital representation. Bridging the gap between human expertise and machine intelligence, this collaborative environment increases decision-making and problem-solving in difficult settings. The Figure 13.6 shows the essential elements for AIDT. In this section, we deliberated each element.

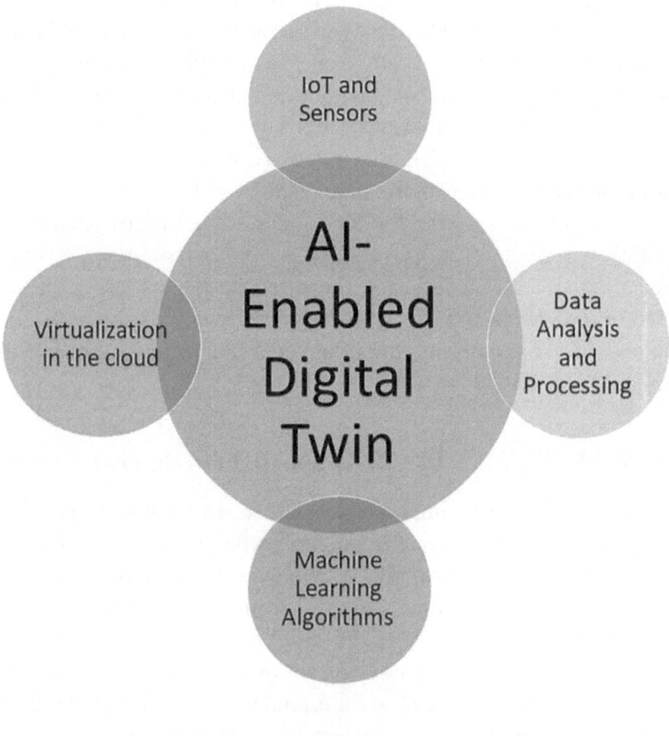

FIGURE 13.6 The Essential Elements for AIDT.

13.3.1 The Internet of Things (IoT) and Sensors

Digital twins powered by smart machines and the web of interconnected devices create a paradigm shift in smart manufacturing and beyond. By combining the power of the Internet of Things (IoT) with sensors, this novel integration can collect detailed data in real-time, laying a solid groundwork for digital twin optimization via AI. The Internet of Things (IoT) is the backbone of artificial intelligence (AI) digital twins, connecting all the many sensors, actuators, and gadgets that make up the data stream. Various factors, including temperature, pressure, vibration, and more, are measured by these Internet of Things (IoT) sensors that are strategically placed into physical systems. The digital twin relies on the data collected by these sensors to provide an accurate representation of the actual object in real-time.

For the Internet of Things (IoT) to work, sensors are essential nodes that must be able to pick up on minute physical details. Their variety enables numerous conditions to be monitored, guaranteeing a thorough comprehension of the system being mirrored in the digital twin. For example, in a production setting, temperature sensors can record changes in heat levels and pressure sensors can measure changes in pressure dynamics. To learn about the mechanical condition of machinery, vibration sensors look for oscillations or abnormalities. The digital twin's precision and reactivity are powered by the complex web of data created by these sensors. Sensors and the IoT really come into their own when it comes to creating and managing ultra-realistic digital twins. These digital representations are dynamic; they change and improve in response to the constant flow of information from the real system.

The digital twin is kept in sync with the physical counterpart's ever-changing conditions by the dynamic nature of data supplied via the Internet of Things. With the help of AI, the already-strong bond between sensors and digital twins can reach new heights. A branch of artificial intelligence known as machine learning algorithms sifts through the mountains of sensor data in search of useful trends, patterns, and insights. Digital twins can mimic the physical system and even foretell its future states and actions because of their analytical capabilities. Important to this integration is the use of AI-enabled predictive analytics. Machine learning-enabled digital twins can forecast the physical system's future performance by poring over past data and looking for trends.

The natural progression for the sensor-driven AI-enabled digital twin ecosystem and the Internet of Things (IoT) is human-machine collaboration. By connecting the actual and virtual worlds, the user interface facilitates interaction and interpretation between the digital representation and human operators. Together, we've built a platform that makes it easy to track, analyze, and act on data from the digital twin, which improves decision-making. The scalability and interoperability of AI-enabled digital twins are enhanced by the incorporation of IoT and sensors. As the size and complexity of systems increase, the capacity to manage a growing amount of data becomes more important, making scalability a must. Adoption of digital twins is possible in a wide variety of settings and sectors because of the interoperability, which guarantees smooth integration with current infrastructures.

13.3.2 Information Analysis and Data Processing

The preservation of the entire system's lifecycle data enables capabilities such as the playback of previous states, study of structural health decline, and intelligent analysis of any historical point. This provides enough information for data analysis and display [9]. Realizing the fantastical qualities of DTs is possible with a better grasp of system mechanisms and data characteristics made possible by the abundance of historical data, which also provides rich sample information for data mining. This allows for secure redundant backup and high-speed data acceptance, which, in turn, provides enough data for intelligent algorithms that analyze the data [33].

To address cloud storage's drawbacks—poor security, slow bandwidth, and significant latency—edge and fog storage options can be introduced. The use of edge and fog computing has the potential to ensure more secure data storage, make data analysis easier on cloud center equipment, and make local data storage a reality. The data that necessitates central processing and data that is not time-sensitive will be sent to the cloud center. The cloud center has constant access to data stored in the edge and fog. To install a system that guarantees data integrity and high database availability in the high-security available database system architecture, it is common practice to use the cloud servers of two or more cloud service providers. Due to the layered storage of the model's enormous data, edge computing is employed to facilitate the data analysis procedure within the sensor network [3]. The technical specifications for a platform that integrates big data are intended to meet the demands in intelligent coal mines sector. In particular, the developed platform incorporates AI-based data analysis tools [31].

Information analysis is a multi-step process that begins with collecting and cleaning up raw data. Preprocessing procedures are used to clean, filter, and standardize the raw sensor data, guaranteeing that the data is consistent and of high quality. Eliminating noise and unnecessary data is an essential first step in laying a solid groundwork for further research. Following this is exploratory data analysis, which seeks to discover correlations, trends, and patterns in the dataset. The analytical tools that analysts and AI algorithms use to learn about the manufacturing system's features include visualizations, descriptive statistics, and others. Here, we lay the framework for future investigations and use that information to build AI models.

13.3.3 Various Algorithms for Machine Learning

An important part of the machine learning model's job is to evaluate a video in the simulation layer in real-time and determine if there are any faults. When no fault is found, the machine learning model keeps silent and does not disrupt the system. If a problem is detected, the operator can intervene in the system by sending a signal to the digital model layer, which will alert it that there is a problem and prevent further time, resources, and money from being wasted. However, this research scenario would benefit from human involvement. A full elimination of human intervention in the system is possible with the aid of additional development and coding [10]. Digital twin applications rely heavily on machine learning to determine if a printed piece is defective. Machine learning algorithms can be taught to examine 3D models in order to detect any flaws in printed items. This helps in defect detection. Structured flaws, surface defects, and other printing mistakes can be detected, using machine learning algorithms that learn the traits and patterns of flawed prints from labeled data. This aids in guaranteeing the printed items' quality and authenticity. Digital twin systems provide massive volumes of data, such as sensor readings, pictures, and other measures, which can be used for automation and efficiency purposes. This data can be processed at scale by machine learning algorithms, which can automate the process of defect discovery. Reduced manual inspection efforts, faster analysis, and the ability to monitor print quality in real-time.

To make intelligent manufacturing a reality, the new generation of intelligent, integrated control systems relies on the intelligent operation layer, which acts as their brain [5]. It achieves deep mining and multi-dimensional analysis of data assets based on enterprise data by making full use of artificial intelligence technology like neural networks and machine learning [34] as well as industrial big data. Therefore, it may offer its services for a variety of applications, including advanced scheduling, data statistics and analysis, quality tracing, energy prediction, and production prediction. Intelligent and scientific decision-making in integrated corporate operations might also be aided by it [35]. Using data collected from a variety of sensors and cameras distributed across various geographical regions, machine learning models draw on the findings of the genetic algorithm. Minimizing the general modeling error as a percentage of the area of forest damaged by the fire and the time when the event occurred is the goal of the evolutionary algorithm, which is designed to optimize all types of "submodels" by default [17].

13.3.4 Virtualization in the Cloud and Modeling

Cloud computing has evolved into a hybrid of numerous technologies because of the exponential growth of computing power. These include utility computing, distributed computing, parallel computing, load balancing, virtualization, hot backup jumbling, and network storage. With the help of cloud virtualization and modeling approaches, artificial intelligence (AI)-enabled digital twins can revolutionize smart manufacturing. Applications spanning from process optimization to predictive maintenance are made possible by this convergence, which is a smart way to use cloud computing to build and optimize digital models of physical systems. When resources like servers, storage, and networks are hosted on the cloud, virtualization allows for the creation of virtual instances of these resources [3]. Digital twins powered by artificial intelligence can take advantage of this virtualization to provide a versatile and scalable simulation and modeling environment.

The storage and processing capacity offered by cloud platforms is ideal for managing the massive datasets and intricate algorithms needed to build and maintain high-fidelity digital twins. As part of the modeling process, digital duplicates of the physical production system are created. Rather than being static copies, these models include dynamic parts that change and adapt based on data collected in real-time. The ability of these models to learn from past data and forecast future states and actions is largely due to machine learning algorithms. When it comes to digital twins, scalability is a major benefit of cloud virtualization. The computing demands of AI algorithms and modeling methodologies can inform the optimal deployment of resources in cloud infrastructure. Cloud computing allows for the dynamic scalability of resources to match the changing demands of processing more data or digital twin complexity. The scalability of AI-enabled digital twins guarantees that they can manage the complexities of large-scale industrial systems without sacrificing performance.

13.4 USE CASES IN SMART MANUFACTURING

Many applications have found usage for digital twins, including process oversight, prediction, and interface with physical assets. Using VR and AR improves interactions with digital twins, and including human factors in digital twins is changing research focused on industry 5.0 [4]. The advent of digital twins has been a watershed moment in smart manufacturing, ushering in an era of dynamic and interconnected frameworks for simulating and improving physical systems. Exploring use cases becomes crucial in this context, providing concrete examples that show how digital twins can be applied and the benefits they provide to smart industrial settings. The following use cases for smart manufacturing are very crucial for smart manufacturing.

13.4.1 Prognostic Maintenance

As a part of smart manufacturing, prognostic maintenance—a subset of predictive maintenance—uses cutting-edge tech to foretell when equipment will break down, so repairs may be made in plenty of time to keep operations running smoothly. Smart manufacturing use cases heavily rely on predictive maintenance to optimize asset management, decrease costs, and increase operational efficiency. Here are a few examples of how prognostic maintenance might be applied in smart manufacturing:

- Using a Unity client user interface and machine learning models for defect identification, additive manufacturing gives operators real-time feedback on product quality by identifying and fixing errors before they're even made [10].
- When it comes to human-robot collaboration, supervised learning-based classification methods including CNNs, LSTMs, HMMs, and SNNs have been documented for application in motion and intent detection and prediction procedures [4].
- Skeleton-based human action recognition is also used for human-robot collaboration [29].
- A versatile multi-sensor method for tracking tool wear has been devised, utilizing vibration, current, and force signals. This method ensures that the overall cutting power is in line with the tool wear curve. To forecast when drilling and milling tools would wear out, Kandilli used data from multiple sensors in a neural network [12].
- The H-DNN design incorporates the prediction flow of 3D physical fields for CCS and consists of two modules: the predictor and the autoencoder. Given that the hyperparameters of the neural network can implicitly depict

the impact of complicated boundary conditions, the ambient elements affect the properties of composite materials and quantifiable loadings such as pressure and displacement [2].

13.4.2 Maximizing Efficiency

Digital twin use cases maximize productivity in smart manufacturing by optimizing processes, reducing downtime, and enhancing overall operational performance through the utilization of modern technology. Achieving this efficiency is greatly facilitated by digital twins, which are virtual representations of physical systems. These twins offer adaptive capabilities, real-time insights, and predictive analytics. Here are some real-world examples of how digital twins have helped smart factories maximize efficiency:

- Entire manufacturing lines or processes can be digitally modeled with the help of digital twins. Manufacturers can enhance the overall efficiency of manufacturing processes by identifying bottlenecks, optimizing workflows, and modeling different situations and configurations [15].
- In smart manufacturing plants, energy utilization is modeled and monitored using digital twins. Digital twins shed light on operations that use a lot of energy by incorporating sensors that measure usage. With this information, manufacturers can optimize equipment utilization during off-peak hours or apply energy-saving measures according to real-time demand—two examples of potential energy efficiency improvements [16].
- With digital twins, supply chain optimization is possible at every stage, not only on the production floor. With the help of logistics, production, and supplier data, manufacturers build digital twins of their supply chain activities [17].
- Efficient demand forecasting, inventory management, and logistics planning are made possible by these digital representations, which provide real-time visibility into the supply chain. Manufacturers may improve operational efficiency, cut down on stockouts, and shorten lead times by optimizing the supply chain with digital twins [19].

13.4.3 Quality Assurance

In smart manufacturing, quality assurance is essential for making sure products live up to high standards and satisfy customers. Because they allow for the simulation and optimization of production conditions, real-time insights, and predictive analytics, digital twins—virtual copies of actual systems or products—play a crucial role in improving quality assurance processes. To illustrate the value of digital twins in smart manufacturing quality assurance, consider the following use cases:

- Artificial intelligence (AI) and the Internet of Things (IoT) technologies show promise for improving operational efficiency and accuracy while also showing their potential in real-world applications. Also, the problems and issues with using machine learning, deep learning, and IoT technologies in tool-embedded TCM are emphasized [12].

- By offering a holistic perspective of the whole production process, digital twins aid in identifying the source of quality concerns. Manufacturers can easily perform corrective actions by tracing back through the digital twin to determine the origin of defects or anomalies. Using this method, quality issues can be resolved more quickly and they won't happen again [16].

13.5 DIFFICULTIES AND POTENTIAL BENEFITS

An AI-enabled digital twin in smart manufacturing can revolutionize the industry, but it is not without its share of problems and advantages. Within the framework of smart manufacturing, let's investigate the pros and cons of using AI-enabled digital twins:

13.5.1 Difficulties

- One of the most important and difficult scenarios involving several robots and human operators in the manufacturing industry. Recognizing the behaviors of both human operators and robots will involve a range of tasks beyond object identification, including position estimation and gesture recognition. Because of this, the system will be able to handle more sophisticated jobs with greater resilience and flexibility, and it will be able to make more complex decisions [1, 13].
- Data that is dependent on physical objects can be brought into harmony with a timetable for decision-making and process optimization with the help of a digital twin. The integration of models and data sets can pique researchers' interest in the practicality and promising future of digital twins and deep learning [2, 16].
- The overall defensive system needs to conduct further quantitative evaluations of the edge-fog-cloud algorithm in relation to other algorithms in order to strengthen the reliability of the results [3].
- A paradigm shifts from one driven by experts to one driven by users is possible with the advent of ever-more-specific modeling tools and the pervasive diffusion of AI [4,6].
- Important technologies for DT technology implementation include industrial information systems, big data, and artificial intelligence. But DT technology is still evolving and active, even though these technologies are developing at a quick pace. One of the key areas for future study is improving how DTs use the findings from related technology [9].
- The multidisciplinary teams that work on design and development projects face a substantial obstacle. These emerge as a result of the diversity of study areas that participate in collaborative efforts; this diversity is both an advantage and a disadvantage, since it can speed up breakthroughs but can cause research to move in various ways due to competing goals [18].

13.5.2 Potential Benefits

- ***Protecting Sensitive Information*—**Personnel with expertise in data analysis, modeling, simulation, and integration are also required for the management and operation of DT systems. It can be difficult to hire competent people or to train current employees. The high implementation cost and complicated design are two other major obstacles that can impede the use of DT technology. Major investments in technological platforms (sensors, software), infrastructure creation, upkeep, and security solutions are necessary for DT solution implementation. Lastly, there is a large operational expenditure required for the expensive DT infrastructure upkeep. Small and medium-sized firms may find it difficult to afford the substantial costs associated with DT system implementation and maintenance [8]. Deployment of DT technology is anticipated to be slowed significantly by its high fixed cost and sophisticated infrastructure.
- ***Compatibility Across Systems*—**With the help of AI-enabled digital twins, producers may easily adjust to changes in demand or product specifications, which, in turn, supports customization. Adaptability is key in ever-changing market conditions. There is a better ability to meet the needs of the market and happier customers [1, 4]. Optimization strategies of AI-enabled DT through deep learning algorithms help to improve self-adaptability in the production of manufacturing industries [5]. E2APPO's well-trained scheduling model outperforms heuristic and metaheuristic algorithms. The scheduling technique that is allowed by DT improves adaptation to complex shop settings and changeable orders, and it provides an ideal balance between time cost and quality [7].
- ***Disparity in Competence*—**More consistent and efficient production processes as a result of a more competent and proficient workforce generally [2, 3]. Lower dependence on traditional wisdom, better uniformity in operational procedures, and improved overall competence [9]. Overall competency levels rose, problem-solving efficiency rose, and the learning curves of less experienced employees shortened. A workforce that can better adjust to changing production processes, more efficient skill development, and individualized learning experiences [18].

13.6 CONCLUSION

Industry 4.0 is driving manufacturing into a new age of efficiency, agility, and creativity, and as we explore the total impact and prospects of AI-enabled digital twins in smart manufacturing, it becomes clear that this technical synergy is a cornerstone of this movement. From predictive maintenance and process optimization to workforce competency development, the combination of artificial intelligence and digital twin technologies offers a myriad of benefits that could revolutionize industrial environments.

Finally, the convergence of AI and digital twins is going to cause a sea change in the smart manufacturing sectors. Working together like this makes manufacturing operations more efficient and resilient to shocks. Artificial intelligence (AI)-enabled digital twins are a promising new development in manufacturing because they can learn from their mistakes, adapt to changing conditions, and ultimately, outperform humans. Companies simply cannot afford to overlook this game-changing technology if they want to remain competitive in today's manufacturing landscape.

REFERENCES

[1] Wang, S., Zhang, J., Wang, P., Law, J., Calinescu, R. and Mihaylova, L. (2024). A deep learning-enhanced digital twin framework for improving safety and reliability in human–robot collaborative manufacturing. *Robotics and Computer-Integrated Manufacturing*, [online] 85, p. 102608. https://doi.org/10.1016/j.rcim.2023.102608.

[2] Xu, X., Wang, G., Yan, H., Zhang, L. and Yao, X. (2023). Deep-learning-enhanced digital twinning of complex composite structures and real-time mechanical interaction. *Composites Science and Technology*, 241, pp. 110139–110139. https://doi.org/10.1016/j.compscitech.2023.110139.

[3] Lv, Z. and Lou, R. (2022). Edge-fog-cloud secure storage with deep-learning-assisted digital twins. *IEEE Internet of Things Magazine*, 5(2), pp. 36–40. https://doi.org/10.1109/IOTM.002.2100145.

[4] Asad, U., Khan, M., Khalid, A. and Lughmani, W. A. (2023). Human-centric digital twins in industry: A comprehensive review of enabling technologies and implementation strategies. *Sensors*, 23(8), p. 3938. https://doi.org/10.3390/s23083938.

[5] Liu, X., Li, J., Wang, H., Jia, W., Yang J. and Guo, Z. (2023). Design of an optimal scheduling control system for smart manufacturing processes in tobacco industry. *IEEE Access*, 11, pp. 33027–33036, 2023, https://doi.org/10.1109/ACCESS.2023.3261883.

[6] Chen, J., Clos, J., Price, D. and Caleb-Solly, P. (2023). Digital twins for human-assistive robot teams in ambient assisted living. TAS '23*: First International Symposium on Trustworthy Autonomous Systems*. https://doi.org/10.1145/3597512.3597520.

[7] Gan, X., Zuo, Y., Zhang, A. and Li, S. (2023). Digital twin-enabled adaptive scheduling strategy based on deep reinforcement learning. *Science China Technological Sciences,* 66, pp. 1937–1951. https://doi.org/10.1007/s11431-022-2413-5

[8] Attaran, M., Attaran, S. and Celik, B. G. (2023). The impact of digital twins on the evolution of intelligent manufacturing and Industry 4.0. *Advances in Computational Intelligence,* 3, 11. https://doi.org/10.1007/s43674-023-00058-y

[9] Yao, J. F., Yang, Y., Wang, X. C. and Zhang, X. P. (2023). Systematic review of digital twin technology and applications. *Visual Computing for Industry, Biomedicine, and Art,* 6, 10. https://doi.org/10.1186/s42492-023-00137-4

[10] Jyeniskhan, N., Keutayeva, A., Kazbek, G., Ali, M. H. and E. Shehab, E. (2023). Integrating machine Learning model and digital twin system for additive manufacturing. *IEEE Access*, 11, pp. 71113–71126. https://doi.org/10.1109/ACCESS.2023.3294486.

[11] Alshathri, S., El-Din Hemdan, E., El-Shafai, W. and Sayed, A. (2023). Digital twin-based automated fault diagnosis in industrial IoT applications. *Computers, Materials & Continua*, 75(1), pp. 183–196. https://doi.org/10.32604/cmc.2023.034048.

[12] Tran, M.-Q., Doan, H.-P., Vu, V. Q. and Vu, L. T. (2023). Machine learning and IoT-based approach for tool condition monitoring: A review and future prospects. *Measurement*, 207, p. 112351. https://doi.org/10.1016/j.measurement.2022.112351.

[13] Yi, S., Liu, S., Xu, X., Wang, X. V., Yan, S. and Wang, L. (2022). A vision-based human-robot collaborative system for digital twin. *Procedia CIRP*, 107, pp. 552–557. https://doi.org/10.1016/j.procir.2022.05.024.

[14] Ma, X., Qi, Q., Cheng, J. and Tao, F. (2022). A consistency method for digital twin model of human-robot collaboration. *Journal of Manufacturing Systems*, 65, pp. 550–563. https://doi.org/10.1016/j.jmsy.2022.10.012.

[15] Zhou, X., Xu, X., Liang, W., Zeng, Z., Shimizu, S., Yang, L. T. and Jin, Q. (2022). Intelligent small object detection for digital twin in smart manufacturing with industrial cyber-physical systems. *IEEE Transactions on Industrial Informatics*, 18(2), pp. 1377–1386. https://doi.org/10.1109/TII.2021.3061419.

[16] Bellavista, P., Giannelli, C., Mamei, M., Mendula, M. and M. Picone, M. (2021). Application-driven network-aware digital twin management in industrial edge environments. *IEEE Transactions on Industrial Informatics*, 17(11), pp. 7791–7801. https://doi.org/10.1109/TII.2021.3067447.

[17] Dashkina, A., Khalyapina, L., Kobicheva, A., Lazovskaya, T., Malykhina, G. and Tarkhov, D. (2020). Neural Network Modeling as a Method for Creating Digital Twins. *Proceedings of the 2nd International Scientific Conference on Innovations in Digital Economy: SPBPU IDE-2020.* https://doi.org/10.1145/3444465.3444535.

[18] Fuller, A., Fan, Z., Day, C. and Barlow, C. (2020). Digital twin: Enabling technologies, challenges and open research. *IEEE Access*, 8, pp. 108952–108971. https://doi.org/10.1109/access.2020.2998358.

[19] Lee, J., Azamfar, M., Singh, J. and Siahpour, S. (2020). Integration of digital twin and deep learning in cyber-physical systems: Towards smart manufacturing. *IET Collaborative Intelligent Manufacturing.* https://doi.org/10.1049/iet-cim.2020.0009.

[20] Alexopoulos, K., Nikolakis, N. and Chryssolouris, G. (2020). Digital twin-driven supervised machine learning for the development of artificial intelligence applications in manufacturing. *International Journal of Computer Integrated Manufacturing*, 33(5), pp. 429–439. https://doi.org/10.1080/0951192x.2020.1747642.

[21] www.marketsandmarkets.com. (n.d.). *Download PDF Brochure – Digital Twin Market Size, Share, Industry Report, Revenue Trends and Growth Drivers.* [online] https://www.marketsandmarkets.com/pdfdownloadNew.asp?id=225269522.

[22] www.3ds.com/fileadmin/PRODUCTS-SERVICES/DELMIA/PDF/Whitepaper/DELMIA-APRISO-Digital-Twin-Whitepaper.pdf

[23] Liu, S., Wang, L. and Wang, X. V. (2021). Sensorless haptic control for human-robot collaborative assembly. *CIRP Journal of Manufacturing Science and Technology*, 32, pp. 132–144. https://doi.org/10.1016/j.cirpj.2020.11.015.

[24] Tao, F., Zhang, H., Liu, A. and Nee, A. Y. C. (2019). Digital twin in industry: State-of-the-art. *IEEE Transactions on Industrial Informatics*, 15(4), pp. 2405–2415. https://doi.org/10.1109/TII.2018.2873186.

[25] Shenglin. (2022). Wongsinglam/semi_labelme: V1.0.0 (v1.0.0). *Zenodo.* https://doi.org/10.5281/zenodo.6393954

[26] Zhang, J., Wang, S., Wang, P., Mihaylova, L. and Law, J. (2022). A vision data repository for human-UR10 robot interactions in manufacturing. *The University of Sheffield. Dataset.* https://doi.org/10.15131/shef.data.16669315.v1

[27] Lin, T. Y. Maire, M., Belongie, S., Hays, J., Perona, P., Ramanan, D., Dollár, P. and Zitnick, C. L. (2014). Microsoft COCO: Common Objects in Context. In: Fleet, D., Pajdla, T., Schiele, B., Tuytelaars, T. (eds.), Computer Vision–ECCV 2014. ECCV 2014. Lecture Notes in Computer Science, vol 8693. Cham: Springer. https://doi.org/10.1007/978-3-319-10602-1_48

[28] Everingham, M., Eslami, S. M. A., Van Gool, L., Williams, C. K. I., Winn, J. and Zisserman, A. (2014). The pascal visual object classes challenge: A retrospective. *International Journal of Computer Vision*, 111(1), pp. 98–136. https://doi.org/10.1007/s11263-014-0733-5.

[29] Dilliraj E. and Ponnusamy, V. (2023). Real-time AI-assisted visual exercise pose correctness during rehabilitation training for musculoskeletal disorder. *Journal of Real-Time Image Processing*, 21(1). https://doi.org/10.1007/s11554-023-01385-6.

[30] Park, K. T., Lee, J., Kim, H. J. and Do Noh, S. (2020). Digital twin-based cyber physical production system architectural framework for personalized production. *The International Journal of Advanced Manufacturing Technology,* 106, pp. 1787–1810. https://doi.org/10.1007/s00170-019-04653-7

[31] Zhang, S., Tang, F., Li, X., Liu, J. and Zhang, B. (2021). A hybrid multi-objective approach for real-time flexible production scheduling and rescheduling under dynamic environment in Industry 4.0 context. *Computers & Operations Research*, 132, p. 105267. https://doi.org/10.1016/j.cor.2021.105267.

[32] Saeid, N. (2019). Industry 5.0—a human-centric solution. *Sustainability,* 11(16), p. 4371. https://doi.org/10.3390/su11164371.

[33] Cooper, J., Noon, M., Jones, C., Kahn, E. and Arbuckle, P. (2013). Big data in life cycle assessment. *Journal of Industrial Ecology*, 17(6), pp. 796–799. https://doi.org/10.1111/jiec.12069.

[34] Freier, P. and Schumann, M. (2021). Decision support systems in the context of cyber-physical systems. *Australasian Journal of Information Systems*, 25. https://doi.org/10.3127/ajis.v25i0.2849.

[35] Sharifi, M. and Taghipour, S. (2021). Optimal production and maintenance scheduling for a degrading multi-failure modes single-machine production environment. *Applied Soft Computing*, 106, p. 107312. https://doi.org/10.1016/j.asoc.2021.107312.

14 Smart Manufacturing
Navigating Challenges, Seizing Opportunities, and Charting Future Directions—A Comprehensive Review

Mrutyunjaya S. Hiremath, Praveen N., Lipsa Subhadarshini, Sujith Kumar Sivanandan, Jayaprada S. Hiremath, and Shantala S. Hiremath

14.1 INTRODUCTION

Industries stand as the backbone of nations' economic prosperity, wielding profound influence over the health and dynamism of economies. The infusion of smart technologies into the fabric of newly developed manufacturing industries has ushered in a transformative era characterized by a remarkable surge in productivity, estimated at a staggering 17–20%, as meticulously documented in a comprehensive study [1]. This surge in efficiency finds its roots in the seamless synergy between enhanced machine utilization and judiciously optimized energy consumption; all orchestrated through the intricate dance of smart manufacturing systems.

In the relentless arena of industrial competition, manufacturers are driven by an unyielding pursuit of innovation, rapid responsiveness to market dynamics, cost-effectiveness, and unwavering reliability—all vital attributes in meeting the ever-evolving demands of consumers. This relentless competition has acted as a catalyst, propelling the paradigm shift toward smart manufacturing—a transformative realm marked by pervasive digitization and the seamless fusion of cyber-physical control, stretching its influence from manufacturing plants to the heart of business outlets [2].

At the epicenter of this revolution lies the formidable edifice of smart manufacturing systems, an embodiment of ingenuity and innovation. These systems stand as a testament to the power of smart devices and sensors, systematically reducing the need for human intervention while empowering a direct conduit for customized manufacturing instructions—transmitted directly from discerning customers via the boundless realm of the Internet [3].

Across the globe, nations have crafted policies and strategies that revolve around the twin pillars of cyber-physical systems and digital manufacturing, igniting the

DOI: 10.1201/9781003473886-14

flames of future industrial prowess. A case in point is Germany's visionary "Industry 4.0 initiative" [4], a bold proclamation that underscores the pivotal role of interconnected machining systems in every facet of production. This visionary concept ingeniously melds the Internet of Things (IoT), Cyber-Physical Systems (CPS), digital manufacturing, and an array of other cutting-edge technologies into a symphony of industrial progress [5]. Likewise, with its "Made in China 2025" and Internet Plus program [6], China has set its sights on elevating its manufacturing sector to unprecedented heights of excellence and innovation.

The grand transition to distributed manufacturing systems marks a decisive departure from traditional control mechanisms, deftly addressing the pressing demands for customization, surges, and troughs in the supply chain, and the production of smaller batches—a realm where traditional models falter. In a telling revelation, McKinsey Global Institute's research has unveiled the astonishing statistic that the manufacturing sector boasts a 60% potential for automation, thus laying bare the vast canvas upon which smart manufacturing technologies can be artistically applied to amplify industrial capacities [7].

Figure 14.1 is a visual compass that guides us through the intricate maze of interconnected, smart manufacturing systems within Industry 4.0. These systems are sinews that link various facets of production, orchestrating a seamless symphony from design to the ultimate customer delivery. The ethereal realm of cloud computing

FIGURE 14.1 Diagram Illustrating Components of Smart Manufacturing [8].

lends its nimble prowess to enable efficient on-demand manufacturing, sustaining the delicate equilibrium of the demand-supply ecosystem with impeccable finesse [9].

Leading global markets have initiated audacious industrial strategies, with Germany's Industry 4.0 as an illustrious exemplar, setting a gold standard for industrial revolutions in the annals of history. These evolving strategies, each a unique chapter in the unfolding narrative of industrial progress, are meticulously chronicled and detailed in Table 14.1, heralding a new era of manufacturing prowess [4].

This study explores the profound significance and untapped potential of smart manufacturing systems, offering a comprehensive understanding of the evolving landscape in this domain. It delves deep into the intricate web of current manufacturing practices and their groundbreaking innovations, focusing on cyber-physical systems and interconnected devices. These innovations are the linchpin of the smart manufacturing revolution, reshaping industries and driving progress [6].

The journey commences with a thoughtful introduction to smart manufacturing systems, setting the stage for a profound exploration of their relevance and impact across various application areas. This introduction unveils the transformative power of smart manufacturing, underscoring its pivotal role in shaping the future of production and industry [6].

As we venture further into the study, the third section unveils the diverse tapestry of technologies interwoven with smart manufacturing and automation. A meticulous examination unfurls the boundless scope of these technologies, revealing their far-reaching implications for enhancing efficiency, precision, and agility within manufacturing processes. This is a journey through the realms of robotics, artificial intelligence, the Internet of Things, and beyond—each a key player in the smart manufacturing ecosystem [6].

TABLE 14.1
Different Countries Declare Strategies for the Next Industrial Development

Country	Declared Strategy
Germany	Industry 4.0 Revolution—Industry 4.0 represents the melding of cutting-edge digital technologies such as the Internet of Things (IoT), Artificial Intelligence (AI), and cloud computing with manufacturing processes, signifying the onset of the Fourth Industrial Revolution.
China	"Made in China 2025" Strategy—This initiative represents China's strategic approach to revolutionizing its industrial sector. It emphasizes a shift toward high-tech fields, including electric vehicles, advanced information technology, telecommunications, robotics, artificial intelligence, renewable energy and transportation systems, innovative materials, and medical equipment.
Japan	Society 5.0—"Society 5.0" refers to a future society integrating advanced digital technologies into every aspect of life to achieve a more interconnected, efficient, and sustainable human-centered society.
United States	Industrial Internet Consortium—The Industrial Internet Consortium is an alliance aimed at accelerating the growth of the Industrial Internet through collaborative partnerships and shared innovations.

In the subsequent section, the spotlight turns toward standardization efforts—a pivotal dimension in smart manufacturing technology. Here, the focus is on the intricate web of standards and protocols that underpin seamless integration and compatibility. These standards serve as the unifying language that enables disparate systems to communicate harmoniously, ensuring that the promise of smart manufacturing is realized across industries and sectors [6].

The journey shifts toward the challenges encountered in implementing smart manufacturing systems—a realm where the complexities and nuances of this transformative journey are laid bare. Insights are offered into effective strategies to navigate these challenges, offering a roadmap to overcome obstacles and unlock the full potential of smart manufacturing [6].

In the fifth section, our gaze extends toward the horizon, where we glimpse the opportunities ahead. Emerging trends come into focus, offering a glimpse of the future and its potential implications across diverse industries. This forward-looking analysis shows how smart manufacturing will continue to evolve and shape the industrial landscape [6].

Finally, the study culminates in the sixth section, where the discoveries and insights acquired throughout this research journey are distilled into a cohesive summary. This section encapsulates the essence of the study, offering a bird's-eye view of the transformative power of smart manufacturing systems and their profound influence on the world of industry and production [6].

In essence, this study is a deep dive into the heart of smart manufacturing—a realm where innovation, technology, and progress converge. It illuminates the path toward a future where efficiency, precision, and adaptability are not mere aspirations but the bedrock upon which industries are built.

14.2 SYSTEM OVERVIEW: UNDERSTANDING THE FOUNDATION

The relentless drive toward process automation is a cornerstone of modern industrial innovation, motivated by a compelling blend of imperatives: to mitigate health risks and amplify productivity. This evolution, witnessed primarily in the manufacturing sector, has traversed a fascinating trajectory from humble beginnings with simple hydraulic and pneumatic systems to today's awe-inspiring array of sophisticated robots [10].

The merits of automation are indeed manifold, offering a treasure trove of advantages that resonate throughout the manufacturing realm. Foremost among these is the tangible uptick in productivity, where automation amplifies the throughput and efficiency of industrial processes. Coupled with this is a concomitant enhancement in quality, with robotic precision reducing errors and ensuring consistently high standards. The allure of automation extends further still, casting a cost-efficient glow upon the manufacturing landscape. Here, savings accrue regarding material costs and the judicious conservation of energy resources. These multifaceted benefits underscore the enduring appeal of automation, propelling it to the forefront of industrial practice [6].

As technology has unfolded its myriad innovations, we've witnessed a seismic shift toward embracing wireless-based automation systems—a transformation that transcends sectoral boundaries. In this epoch, the marriage of automation and the

Internet of Things (IoT) has allied profound consequences. The IoT's allure lies in its capacity to simplify operations across diverse domains, transcending manufacturing to permeate the power and energy sectors with its transformative potential [11].

The infusion of the Internet into control systems yields a plenitude of additional benefits, each contributing to a more agile and responsive industrial ecosystem. Remote sensing, empowered by IoT, extends the reach of monitoring and control, enabling operators to gain real-time insights and take swift action. Distance control ushers in an era where machines and processes can be manipulated from afar, reducing human exposure to hazardous environments. Easier data acquisition provides information for analysis, decision-making, and process optimization. Flexibility—a coveted attribute—is substantially bolstered, empowering automation systems to adapt seamlessly to evolving demands and conditions [11].

With its expansive scope and continually emerging technologies, the Internet of Things holds the potential to revolutionize a diverse array of systems. It accomplishes this by simplifying operations and significantly reducing the need for direct human intervention—all facilitated by a global networking infrastructure. The concept of smart manufacturing, firmly anchored in the IoT paradigm, has the power to reshape industrial landscapes. It leverages IoT features to usher in a new era characterized by interconnected systems and intelligent automation. Technologies like microgrids, smart grids, advanced metering systems, centralized load dispatch, control systems, and distribution system automation are emblematic of this paradigm shift [11].

Within the power system transmission and distribution, the application of smart technology in automation and control represents a vibrant arena of ongoing research and development. Utility companies and local governments are embarking on pioneering pilot projects that herald the implementation of these advanced technologies. These projects serve as vanguards, forging a path toward the seamless integration of smart technology into the broader landscape of automation and control systems. As these initiatives gather momentum, they promise to enhance power systems' reliability, efficiency, and sustainability, ushering in a future where intelligent automation takes center stage [11].

In summary, the march toward process automation is not merely an industrial imperative but a transformative journey propelled by a convergence of technological marvels. With its roots in simple systems and the complex world of robotics, automation offers many benefits, from heightened productivity to elevated quality and resource conservation. The integration of the Internet of Things ushers in a new era of interconnectedness, simplifying operations across sectors and empowering industries to adapt, respond, and thrive in an ever-evolving landscape. In particular, the power and energy sectors stand to reap substantial rewards from this digital transformation as they explore the frontiers of smart technology in automation and control. The ongoing pilot projects in this realm herald a promising future where automation and intelligence coalesce to redefine industrial landscapes.

14.2.1 Framework of Smart Manufacturing Systems

The Smart Manufacturing System (SMS) represents a profound revolution in manufacturing, transcending traditional practices to embrace the full potential of digital transformation. At its core, SMS is a multifaceted endeavor, underpinned by a rich tapestry of technologies and principles, each contributing to the overarching goal of elevating productivity and competitiveness within the manufacturing sector [6].

One of the foundational pillars of SMS is interoperability, enabling seamless communication and cooperation among various components and systems within a manufacturing environment. This harmonious interaction between machines, sensors, and processes forms the bedrock upon which the digitized manufacturing landscape is built. Through interoperability, disparate elements unite to operate as a synchronized whole, fostering efficiency and synergy.

Real-time control and monitoring are integral facets of SMS, providing manufacturers unprecedented visibility and command over their production processes. These capabilities empower manufacturers to make informed decisions swiftly, respond to deviations in real time, and optimize operations for peak efficiency. The ability to monitor every facet of production in real time is akin to having a digital nerve center that offers insights and control like never before.

Flexibility in production is a hallmark of SMS, allowing manufacturers to adapt to dynamic market conditions swiftly. This adaptability is crucial in a fast-paced global economy where consumer preferences, market demands, and external factors can change rapidly. The flexibility to reconfigure production lines, shift priorities, and accommodate customization requests is a key driver of competitiveness.

In today's hyper-connected world, rapid response to market changes is not a luxury but a necessity. SMS equips manufacturers with the agility to swiftly pivot and recalibrate production in response to market shifts, ensuring they stay ahead of the curve. This capability enables manufacturers to seize emerging opportunities and navigate challenges with finesse.

Advanced sensors are the sensory organs of SMS, providing a wealth of data that serves as the lifeblood of informed decision-making. These sensors can monitor everything from temperature and humidity to pressure and machine performance, delivering constant data for analysis and optimization. This data-driven approach enables manufacturers to proactively address issues, predict maintenance needs, and enhance efficiency.

Big data analytics is the brainpower behind SMS, processing the vast amounts of data generated by sensors and production processes. Manufacturers gain invaluable insights into their operations through sophisticated algorithms and data analysis. These insights inform continuous improvement efforts, uncover hidden patterns, and support data-driven decision-making.

SMS operates in two distinctive modes, each tailored to specific needs and objectives. Production engineers play an active role in the semi-autonomous mode, setting goals and parameters. This mode fosters collaboration between human expertise and automated systems, leveraging both strengths. In contrast, the fully autonomous mode delegates decision-making to the system, independently determining and implementing optimal operating parameters across interconnected production units. This autonomy minimizes human intervention and enhances efficiency [12].

Manufacturers must prioritize several key areas in a highly competitive landscape to maintain their edge. Cost-effectiveness remains a top concern, requiring efficient resource allocation and cost optimization. Achieving optimal manufacturing and delivery times is another critical goal, as delays can impact customer satisfaction and market competitiveness. Product quality is non-negotiable, as subpar products can tarnish reputations and lead to costly recalls. Customization—the ability to tailor products to individual customer needs—is an increasingly important factor in a consumer-centric market [3].

An equally vital aspect of smart manufacturing is the capacity to adapt and improve continually. This adaptability hinges on information and the ability to respond to changing environmental factors. It's essential to recognize that implementing smart manufacturing involves selecting the right mix of technologies and strategies to transition from existing systems to smart ones. This process is no small feat and often involves navigating complex terrain. Strategies like the Supply Chain Readiness Level and Manufacturing Enterprise Solutions Association (MESA) Manufacturing Transformation Strategy are commonly employed and rooted in ISA-95 methods. These strategies go beyond focusing solely on specific technologies or manufacturing execution systems. Instead, they emphasize the foundational role of information and communication technology in orchestrating the transition to a smart system [13].

Incorporating information and communication technologies (ICT) is a strategic linchpin of smart manufacturing, exemplified by the Industry 4.0 model. This approach entails transforming various services within the industry into smart services by integrating efficient data and process management. By utilizing the Internet of Things (IoT), systems and services are interlinked, ushering in a new era of interconnectedness. Developing data analysis software with integrated security systems is an indispensable component of this transformation, as it safeguards the integrity of incoming data while facilitating in-depth analysis [6].

Figure 14.2, presented by the National Institute of Standards and Technology (NIST), provides a visual map for a holistic perspective of the smart manufacturing ecosystem. This illustration vividly portrays the intricate interplay among different domains of smart manufacturing and their functions. This ecosystem delineates the connections between an enterprise's product, production process, and business aspects, each with its lifecycle. The components of the NIST smart manufacturing system, portrayed in Figure 14.2, are further elucidated in Table 14.2, providing a comprehensive guide to understanding the various elements and their roles [6].

The wave of digital transformation is reshaping industries worldwide, and smart manufacturing is at its vanguard. Initiatives such as Germany's Industry 4.0, China's Made in China 2025, the USA's Industrial Internet, and Japan's Society 5.0 are pivotal national strategies in technology and manufacturing [14].

These initiatives may differ in their implementation methods, target industries, and projected timelines. However, they share a common goal: to harness the potential of smart and digital technologies to elevate global manufacturing systems. The amalgamation of digital transformation and smart technologies is forging the foundation of next-generation industries. As these strategies evolve, they promise to usher in an era of unparalleled efficiency, innovation, and competitiveness within the manufacturing sector.

In 2011, Germany ushered in a transformative era in the industrial landscape with the inception of Industry 4.0—a visionary movement that continues to shape the future of manufacturing. Industry 4.0 is a comprehensive paradigm shift around six

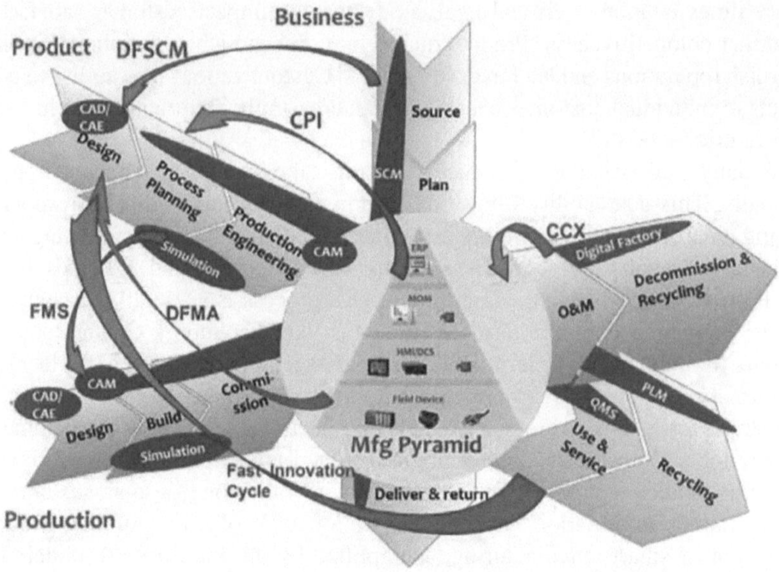

FIGURE 14.2 NIST's Model of the Smart Manufacturing Ecosystem [8].

TABLE 14.2
Elements Associated with NIST's Intelligent Manufacturing Framework

Components	Description
Comprehensive Product Lifecycle Oversight	Overseeing the product's entire lifecycle, from its initial conception, design, and manufacturing stages to its eventual disposal.
Logistics and Supply Chain Coordination	Coordinating the movement of materials, finished goods, and associated information among suppliers, distributors, and final consumers.
Supply Chain-Focused Design Strategy	Creating products with a design that optimizes and enhances the efficiency of the supply chain.
Persistent Process Enhancement	An ongoing cycle of reengineering and improving engineering and management systems is in progress.
Ongoing System Optimization	A continual process of diagnosing, forecasting, and enhancing the performance of production systems.
Design Optimization for Manufacturing and Assembly	Designing for simplified manufacturing of components and straightforward assembly of the product.
Adaptive/Modular Manufacturing System	Adaptable machining that allows for various production types to be reconfigured without overhauling the entire process.
Production System Hierarchy	The tiered organizational framework of a current manufacturing system.
Rapid Product Development Cycle	Enhancing the cycle of introducing new products by predicting trends, collecting data from product use, and incorporating this feedback into the initial stages of product development.

fundamental pillars: the digital economy and society, sustainability, innovative workplaces, well-being, intelligent mobility, and civil security [15]. At its core, Industry 4.0 aspires to intertwine traditional manufacturing with cutting-edge information technology and services seamlessly. This integration hinges on robust internet connectivity and highly efficient algorithms [16]. Central to its operation is utilizing the Internet of Things (IoT)—a powerful tool that enhances supply chain management and emphasizes producing high-value-added products. Simultaneously, it facilitates the global distribution of manufacturing modules and machine tools, revolutionizing how industries operate and innovate [17].

Figure 14.3 visually represents Industry 4.0's architectural framework through the Reference Architectural Model Industry 4.0 (RAMI4.0) [18]. This model presents a multidimensional view of Industry 4.0's components, encompassing key layers such as assets, integration, communication, information, functionality, and business. Furthermore, it elucidates the hierarchy levels, from product and field devices to the enterprise and the interconnected world. This intricate visualization elucidates the seamless flow of information and work processes that underpin the Industry 4.0 ecosystem.

Meanwhile, China's ambitious initiative, "Made in China 2025," launched in 2015, mirrors the spirit of Germany's Industry 4.0. This endeavor is dedicated to establishing China as a global manufacturing powerhouse, leveraging digital manufacturing technologies to enhance international competitiveness [19]. China's manufacturing journey can be divided into three phases: an Incubation period, until 1991, marked by structural reforms and the establishment of special economic zones. A Navigation period, until 2001, was characterized by infrastructural development, WTO accession, and finally, the Dynamic years. During the Dynamic years, China focused on developing a global manufacturing supply chain, positioning itself as a prominent manufacturing nation and the world's second-largest economy. "Made in China 2025" sets audacious goals, envisioning China as a significant producer of mobile

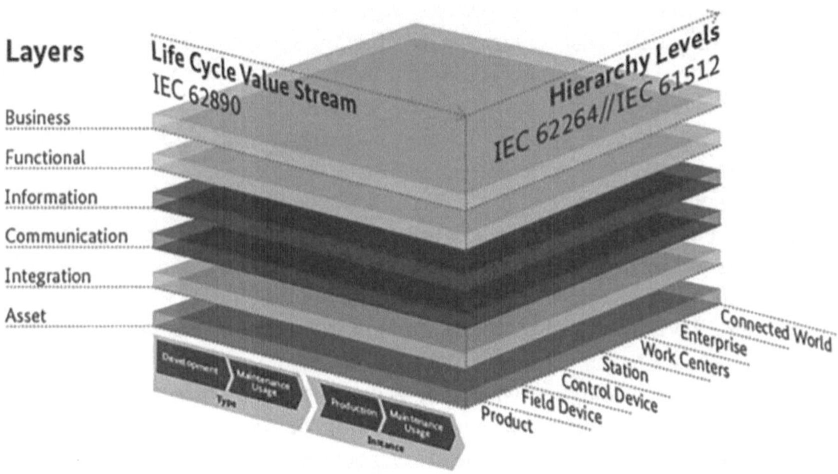

FIGURE 14.3 RAMI4.0 — Reference Architecture Model for Industry 4.0 [11].

phone chips, industrial robots, and renewable energy equipment for the global market by 2025. The initiative seeks to transition from labor-intensive manufacturing to more automated and technologically advanced production systems [20].

In parallel, Japan unveiled "Society 5.0" in January 2016 [21], representing a forward-thinking vision that marks the fifth stage of societal evolution, following hunting, agrarian, industrial, and information societies [22]. Society 5.0 is predicated on seamlessly integrating Cyber-Physical Systems (CPS) with Information and Computer Technology (ICT). It envisions AI-based robots as pivotal contributors to value creation [17]. This transformative strategy encompasses various service platforms, including smart manufacturing, intelligent transportation, energy value chains, regional inclusive care systems, and infrastructure maintenance. The overarching objective is to construct a resilient and innovative society that benefits Japan and contributes significantly to global progress. Society 5.0 epitomizes the fusion of advanced technology, human ingenuity, and societal well-being in an interconnected world.

General Electric's pioneering introduction of the Industrial Internet Platform (IIP) in the United States heralds a profound shift in the industrial landscape, characterized by the seamless integration of the Internet into the fabric of industrial production and control mechanisms [17]. IIP, delivered as a Software as a Service (SaaS) model, represents the vanguard of this transformation, serving as a linchpin that orchestrates the intricate dance between the physical and cyber components of the industrial sphere. Its current mission is a testament to its visionary nature, primarily focusing on managing intelligent products and detecting and pinpointing faults within the industrial ecosystem [16].

In a parallel narrative of innovation, Wang et al. [23] put forth a revolutionary concept—an Industrial Operating System (Ind-OS). This visionary system, envisioned as the steward of the entire manufacturing apparatus, would operate in tandem with IIP, ushering in a new era of industrial orchestration. The grand vision for Ind-OS extends beyond mere oversight; it aims to serve as the central nervous system that seamlessly integrates various key systems, from the hallowed halls of Enterprise Resource Planning (ERP) and Enterprise Information Systems (EIS) to the corridors of Human Resource Management (HRM), Customer Relation Management (CRM), and the Manufacturing Executive System (MES). Ind-OS aspires to be the maestro conducting a symphony of systems, harmonizing their efforts to achieve unprecedented efficiency and cohesion within the industrial realm.

The framework of the Industrial Internet is articulated through a comprehensive architecture, known as the 5C model. At its core lies the foundational Connection layer, dedicated to meticulously monitoring conditions within the industrial ecosystem. Above it, the Conversion layer deftly translates the raw data harvested from the Connection layer into actionable information, poised to unleash its potential in the subsequent layers. The Cyber layer, the third stratum in this architectural edifice, serves as the fulcrum upon which data is harnessed to facilitate informed decision-making. The Cognition layer, populated by diligent technicians, stands poised to decipher and interpret the data, bridging the gap between raw information and meaningful insights. At the pinnacle of this structure lies the Configure layer, where strategic decision-makers wield the power to shape the course of industrial operations. In

TABLE 14.3
Traits of Recently Developed Intelligent Technologies

Country of Origin	Industry 4.0 Germany	Made in China 2025 China	Society 5.0 Japan	Industrial Internet USA
Key Area of Concentration	Global Small and Medium Enterprises	Key Sectors in the Chinese Economy	Development of a Smart Society in Japan and Globally	Global Industrial Sectors
Significant Contributions to the Field	Global Digital Transformation in Manufacturing Industries	Establishing China as the Foremost Global Manufacturing Hub	Digital Transformation Across 12 Key Sectors of Society	Industry Digitalization via Internet Integration
Presented	2011	2015	2016	2012
Fundamental Technology	Web and Connected Devices	Web and Connected Devices	Web and Connected Devices	Web and Connected Devices

this layer, the insights garnered from the Cognition layer converge to inform strategic decisions, defining the path forward for the industrial enterprise [24].

Table 14.3 emerges as an illuminating tableau, offering a comparative overview that unveils these emerging technologies' intricate relationships and distinctive characteristics within the expansive canvas of the Industrial Internet. It serves as a compass, guiding stakeholders and enthusiasts through the dynamic landscape of technological innovation, presenting a holistic view of the transformative forces at play in the industry world.

14.2.2 Application of IoT in Process Industries

The proliferation of the Internet of Things (IoT) has marked a transformative epoch in industrial automation, particularly gaining traction in developed nations where its adoption has become increasingly pervasive. One compelling facet of this technological wave is the advent of Automatic Guided Vehicles (AGVs), representing a remarkable fusion of robotics and logistics. These AGVs serve as a testament to the boundless potential of IoT, revolutionizing the landscape of uncrewed transportation and logistics with their remarkable capabilities, including self-loading and unloading of goods.

At the heart of AGV functionality lies their ability to navigate autonomously—a feat made possible through sophisticated technologies. These include floor markings, vision cameras, radio waves, or lasers, each serving as a navigational beacon guiding AGVs through their intricate choreography within industrial facilities [25]. However, their repertoire extends beyond the realm of goods transport. AGVs have gracefully expanded their roles to encompass diverse tasks, from the noble pursuit of cleanliness to the gracious role of assisting with the transportation of people. In a testament to the power of centralized management, these multifaceted AGVs often find themselves under the watchful eye of cloud-based servers, orchestrating their every move with precision and efficiency [26].

The grand tapestry of industrial automation has witnessed a crescendo with the emergence of Industry 4.0—a watershed moment signifying the dawn of a new era in robotic industrial automation. Characterized by the harmonious integration of cyber-physical systems and the transformative capabilities of IoT technologies, Industry 4.0 stands as a testament to human ingenuity and technological progress [27]. Within this intricate ecosystem, the IoT architecture reigns supreme, its foundation resting on four foundational layers, each playing a pivotal role in the orchestration of data and information.

At the bedrock of this architectural marvel lies the first layer, where sensors and actuators are integrated seamlessly into hardware, serving as the vanguard of data collection. Here, the physical world interfaces with the digital realm, capturing the essence of the industrial landscape with precision and accuracy. Ascending the technological hierarchy, we encounter the networking layer—a digital thoroughfare through which data flows seamlessly from sensors to control units and vice versa. This neural network layer ensures the uninterrupted transmission of critical information, orchestrating the symphony of industrial processes.

The ascent continues as we ascend to the service and interface layers, serving as the portals through which human interaction with control units is facilitated. Here, users interface with the digital infrastructure, accessing services tailored to their needs and requirements. It is the bridge that connects the human intellect with the machinery of industry, fostering a dynamic and responsive ecosystem [27].

Table 14.4 serves as a guiding compass, delineating the intricate design considerations that underpin the successful implementation of IoT in industrial contexts. It offers valuable insights into the multifaceted decision-making processes that drive the integration of IoT technologies into the industrial fabric. Meanwhile, Figure 14.4 stands as a visual testament to the intricate working mechanisms that underlie the IoT architecture within industries, showcasing the journey of information from the physical world through sensors and machine learning tools, onward to smart-perception devices before being entrusted to the network layer for transmission to the application layer. Within this application layer, the magic happens, where data transforms into actionable insights, and the IoT-integrated system responds with precision to the needs and demands of the industry [27].

Figure 14.4 serves as a visual gateway into the intricate architecture of the Internet of Things (IoT), unveiling the complex web of relationships that bind physical

TABLE 14.4
Design Considerations for IoT Industries

Goals	Explanation
Power	Functional constraints due to restricted power availability.
Response Time Delay	Duration needed for data transfer.
Data Transmission Capacity	The maximum data capacity the system can carry.
Expansion Capability	The maximum number of devices the system can support.
Network Structure	The sequence of communication where one node communicates with another node.
Security and Safety Measures	The degree of security within the system.

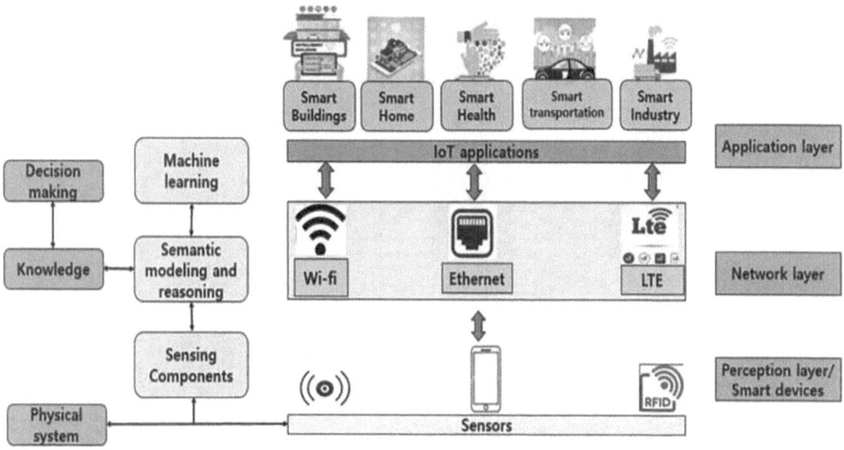

FIGURE 14.4 Schematic Representation of IoT Architecture [28].

systems, often referred to as "things," with the tangible entities they interact with within the physical world. These interactions are a symphony orchestrated by an ensemble of sensors, each attuned to a specific facet of the environment, be it temperature, humidity, speed, or pressure. These sensors serve as the sensory organs of the IoT ecosystem, capturing the nuances of the physical realm with precision and conveying this data to the digital realm for analysis and action. In response to the insights from this data, the IoT deploys its arsenal of actuators and controllers, carrying out commands and orchestrating actions by the system's design and deployment imperatives. It is a seamless fusion of the physical and digital realms—a dance of data and action underpinned by the elegant simplicity of sensors and the dynamic agency of actuators and controllers [29].

IoT is not confined to the industrial sphere alone; its tendrils extend far and wide, permeating diverse domains and redefining the boundaries of human interaction with the world. From the cozy confines of smart homes and buildings, where temperature and lighting adjust in harmony with human presence, to the domain of personal health, where wearable devices monitor vital signs and transmit data for analysis and intervention, the reach of IoT knows no bounds. In transportation, IoT fuels the intelligence behind traffic management systems, optimizing routes and mitigating congestion, while in industries, it orchestrates a symphony of machinery, ensuring seamless operations and predictive maintenance [29].

The genesis of IoT can be traced back to a pivotal moment in the annals of technological history when the world witnessed a shift in the balance of connected entities. As Figure 14.5 illustrates, this transformation occurred between 2008 and 2009, when the number of connected devices surpassed the world's human population. However, the concept of the "Internet of Things" was bestowed with its name nearly a decade earlier, in 1999, by the visionary thinker Kevin Ashton [28]. Cisco Technology further illuminates the evolution of IoT through an insightful infographic [30]. This visual masterpiece encapsulates the essence of IoT in action—a

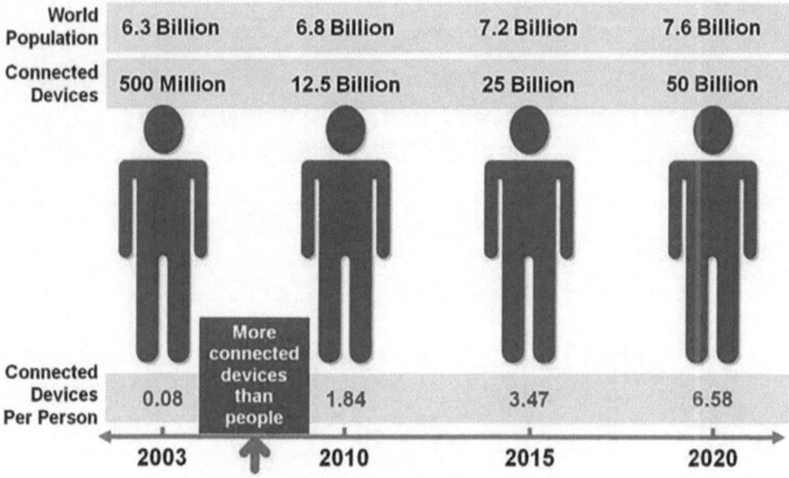

FIGURE 14.5 Number of Connected Devices per Person [30].

seamlessly interconnected system where an alarm clock, train schedule, traffic management system, coffee maker, and car engage in a harmonious exchange of real-time information. In this orchestrated symphony, the alarm clock awakens its owner based on the train schedule, which, in turn, adjusts its timings based on traffic conditions. At the same time, the coffee maker commences brewing, ensuring that the commuter starts the day seamlessly.

This interconnectedness, manifesting through the intricate choreography of devices, appliances, and systems, stands as a testament to the interoperability that defines IoT. These entities operate independently, yet their actions are choreographed in concert, driven by real-time information and a shared mission to enhance efficiency and convenience. The remarkable increase in connected devices per person, from a mere 0.08 in 2003 to 3.47 in 2015, foreshadows the exponential growth of IoT. Predictions point to a staggering 6.58 connected devices per person by 2020, underscoring the relentless march of IoT into our daily lives [31].

At its core, IoT aspires to extend the benevolent embrace of the Internet to the physical world, offering a pantheon of benefits that include consistent connectivity, seamless data exchange, and remote controllability. It is the promise of an interconnected world where the digital and physical realms harmonize, creating a symphony of efficiency, convenience, and innovation [29].

14.3 NAVIGATING THE LANDSCAPE OF SMART MANUFACTURING STANDARDS

Smart manufacturing technology represents the nexus of information technology, industrial manufacturing techniques, and human ingenuity, heralding a transformative era in the landscape of manufacturing systems. In this epoch of innovation, several seminal publications have served as beacons, illuminating the path and shaping the contours of this dynamic domain.

One such influential publication is the "Current Standards Landscape for Smart Manufacturing Systems," unveiled by the esteemed National Institute of Standards and Technology (NIST) in February 2016 [8]. This comprehensive document is a testament to the meticulous groundwork laid by NIST in elucidating the evolving standards landscape within smart manufacturing. It serves as a compass, navigating stakeholders through the intricate web of standards, offering insights into the prevailing standards landscape, and providing a foundation for smart manufacturing systems to flourish. With meticulous research and expert guidance, this publication has become an indispensable resource for industry professionals and policymakers, charting the course for the global adoption of smart manufacturing systems.

Germany, a vanguard in the smart manufacturing revolution, has played a pivotal role in shaping the trajectory of this technological marvel. In 2015, the German Commission for Electrical, Electronic & Information Technologies of DIN and VDE unveiled the "German Standardization Roadmap Industry 4.0" [32]. This roadmap represents a seminal contribution to the field, encapsulating the wisdom and vision of German experts in standards development. It outlines a strategic path toward harmonizing standards, fostering an environment where smart manufacturing can thrive. Germany's leadership in this endeavor has reverberated globally, inspiring nations and industries to embark on their journeys towards smart manufacturing excellence.

China, another formidable force in the manufacturing world, has also left an indelible mark on the smart manufacturing landscape. In 2015, the Ministry of Industry and Information Technology of China (MIIT) and the Standardization Administration of China (SAC) jointly published the "National Smart Manufacturing Standards Architecture Construction Guidance" [33]. This authoritative document provides invaluable guidance for constructing a robust standards architecture, laying the groundwork for the seamless integration and deployment of nationwide smart manufacturing systems. China's commitment to standardization in smart manufacturing reflects its vision of becoming a global leader in this transformative field.

These seminal publications, born from the collective wisdom of institutions and experts, hold the power to shape the destiny of manufacturing systems worldwide. They provide the knowledge, framework, and roadmap necessary to unlock the full potential of smart manufacturing, fostering a future where innovation knows no bounds and productivity soars to new heights. In the tapestry of smart manufacturing's evolution, these publications stand as foundational threads, weaving together the past, present, and future of a revolution that promises to redefine the very essence of manufacturing.

Several eminent organizations have stepped forward as custodians of order and uniformity in the grand tapestry of standardization that envelops smart manufacturing systems. With their unwavering commitment to establishing and upholding standards, these organizations have played a pivotal role in the trajectory of global smart manufacturing systems.

The International Electrotechnical Commission (IEC) is among the foremost guardians of standardization. This august institution has diligently cultivated a rich repository of standards that span the diverse spectrum of smart manufacturing. Its contributions extend to industrial automation, electrical engineering, and information technology. Through a meticulous process of consensus-building and technical expertise, the IEC has bestowed a framework that ensures interoperability and harmonious coexistence within the intricate fabric of smart manufacturing systems.

The International Society of Automation (ISA) is another stalwart in the standardization arena. With a steadfast focus on industrial automation and control systems, ISA has been instrumental in devising standards that underpin the seamless integration of technology and manufacturing prowess. These standards enhance the efficiency of manufacturing processes and bolster the reliability and safety of smart manufacturing systems, ushering in an era of unparalleled performance and security.

The International Standards Organization (ISO), renowned for its global reach and impact, has left an indelible mark on the standardization landscape of smart manufacturing systems. ISO's imprimatur extends to quality management, data exchange, and information security. Its standards serve as a lodestar, guiding organizations and nations in their quest for excellence in smart manufacturing. ISO's imprimatur signifies a commitment to best practices, fostering a world where smart manufacturing systems transcend borders and boundaries.

Within the United States, the American National Standards Institute (ANSI) has assumed a pivotal role in shepherding standards development. ANSI's influence radiates across diverse sectors, including manufacturing, technology, and engineering. It champions the cause of interoperability and safety, ensuring smart manufacturing systems adhere to rigorous standards that safeguard industry and society.

14.4 EXPLORING THE CHARACTERISTICS AND CHALLENGES

Smart manufacturing systems are designed to address the numerous challenges and complexities of the current industries. However, the implementation of these systems is not without its own set of challenges. Key issues include security vulnerabilities, difficulties in system integration, uncertain return on investment for new technologies, and financial constraints associated with establishing new smart manufacturing systems or upgrading existing ones with such technology. In the following sections, we delve into these challenges smart manufacturing systems face and potential solutions.

14.4.1 Addressing Security Challenges in the Era of Smart Manufacturing

Integrating networked systems within a smart manufacturing environment is fundamental to the seamless flow of information, real-time monitoring, and efficient coordination. However, this increased connectivity also creates a heightened need for robust security measures to safeguard the system, data, and manufacturing processes. Let's delve deeper into the challenges and solutions related to security in smart manufacturing systems:

1. **Cybersecurity Threats:**
    - ***Challenge:*** The interconnected nature of smart manufacturing systems exposes them to various cybersecurity threats, including malware, ransomware, and data breaches.

- ***Solution:*** Employ comprehensive cybersecurity strategies, encompassing intrusion detection systems, firewalls, regular security audits, and continuous monitoring. Implement threat intelligence and stay updated on emerging threats [34].

2. **Data Privacy and Compliance:**
   - ***Challenge***: Smart manufacturing systems handle sensitive data, including proprietary designs, production schedules, and customer information. Ensuring compliance with data privacy regulations like GDPR and CCPA is critical.
   - ***Solution***: Implement data encryption, access controls, and anonymization techniques. Develop and enforce strict data privacy policies and procedures and conduct regular compliance audits.
3. **Insider Threats:**
   - ***Challenge***: Threats can emanate within the organization, including disgruntled employees or unintentional data breaches.
   - ***Solution***: Implement user access controls and privileged access management. Conduct employee training on cybersecurity best practices and create a culture of security awareness.
4. **Supply Chain Vulnerabilities:**
   - ***Challenge***: Smart manufacturing systems often rely on a global supply chain, which can introduce vulnerabilities through third-party components and suppliers.
   - ***Solution***: Assess and audit the security practices of suppliers and third-party vendors. Establish stringent supplier security requirements and conduct periodic security assessments.
5. **Legacy Systems and Updates:**
   - ***Challenge***: Integrating smart technology into existing manufacturing systems may be complex, especially if legacy systems are involved. Legacy systems may have vulnerabilities and lack support.
   - ***Solution***: Gradually phase out legacy systems where possible and ensure that remaining legacy components are regularly patched and updated. Implement secure gateways and isolation for legacy systems.

In the evolving landscape of smart manufacturing, security is not a one-time consideration but an ongoing commitment. By proactively addressing security challenges and implementing robust solutions, organizations can harness the benefits of smart manufacturing while safeguarding their operations from potential threats and vulnerabilities [34].

14.4.2 System Integration in Smart Manufacturing

Integrating new smart manufacturing technologies with existing machinery is a common challenge organizations face when transitioning to modern manufacturing systems. This challenge can be particularly daunting due to compatibility issues between older and newer equipment. Here, we explore the challenges associated with integration and propose solutions to ensure a smooth transition to smart manufacturing:

1. **Legacy Equipment:**
    - ***Challenge***: Many manufacturing facilities have legacy machinery that predates the era of smart manufacturing. These machines may lack digital interfaces and operate on outdated communication protocols.
    - ***Solution***: Implement retrofitting solutions that bridge the gap between legacy machinery and modern systems. This may involve adding sensor arrays and communication modules to older equipment to make them compatible with digital networks.
2. **Communication Protocols:**
    - ***Challenge***: Older machines often use proprietary or outdated communication protocols that do not align with the interoperability requirements of smart manufacturing.
    - ***Solution***: Invest in intermediate legacy equipment and modern systems gateway devices. These gateways can translate data between different communication protocols, enabling seamless integration.
3. **Data Standardization:**
    - ***Challenge***: Data generated by legacy machines may not adhere to standardized formats, making it difficult to aggregate and analyze information from heterogeneous sources.
    - ***Solution***: Implement data standardization processes that transform diverse data formats into a common structure. This enables unified data analytics and decision-making across the manufacturing environment.
4. **Scalability:**
    - ***Challenge***: The scalability of existing equipment may be limited, making it challenging to expand manufacturing operations to accommodate new technologies and increased production demands.
    - ***Solution***: Consider phased upgrades or modular expansions that allow for integrating new equipment while gradually replacing or augmenting older machines. This approach minimizes disruptions to ongoing operations.
5. **Interoperability:**
    - ***Challenge***: Ensuring that new and old equipment can effectively communicate and collaborate is crucial for the success of smart manufacturing.
    - ***Solution***: Prioritize interoperability by selecting equipment and systems that adhere to industry standards and communication protocols. Implement middleware solutions that facilitate seamless data exchange between diverse components.

Organizations can effectively integrate new smart manufacturing technologies with existing machinery by addressing these challenges and implementing the suggested solutions. This streamlines the transition to smart manufacturing and maximizes the utilization of valuable assets while improving overall manufacturing efficiency [35].

14.4.3 Interoperability Challenges in Smart Manufacturing

Interoperability is a critical aspect of smart manufacturing systems, as it ensures the seamless exchange of data and information between various components, systems, and devices. Let's delve deeper into the concept of interoperability and its four recognized levels within the context of Industry 4.0 (I4.0) while also exploring the challenges and solutions associated with achieving effective interoperability:

1. **Communication Protocol Alignment:**
   - ***Challenge***: Discrepancies in communication protocols can hinder interoperability.
   - ***Solution***: Standardize communication protocols and ensure all systems adhere to these standards to facilitate smooth data exchange.
2. **Bandwidth and Frequency Differences:**
   - ***Challenge***: Variations in communication bandwidth and operational frequency can limit interoperability.
   - ***Solution***: Invest in communication infrastructure that supports the required bandwidth and frequency for data exchange.
3. **Communication Modes:**
   - ***Challenge***: Differences in communication modes (e.g., wired vs. wireless) can pose challenges.
   - ***Solution***: Select communication modes that align with the requirements of the manufacturing environment and ensure compatibility.
4. **Hardware Capabilities:**
   - ***Challenge***: Varied hardware capabilities across systems can affect interoperability.
   - ***Solution***: Standardize hardware specifications or implement middleware solutions that bridge hardware disparities.
5. **Standards Adherence:**
   - ***Challenge***: Lack of adherence to industry standards can hinder interoperability.
   - ***Solution***: Promote and enforce industry standards to ensure consistent practices and protocols.

Achieving effective interoperability in smart manufacturing requires a concerted effort to address these challenges. By focusing on aligning communication protocols, standardization, and technical compatibility, organizations can unlock the full potential of smart manufacturing systems [11, 36] and realize the benefits of enhanced data exchange, collaboration, and decision-making across the manufacturing ecosystem.

14.4.4 Ensuring Safety in Human-Robot Collaboration

The advent of collaborative robots, often called cobots, has ushered in a new era in manufacturing—one characterized by safe and productive human-machine collaboration within shared workspaces. Creating a safe and efficient collaborative workspace where humans and cobots can coexist harmoniously requires attention to the following aspects:

1. **Safety Standards Compliance:**
   - ***Importance***: Adherence to industry-specific safety standards and guidelines is vital.

- ***Measures***: Ensure that cobots meet safety standards, including safety-rated sensors, speed and force limitations, and compliance with ISO and ANSI safety standards.

2. **Sensory Systems:**
   - ***Importance***: Cobots should be equipped with advanced sensory systems for real-time environment monitoring and human presence detection.
   - ***Measures***: Employ sensors, cameras, and other technologies to enable cobots to detect and respond to human proximity and movements.
3. **Training and Education:**
   - ***Importance***: Proper training and education programs for human workers and robot operators are essential.
   - ***Measures***: Develop training modules that cover safe operating procedures, emergency protocols, and the fundamentals of cobot-human collaboration.
4. **Risk Mitigation Strategies:**
   - ***Importance***: Implement risk mitigation strategies such as reduced operating speeds, protective barriers, or automated shutdown mechanisms.
   - ***Measures***: Configure cobots to slow down or stop when humans approach, ensuring safe coexistence.

In conclusion, collaborative robots represent a transformative development in the manufacturing industry, where humans and machines can work together synergistically. Prioritizing occupational health and safety, conducting thorough risk assessments, complying with safety standards, and implementing robust safety measures are crucial steps in ensuring the successful integration of cobots into the modern industrial workspace [37].

14.4.5 Multilingual Capabilities in Smart Manufacturing Systems

In today's interconnected and globalized manufacturing landscape, the need for smart manufacturing systems to transcend language barriers is paramount. To embody the concept of "smart," these systems must possess multilingual capabilities that enable them to understand and interpret instructions provided in human languages and translate them into machine language. This fusion of linguistic understanding and technological prowess represents a significant leap in the evolution of manufacturing processes [38]. Here, we explore the significance of multilingual capabilities in smart manufacturing systems, the role of AI and advanced technologies, and the benefits of human-machine interaction through natural language inputs.

14.4.5.1 The Significance of Multilingual Capabilities

Multilingual capabilities empower smart manufacturing systems to communicate and collaborate seamlessly in a linguistically diverse environment. This functionality is particularly crucial in industries with a global footprint, where operators and stakeholders may speak different languages. The ability to interpret instructions in human language opens the door to more intuitive and efficient human-machine interaction.

14.4.5.2 Integrating AI and Advanced Technologies

Integrating Artificial Intelligence (AI) and advanced technologies is the linchpin of enabling multilingual capabilities in smart manufacturing systems. AI-driven natural language processing (NLP) algorithms and machine learning empower these systems to effectively comprehend and respond to human language inputs. Here's how it works:

1. ***Voice Commands***: Smart manufacturing systems equipped with voice recognition technology can receive and process spoken instructions from operators. NLP algorithms analyze the voice command, extract meaning, and convert it into actionable tasks for machines.
2. ***Text Inputs***: Operators can provide instructions or input data in written form, either through touchscreens, keyboards, or mobile devices. NLP algorithms analyze the text, extract relevant information, and initiate corresponding manufacturing operations.
3. ***Translation Capabilities***: In multilingual environments, smart manufacturing systems can translate instructions between languages, ensuring clear communication between operators of different linguistic backgrounds.

14.4.5.3 Meeting the Evolving Needs of Industry 4.0

Adopting multilingual capabilities aligns with the evolving needs of Industry 4.0, where smart technologies are reshaping manufacturing paradigms. These capabilities bridge linguistic divides, foster greater human-machine collaboration, and enhance the adaptability of manufacturing systems to a diverse and technologically advanced industrial environment.

In conclusion, equipping smart manufacturing systems with multilingual capabilities is a strategic imperative for manufacturers aiming to thrive in a globalized world. By integrating AI, NLP, and advanced technologies, these systems can transcend language barriers, facilitate efficient communication, and empower operators to interact with machines using their preferred language inputs. This evolution represents a pivotal step toward achieving the true potential of smart manufacturing in the modern era.

14.4.6 Return on Investment Analysis for New Technology Adoption

Integrating advanced technology into an existing manufacturing system represents a significant decision for any organization. This transition requires a thorough technological assessment and a comprehensive financial analysis to evaluate the feasibility and potential benefits. In this context, assessing the return on investment (ROI) becomes a critical factor that guides decision-makers. Explore the essential financial analysis and ROI assessment components when transitioning to advanced manufacturing technology [35].

14.4.7 Cost-Benefit Analysis

One of the primary components of financial analysis is a detailed cost-benefit analysis. This involves identifying and quantifying all costs associated with adopting new technology. These costs can include the following:

1. ***Acquisition Costs***: The initial cost of purchasing and installing the new technology, including hardware, software, and infrastructure upgrades.
2. ***Training and Skill Development***: Expenses related to training the workforce to operate and maintain the advanced systems.
3. ***Downtime Costs***: Production losses or disruptions during the transition period when the existing system is offline or undergoing modifications.
4. ***Maintenance and Support***: Ongoing costs for servicing, maintaining, and updating the technology.
5. ***Energy and Resource Costs***: Changes in energy consumption and resource utilization resulting from the new technology.

Quantify The Potential Gains:

1. ***Increased Productivity***: Assess how the new technology enhances production efficiency, reduces lead times, and increases output.
2. ***Improved Quality***: Determine the impact on product quality, including reduced defects and waste.
3. ***Enhanced Customization***: Consider the ability to tailor products to customer demands, potentially opening new markets.
4. ***Cost Savings***: Identify areas where the new technology reduces operational costs, such as labor, materials, or energy.

14.4.8 ROI Calculation

Calculating the ROI becomes essential once the costs and benefits are identified and quantified. The ROI formula is straightforward:

$$ROI = \left({Net\ Profit\ from\ Investment} \middle/ {Total\ Investment\ Cost} \right) \times 100$$

In this formula:

- Net Profit from Investment includes all the benefits gained from the technology minus the total costs incurred.
- Total Investment Cost encompasses the initial acquisition cost, training, downtime, maintenance, and other expenses related to the technology transition.

A positive ROI indicates that the investment will likely generate profits over time. Decision-makers typically compare this ROI to predetermined thresholds or expected rates of return to determine whether the investment aligns with the organization's financial goals.

14.4.9 Time to Recoup Investment

In addition to ROI, organizations consider the time required to recoup the initial investment through increased revenues or efficiency gains. This metric is crucial because it indicates how long the investment will take to deliver a positive return.

14.4.10 Risk Assessment

Financial analysis should also encompass risk assessment, identifying potential risks and uncertainties associated with the technology transition. These risks could include market volatility, regulation changes, or unexpected technical issues.

In conclusion, a comprehensive financial analysis and ROI assessment are vital to the decision-making process when transitioning to advanced manufacturing technology. It ensures that the transition is both technologically feasible and financially sound. By quantifying costs, benefits, and ROI, organizations can make informed decisions that align with their strategic objectives and long-term sustainability.

14.5 PROSPECTS AND EMERGING TRENDS IN SMART MANUFACTURING

Smart manufacturing systems have garnered significant attention and are experiencing widespread adoption globally, driven by their potential to unlock market opportunities and enhance operational efficiency. Various studies and reports provide insights into the growing adoption of these intelligent manufacturing solutions across different industries, highlighting their significance [39].

14.5.1 SMEs Embrace Intelligent Manufacturing

Research conducted by Zhengxin et al. [39] underscores the increasing interest of small and medium-sized enterprises (SMEs) in embracing intelligent manufacturing. Historically, large corporations dominated the adoption of advanced manufacturing technologies, but SMEs now recognize the benefits and feasibility of intelligent manufacturing systems.

14.5.2 Automation Potential in Key Sectors

A report by McKinsey & Company, as presented by Manyika [7], sheds light on the vast automation potential in various sectors. The report suggests that industries such as accommodation and food services could witness automation levels of up to 73%. Similarly, sectors like manufacturing, transportation, and warehousing have the potential for automation rates of 60%. These findings emphasize the strategic focus on implementing cutting-edge technology to boost productivity and overall performance in these sectors.

14.5.3 Intent to Upgrade Manufacturing Systems

A 2017 WBR [35] survey reveals industry participants' intentions regarding upgrading their manufacturing systems. Figure 14.6 shows that the survey findings indicate

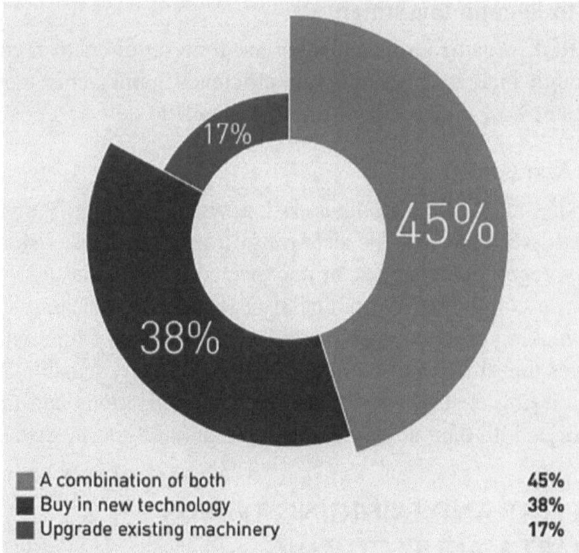

FIGURE 14.6 Survey Report by WBR Digital on Adoption of Smart Manufacturing [33].

that 38% of respondents planned to install new technology, 45% aimed to upgrade their existing machinery with smart manufacturing systems, and 17% considered combining both approaches. This indicates a substantial willingness to invest in and transition to smarter manufacturing technologies.

14.5.4 Disparity between Current and Future Systems

Qin et al. [40] analyzed the disparities between current manufacturing systems and the capabilities offered by smart manufacturing or Industry 4.0 technologies. Figure 14.7 illustrates that several critical components and functionalities of smart manufacturing technologies are absent even in the most advanced reconfigurable manufacturing systems. These include self-configuration, self-optimization, early awareness, decision-making, and predictive maintenance. The analysis underscores the significant room for growth and the compelling opportunities for adopting advanced manufacturing systems.

14.5.5 Enhancing Industrial Automation

The increasing complexity of smart manufacturing systems has led to intricate interactions and interrelations between machines and subsystems, shaping the architecture of industrial automation [42]. Automation technologies have expanded the scope for sharing information and data across interconnected systems. This advancement has notably improved process management, machine interoperability, and handling exceptions effectively. These advancements further motivate the adoption of smart manufacturing systems [41].

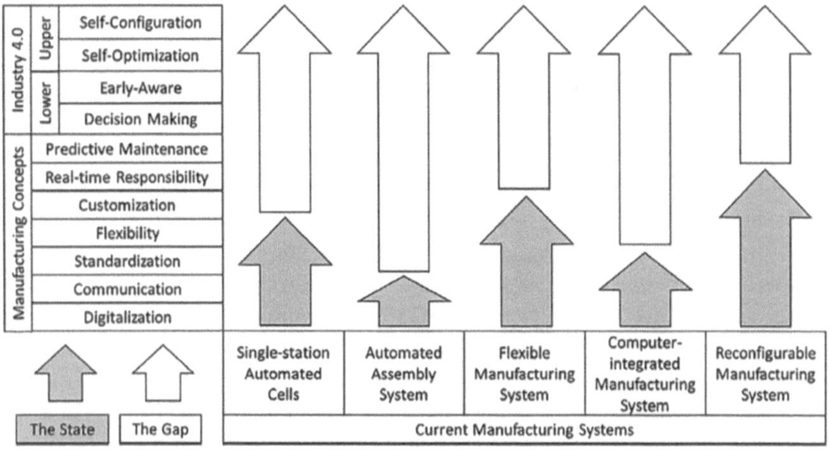

FIGURE 14.7 Gap in Technology: Current Manufacturing Systems vs. Industry 4.0 [41].

14.5.6 Transition to Intelligent Manufacturing

Adopting automated and intelligent systems signifies a pivotal transition to the next manufacturing level, often called smart or intelligent manufacturing. These systems leverage end-user data and information through cloud computing, enabling the rapid production of customized products [43]. By aligning with current market demands, optimizing production costs, offering design flexibility, and expediting product delivery, these technologies cater to the evolving needs of industries and consumers.

In conclusion, the increasing adoption of smart manufacturing systems across diverse industries reflects their growing importance in enhancing operational efficiency, meeting market demands, and positioning organizations for future success. As technology continues to evolve, smart manufacturing is expected to play a central role in driving innovation and competitiveness in the industrial landscape.

14.6 CONCLUSION

Smart manufacturing systems are increasingly pivotal in advancing technology in today's industrial landscape. These systems enhance operational efficiency and productivity and significantly impact the global economy. The rise of the Internet of Things (IoT) and the Industrial Internet of Things (IIoT) is crucial in elevating manufacturing systems with smart technologies. Research indicates that numerous industries are eager to upgrade to smart manufacturing systems. A primary challenge in this transition is ensuring compatibility between existing machinery and new technological advancements.

Smart devices, such as IoT and Cyber-Physical Systems (CPS), have emerged as universal paradigms capable of transforming industries with their sensing, identification, remote control, and automated capabilities. Initiatives like Industry 4.0, Society 5.0, Made in China 2025, and the Industrial Internet are all underpinned by Internet technology and interconnected devices. They focus on process control with minimal human intervention and intelligent decision-making, significantly impacting the global market. This chapter has discussed the concept of smart manufacturing, its associated technologies, new paradigms, and the challenges and opportunities they present.

The interlinking of physical objects through the Internet or IoT has played a vital role in enhancing and upgrading existing systems and processes. Internet technology has revolutionized SCADA systems in automation by introducing remote sensing and control capabilities. Integrating IoT in various sectors, such as public transportation, power grid operations, healthcare, traffic coordination, asset tracking, fleet management, and self-driving vehicles like AGVs, has improved lifestyles and boosted socio-economic status globally. IoT's role in minimizing human involvement in hazardous environments and increasing productivity is particularly noteworthy.

Developing associated technologies like artificial intelligence, cyber-physical systems, big data processing, augmented and virtual reality, IoT, and robotics is essential to successfully implementing smart manufacturing technology. There is a significant gap between the conceptual smart manufacturing system and existing manufacturing practices. Furthermore, to facilitate effective communication between machines in industries, implementing the latest IPv6 technology is necessary, as it supports a higher number of interconnected devices.

REFERENCES

[1] N. Tuptuk, S. Hailes, Security of smart manufacturing systems, Journal of Manufacturing Systems 47 (2018) 93–106.

[2] L. Wang, A.J. Shih, Challenges in smart manufacturing, Journal of Manufacturing Systems 40 (2016) 1.

[3] H.S. Kang, J.Y. Lee, S. Choi, H. Kim, J.H. Park, J.Y. Son, B.H. Kim, S.D. Noh, Smart manufacturing: Past research, present findings, and future directions. International Journal of Precision Engineering and Manufacturing-Green Technology 3 (2016) 111–128. https://doi.org/10.1007/s40684-016-0015-5.

[4] C. Arnold, D. Kiel, K.-I. Voigt, How the industrial internet of things changes business models in different manufacturing industries, International Journal of Innovation Management 20 (08) (2016) 1640015.

[5] A.W. Scheer, Industry 4.0: From vision to implementation, AWS Institute for Digital Products and Processes gGmbH, Whitepaper 9 (September 2015).

[6] B. Lydon, Industry 4.0: Intelligent and flexible production, ISA InTech Magazine [online] (2016) [available online https://www.isa.org/intech-home/2016/may-june/features/industry-4-0-intelligent-and-flexible-production].

[7] J. Manyika, A future that works: Automation employment and productivity (2017) [available online https://www.mckinsey.com/featured-insights/digital-disruption/harnessing-automation-for-a-future-that-works/de-DE].

[8] Y. Lu, K.C. Morris, S. Frechette, Current standards landscape for smart manufacturing systems, National Institute of Standards and Technology, NISTIR 8107 (2016) 39.

[9] S. Phuyal, D. Bista, J. Izykowski, R. Bista, Design and implementation of cost efficient SCADA system for industrial automation, International Journal of Engineering and Manufacturing (IJEM) 10 (2) (2020) 15–28, doi:10.5815/ijem.2020.02.02.

[10] M.P. Groover, Fundamentals of Modern Manufacturing, John Wiley & Sons, Inc, 2010.

[11] Y. Lu, Industry 4.0: A survey on technologies, applications, and open research issues, Journal of Industrial Information Integration 6 (2017) 1–10.
[12] P. Alavian, Y. Eun, S.M. Meerkov, L. Zhang, Smart production systems: Automating decision-making in a manufacturing environment, International Journal of Production Research 58 (3) (2020) 828–845.
[13] K. Jung, B. Kulvatunyou, S. Choi, M.P. Brundage, An overview of a smart manufacturing system readiness assessment, In IFIP International Conference on Advances in Production Management Systems, Springer, 2016, pp. 705–712.
[14] L. Li, China's manufacturing locus in 2025: With a comparison of "made-in-China 2025" and "Industry 4.0", Technological Forecasting and Social Change 135 (2018) 66–74.
[15] M.A.K. Bahrain, M.F. Othman, N.N. Azli, M.F. Talib, Industry 4.0: A review on industrial automation and robotics, Jurnal Teknologi 78 (6–13) (2016) 137–143.
[16] M. Rüßmann, M. Lorenz, P. Gerbert, M. Waldner, P. Engel, M. Harnisch, J. Justus, Industry 4.0: The future of productivity and growth in manufacturing industries, Boston Consulting Group 9 (1) (2015) 54–89.
[17] "Toward realization of the new economy and society," Keidanren (Japan Business Federation) [online], (Available: www.keidanren.or.jp/en/policy/2016/029_outline.pdf), April 19, 2016.
[18] P. Adolphs, U. Epple, Reference architecture model industrie 4.0 (rami4. 0), ZVEI and VDI, Status Report (2015).
[19] L. Li, The path to Made-in-China: How this was done and prospects, International Journal of Production Economics 146 (1) (2013) 4–13.
[20] J. Wubbeke, M. Meissner, M.J. Zenglein, J. Ives, B. Conrad, Made in China 2025: The making of a high-tech superpower and consequences for industrial countries, Mercator Institute for China Studies 17 (2016) 2009–2017.
[21] M. Fukuyama, Society 5.0: Aiming for a new human-centered society, Japan Spotlight 1 (2018) 47–50.
[22] K. Fukuda, Science, technology and innovation ecosystem transformation toward Society 5.0, International Journal of Production Economics (2019) 107460.
[23] J. Wang, C. Xu, J. Zhang, J. Bao, R. Zhong, A collaborative architecture of the industrial internet platform for manufacturing systems, Robotics and Computer-Integrated Manufacturing 61 (2020) 101854.
[24] J.-Q. Li, F.R. Yu, G. Deng, C. Luo, Z. Ming, Q. Yan, Industrial internet: A survey on the enabling technologies, applications, and challenges, IEEE Communications Surveys & Tutorials 19 (3) (2017) 1504–1526.
[25] N. Gupta, A. Singhal, J.K. Rai, R. Kumar, Path Tracking of Automated Guided Vehicle, in Seventh International Conference on Contemporary Computing (IC3), Noida, 2014.
[26] L. Schulze, A. Wtillner, The Approach of Automated Guided Vehicle Systems, in IEEE International Conference on Service Operations and Logistics, and Informatics, Shanghai, 2006.
[27] X.T.R. Kong, X. Yang, G.Q. Huang, H. Luo, The Impact of Industrial Wearable System on Industry 4.0, in IEEE 15th International Conference on Networking, Sensing and Control (ICNSC), Zhuhai, 2018.
[28] K. Ashton, That 'Internet of Things' thing, RFID Journal (2009) [available online www.rfidjournal.com/articles/view?4986]. Accessed on 25 June 2019.
[29] L. Gao, X. Bai, A unified perspective on influencing consumer acceptance of Internet of things technology, Asia Pacific Journal of Marketing and Logistics 26 (2) (2014) 211–231.
[30] D. Evans, The Internet of Things [Infographic], Cisco Internet Business Solutions Group [available online: https://blogs.cisco.com/diversity/the-internet-of-things-infographic] 2011.

[31] D. Evans, The Internet of Things how the next evolution of the internet is changing everything, Cisco IBSG (2011) [available online http://www.cisco.com/web/about/ac79/docs/innov/IoT_IBSG_0411FINAL.pdf]

[32] L. Adolph, T. Anlahr, H. Bedenbender, German standardization roadmap: Industry 4.0, in Version 2, DIN eV, Berlin, 2016.

[33] Q. Li, H. Jiang, Q. Tang, Y. Chen, J. Li, J. Zhou, Smart manufacturing standardization: Reference model and standards framework, in OTM Confederated International Conferences On the Move to Meaningful Internet Systems, Springer, 2016, pp. 16–25.

[34] K.-D. Thoben, S. Wiesner, T. Wuest, "Industrie 4.0" and smart manufacturing review of research issues and application examples, International Journal of Automation Technology 11 (1) (2017) 4–16.

[35] W. Digital, Outsmarting the competition: Smart manufacturing report, Smart Manufacturing Leaders Summit, 2018 [available online www.manufacturing.wbresearch.com].

[36] D. Chen, G. Doumeingts, F. Vernadat, Architectures for enterprise integration and interoperability: Past, present and future, Computers in Industry 59 (7) (2008) 647–659.

[37] L. Gualtieri, I. Palomba, E. J. Wehrle, and R. Vidoni, The opportunities and challenges of SME manufacturing automation: Safety and ergonomics in human-robot collaboration, in Industry 4.0 for SMEs, Springer, 2020, pp. 105–144.

[38] J. Uhlemann, R. Costa, J.-C. Charpentier, Product design and engineering—past, present, future trends in teaching, research, and practices: Academic and industry points of view, Current Opinion in Chemical Engineering 27 (2020) 10–21.

[39] Z. Wang, M. Shou, S. Wang, R. Dai, K. Wang, An empirical study on the key factors of intelligent upgrade of small and medium-sized enterprises in China, Sustainability 11 (3) (2019) 619.

[40] J. Qin, Y. Liu, R. Grosvenor, A categorical manufacturing framework for Industry 4.0 and beyond, Procedia CIRP 52 (2016) 173–178.

[41] T. Tommila, O. Venta, K. Koskinen, Next generation industrial automation-needs and opportunities, Automation Technology Review 2001 (2001) 34–41.

[42] A.I. Omer, M.M. Taleb, Architecture of industrial automation systems, European Scientific Journal 10 (3) (January 2014).

[43] A. Alcina, Translation technologies: Scope, tools and resources, Target. International Journal of Translation Studies 20 (1) (2008) 79–102.

15 Industry 4.0 in Manufacturing, Communication, Transportation, Healthcare

Priyanga Subbiah, Krishnaraj Nagappan, Tamil Selvi A., Poornima, and Kiran Bellam

15.1 INTRODUCTION

15.1.1 MANUFACTURING

"Industry 4.0," sometimes called the "Fourth Industrial Revolution," describes the integration of state-of-the-art technology with higher degrees of automation in relation to production processes. Modern manufacturing is defined by the integration of digital and physical systems, the advent of fully automated processes, and the use of analytics in long-term planning. From a manufacturing sector point of view, the following are some of the most important features of the fourth industrial revolution:

- **Internet of Things (IoT):** In order to collect and analyse data in real time, Industry 4.0 makes use of the IoT to link together machines, sensors, and other physical objects that can be connected to the internet. Because of this interconnectivity, manufacturing processes can be monitored, controlled, and optimized more effectively.
- **Big Data and Analytics:** Advanced analytics methods are used to make sense of the massive amounts of data produced by IoT gadgets. Manufacturers can increase output and cut costs by using this data-driven strategy to better understand operational inefficiencies, quality issues, and predictive maintenance.
- **Cyber-Physical Systems (CPS):** The term "Industry 4.0" pertains to the coming together of digital technology and traditional infrastructures. Connected manufacturing systems (CPS) refer to a class of sophisticated, interconnected machinery that integrate sensors, actuators, and controls with software systems to enable intelligent operations. These automated systems have the capability to monitor and control physiological processes

DOI: 10.1201/9781003473886-15

without the need for human intervention, resulting in increased flexibility and adaptability of operations.

- **Machine Learning (ML) (AI) and Artificial Intelligence:** Data analysis, prediction, and automated decision-making are all made possible by AI and ML algorithms. They can enhance manufacturing processes by optimizing production schedules, anticipating maintenance needs, spotting anomalies, and so on.
- **Additive Manufacturing (3D Printing):** Industry 4.0 includes additive manufacturing technologies, such as 3D printing. This enables on-demand production, customization, and rapid prototyping, reducing costs and lead times while allowing greater design flexibility.
- **Human-Machine Interaction:** The value of human expertise and the synergy between humans and machines are recognized in the Fourth Industrial Revolution. It promotes the use of AR, VR, and wearable tech to improve training, maintenance, and human-machine interface issues.
- **Supply Chain Integration:** Supply chain integration via digital platforms and real-time information exchange is at the heart of the Fourth Industrial Revolution, sometimes called Industry 4.0. This paves the way for easier communication, better supply management, and more agility in meeting customer requests.

15.1.2 Communication

A major focus of the Fourth Industrial Revolution (4IR) and the concept of Industry 4.0 is cyber-physical systems (CPS). This chapter delves into the recent developments in automated systems and data transfer as well as their meteoric rise in popularity within manufacturing. Some of the most crucial aspects of communication in the era of Industry 4.0 are summarized here.

- The incorporation of Industries 4.0 ideas relies heavily on cyber-physical systems like sensors, actuators, and controllers.
- Communication technologies such as wireless connectivity and standardized communication interfaces are used to connect devices and enable them to communicate more efficiently.
- Open, secure and near real-time communication systems are used to exchange data, analyze, and control intelligent actions for various processes in manufacturing.
- In the Industry 4.0 framework, the exchange of data among devices and systems is facilitated by communication protocols such as OPC UA, MQTT, and DDS.
- Through the help of wireless connectivity and sensors, machines in factories can now keep tabs on the entire production process, map it out, and make decisions on their own—all thanks to Industry 4.0.
- This technology uses CPS to share and analyze data and control intelligent actions for various processes in manufacturing.

15.1.3 Transportation

Industry 4.0, also known as the Fourth Industrial Revolution, is characterized by the amalgamation of analogue and digital procedures. The implementation of Industries 4.0 is driving the evolution of transportation and logistics organization and execution. Industries 4.0 is anticipated to have an impact on the transportation sector in various ways.

Increased Efficiency: Transport companies can benefit from automation, AI, and the Internet of Things, all components of the Industry 4.0 framework.

Improved Safety: Advanced sensors and analytics can help transport companies identify potential safety risks and take proactive measures to avoid accidents.

Improved Customer Experience: With the help of Industry 4.0 technologies, transport companies can provide their customers with information about their shipments in real time, thus improving transparency and customer satisfaction.

Greater Sustainability: Technologies included in the Industry 4.0 effort might help the transportation industry reduce its environmental impact in a number of ways, including by improving fuel efficiency, cutting down on waste, and improving route planning. Through the combination of digital and physical technologies, Industry 4.0 is revolutionizing the transportation sector. It enhances efficiency, safety, customer experience, and sustainability.

Augmented Reality: The use of technology to enhance the patient experience and improve clinical outcomes.

Healthcare 4.0 is a conceptual framework for intelligent and interconnected healthcare that encompasses the augmentation of automation. One of the most notable and significant distinctions between Healthcare 4.0 and preceding healthcare models is the integration of digital technologies with the aim of enhancing patient outcomes and diminishing expenses.

- The terms “Hospital 4.0” and “Medicine 4.0” are becoming increasingly popular in the healthcare sector.
- In Healthcare 4.0, data drives digital health technologies like telemedicine, mHealth, Wi-Fi health, eHealth, and mHealth.
- To effectively prepare for Industry 4.0, healthcare organizations may consider adopting a patient-centric approach to care, alongside embracing digital transformation and investing in novel technologies. Ultimately, Industry 4.0 is transforming the healthcare industry and healthcare organizations must adapt to remain competitive and provide the best possible care to patients.

15.1.4 Healthcare 4.0

In Healthcare 4.0, artificial intelligence (AI) may help doctors more accurately diagnose patients and provide more effective care. The utilization of AI algorithms for the analysis of medical images such as X-rays and MRIs can facilitate the early detection and treatment of diseases and abnormalities. Medical practitioners will be able to foresee potential health effects, thanks to artificial intelligence's (AI) capacity to sift through mountains of data in search of patterns. This, in turn, can facilitate the development of customized treatment plans for patients.

- AI can improve the accuracy and speed of disease diagnosis, reduce errors, and improve performance in predicting and detecting different diseases.
- For example, AI can detect dangerous tumours in medical images, allowing pathologists to diagnose the disease in less time.

Predictive models built with AI can help doctors zero in on patients most likely to experience a particular health problem early on, which is crucial for optimal treatment and prevention outcomes.

15.2 LITERATURE SURVEY

The research utilized interpretive structural modeling to establish their interrelation. The research employed the FMICM-AR methodology to investigate the determinants that impact the driving and dependent variables. A hierarchical arrangement of competencies was presented through a structural model, which also assigned priority to them [1].

Cutting-edge tech propels manufacturing toward smart systems that boost flexibility and efficiency. A conceptual model measures operational performance, validated by experiments. Smart manufacturing aids Industry 4.0 transition and tackles industry challenges [2].

Organizations globally prioritize Industry 4.0 adoption, but limited empirical studies exist on key technological factors. This article fills the research gap by identifying and prioritizing factors in an emerging economy [3].

Comparing blockchain-based IoT healthcare Industry 4.0 systems can be difficult due to the complexity of assessing their security and privacy properties. For accurate ranking and weighting, this chapter presents the S-FWZIC method and a unified MCDM framework [4].

Industry 4.0 integrates modern technologies, with AI as a key component enabling autonomous tasks and predictive maintenance. This article surveys AI and XAI methods in Industry 4.0, exploring their applications, opportunities, challenges, and future research directions [5].

This study analyzes Industry 4.0 adoption in Latin American organizations using TAM, GITAM, and TPB models. Results highlight influential factors and their implications for developing countries [6].

Industry 4.0 promotes networked manufacturing by connecting existing machinery to the Internet of Things. The benefits outweigh the costs of retrofitting, but there are challenges to be overcome. With a focus on technical challenges and enabling technologies, this chapter provides an overview of brownfield development while also serving as a guide for intelligent factory transformation sector 5.0 [7].

Industry 4.0 transforms industries and places humans at the center of digital transformation. This study identifies the influence of intangible factors across different maturity levels, resulting in a conceptual framework for nurturing intellectual capital. Further research is needed to validate the framework across various sectors and assess its applicability in manufacturing companies [8].

Technological advancements have transformed SCADA systems in smart factories. This chapter explores SCADA systems in the steel industry, addressing interoperability and technology adoption challenges [9].

Industry 4.0, also known as I4.0, denotes the Fourth Industrial Revolution, which entails the assimilation of sophisticated technologies, including digitalization and artificial intelligence, into industrial systems. This integration brings about substantial changes in the operational mechanisms of industries. The present study offers an examination of the effects of Industry 4.0 on the electrical utility sector, catering to both scholars and professionals alike [10].

15.3 SMART COMMUNICATION: EMPOWERING CONNECTIVITY IN INDUSTRY 4.0

The conception of smart communication has arisen as a revolutionary force in the age of Industry 4.0, enabling connectivity across multiple sectors. This explanation emphasizes the value of smart communication in Industry 4.0 and how it improves connectivity while fostering innovation. This integration is having a transformative impact on various industries, including manufacturing, communication, transportation, and healthcare. Between machines, systems, and people, seamless connectivity and real-time data exchange are made possible by smart communication, which serves as a catalyst [11]. These gadgets collect real-time data from production lines, supply networks, or infrastructure, thanks to sensors and connection capabilities. Following analysis and use of the data, operations are optimized, decision-making is improved, and predictive maintenance is made possible, increasing productivity and decreasing downtime. Industry 4.0's smart communication is strengthened even further by the use of big data analytics. Organizations are able to obtain important insights, spot patterns, and make data-driven decisions by analyzing vast amounts of data that have been gathered from various sources. Through proactive response made possible by real-time data analytics, businesses may optimize processes, reduce risks, and boost overall performance. Furthermore, AI-based solutions are essential for smart communication. Chabot's and virtual assistants with AI capabilities improve consumer interactions, offer tailored advice, and speed up communication procedures. Intelligent data processing and analysis are made possible by machine learning algorithms, which also promote automation and boost effectiveness in a variety of fields. Transportation systems are significantly impacted by intelligent communication as well. Intelligent traffic systems optimize traffic flow, ease congestion, and improve safety by using real-time data from sensors and linked vehicles. Intelligent communication networks enable autonomous cars to navigate and communicate with one another, increasing transportation effectiveness and lowering environmental impact.

Smart communication makes it possible to give individualized and accurate medical treatments in the field of healthcare. The collecting and analysis of patient data is made easier by big data analytics and IoT devices, which improves diagnosis and treatment choices. Access to healthcare services in isolated or underserved locations is ensured by telemedicine, which enables remote monitoring and consultation. Industry 4.0's fundamental enabler of connectedness, effectiveness, and innovation across industries is smart communication. Organizations may harness the power of real-time data, optimize operations, and arrive at wise judgments by utilizing IoT, big data analytics, and AI. Smart communication will be essential in unlocking new opportunities, changing sectors, and propelling the subsequent wave of technical improvements as Industry 4.0 continues to develop.

15.3.1 Communication Revolution: Industry 4.0's Influence on the Way We Connect

Industry 4.0 has given rise to "smart factories" in the manufacturing sector, which use automation, sensors, and IoT devices to build intelligent and linked production processes. It is now possible to communicate seamlessly across machines, systems, and people, which allows for real-time data transmission, effective coordination, and improved decision-making. This advancement in factory communication has improved quality control, decreased downtime, and increased production. Additionally, communication between organizations and individuals has changed because of Industry 4.0. A hyper-connected society has been made possible by the growing use of cellphones, social media websites, and online communication tools. Through numerous digital channels, people may now interact with one another globally, share information immediately, and communicate in real time. This has sped innovation, facilitated worldwide collaboration, and increased networking and knowledge-sharing opportunities. The emergence of Industry 4.0 has sparked a communication revolution that has radically changed how we communicate and engage with one another in different spheres of our life. This overview examines the enormous impact of Industry 4.0 on communication while highlighting the important developments and changes that have transformed our connectivity [12]. These technologies have changed how people communicate, allowing for quicker, more effective, and more integrated ways to engage with one another.

Industry 4.0 has ushered in a new era of interconnected and intelligent transportation systems in the transportation sector. Real-time monitoring of traffic conditions, route optimization, and safety improvements are made possible by the integration of IoT devices, sensors, and data analytics. Vehicles, traffic management systems, and infrastructure are all connected through communication networks, allowing for smooth information sharing and coordination for effective transportation operations. Additionally, thanks to Industry 4.0, the healthcare sector has witnessed a communication revolution. Connected medical devices, wearable technology, and telemedicine systems have transformed communication between patients and doctors as well as remote healthcare monitoring. Early detection, individualized treatments, and better patient outcomes are made possible by the collection and analysis of real-time data. The delivery of healthcare services has been made possible by communication technologies, particularly in distant places with little access to medical facilities. Industry 4.0's revolution in communication has not been without difficulties. In this networked environment, data security, privacy, and ethical concerns have become crucial factors to take into account. In the digital age, it is more important than ever to protect personal information, ensure secure communication routes, and fix potential weaknesses. Industry 4.0 has sparked a communication revolution that is changing how we communicate and engage with one another. The fusion of cutting-edge technologies has revolutionized transportation and healthcare, facilitated seamless communication, improved manufacturing processes, and enabled global connectedness. Emerging technologies will continue to influence the communication environment as Industry 4.0 develops, bringing with them new opportunities and difficulties for connectivity and cooperation.

15.3.2 Intelligent Communication: Harnessing Industry 4.0 Technologies for Enhanced Connectivity

With the help of innovative technology, intelligent communication has emerged as a crucial element of Industry 4.0, revolutionizing how people communicate and enhancing connection. In order to promote improved connectivity and collaboration, this description analyzes the idea of intelligent communication and how it makes use of Industry 4.0 technology. The development of intelligent communication systems that can adapt, learn, and optimize communication procedures is being facilitated by these technologies, which are revolutionizing conventional communication techniques. The use of IoT devices and sensors is at the core of intelligent communication. These interconnected gadgets make it possible to gather enormous amounts of real-time data from many different sources, including machinery, equipment, and environmental conditions. This information lays the groundwork for intelligent communication systems, allowing for well-informed choice-making, insightful analysis, and seamless connectivity between objects, systems, and people. AI is essential to intelligent communication because it gives computers and other systems the ability to analyses and interpret data, comprehend context, and make wise judgments [13]. Big data analytics process and analyzes enormous amounts of data in real-time, further enhancing intelligent communication. Organizations can obtain important insights, make data-driven decisions, and improve communication strategies by seeing patterns, trends, and correlations in the data. Real-time analytics allow for proactive communication, allowing businesses to react quickly to shifting conditions and changing client needs. With the digital twin idea, intelligent communication is revolutionizing manufacturing industry production processes. By simulating digital twins of physical assets, real-time monitoring, analysis, and optimization are made possible. This technology makes it easier for physical assets and their digital equivalents to communicate, enabling predictive maintenance, streamlining processes, and boosting overall effectiveness.

15.3.3 The Future of Communication: Industry 4.0's Transformational Impact

The revolutionary influence of Industry 4.0 is reshaping communication and changing the way people connect, communicate, and collaborate. This summary explores the significant changes wrought by Industry 4.0 and demonstrates how they have completely altered the environment of communication. Industry 4.0 is the combination of revolutionary technologies, including automation, big data analytics, AI, and the IoT. With the help of these technologies, communication methods are being redefined, ushering in a time of smarter, quicker, and more effective connectedness.

The IoT is one of the main forces behind the change of communication in Industry 4.0. Systems, sensors, and linked devices are seamlessly integrated to form a network that allows for real-time data transfer and communication. This connectivity goes beyond conventional gadgets to encompass commonplace items, ushering in a new era of smart infrastructure, smart cities, and smart households [14]. The future of communication is heavily reliant on AI, since it makes intelligent automation, natural language processing, and personalized experiences possible. By enabling effective and context-aware communication, chabots, virtual assistants, and AI-powered algorithms improve consumer interactions and streamline company operations. AI-driven analytics offer insightful knowledge into customer behavior, allowing businesses to customize their communication plans and send individualized, focused messages.

Furthermore, by processing and analyzing massive amounts of data in real-time, big data analytics revolutionizes communication. This makes it possible for businesses to get useful insights, make data-driven choices, and spot new trends and patterns. Real-time analytics enables firms to respond quickly to market developments, allowing them to modify and improve their communication strategy as necessary. The concept of Industry 4.0 encompasses the integration of smart cities and intelligent infrastructure within the realm of communication in the future. The integration of advanced communication networks, sensors, and artificial intelligence systems enables the achievement of real-time resource monitoring, analysis, and optimization. As a result, there is better traffic control, more effective use of energy, increased public safety, and an overall improvement in quality of life.

15.4 THE DIGITAL TRANSFORMATION OF HEALTHCARE

Provider access and management of healthcare services are being transformed by the digital revolution in healthcare. The healthcare industry is currently undergoing a significant shift toward a more patient-focused, efficient, and data-oriented approach, facilitated by the implementation of advanced technologies and the incorporation of digital innovations. Every element of healthcare is being transformed by these technologies, from patient care and diagnostics to research and management. Healthcare providers can improve the quality and accessibility of care through the digital transformation. Information about patients may be shared easily thanks to electronic health records, which improves professional coordination and continuity of care. In remote or underdeveloped locations, telemedicine provides remote consultations, allowing patients to obtain healthcare services from the comfort of their homes. Individuals are empowered to measure vital signs, monitor their health, and successfully manage chronic illnesses, thanks to wearable technology and mobile health applications. These tools can analyze enormous volumes of medical data, spot patterns, and help medical personnel make precise diagnoses, forecast the success of treatments, and customize patient care. Big data analytics are also essential for population health management and medical research. Researchers can learn more about illness patterns, spot trends, and create tailored interventions by analyzing massive databases. Healthcare organizations may identify high-risk individuals, put preventive measures in place, and allocate resources more efficiently by using population health management. Although the digitalization of healthcare has many advantages, there are drawbacks as well [15]. Among the most important factors are patient data protection and privacy, system interoperability, and the requirement that healthcare workers learn new digital skills. The digitalization of healthcare is transforming the sector, enhancing patient care, and spurring innovation. Healthcare providers may improve efficiency, accessibility, and quality of care by embracing digital solutions and utilizing cutting-edge technologies, which will eventually improve health outcomes for both people and communities.

15.4.1 Harnessing the Power of Industry 4.0 in Healthcare

The healthcare industry is experiencing a significant transformation with the implementation of Industry 4.0 technology. This revolution in healthcare is opening up new possibilities for bettering patient care, increasing operational effectiveness, and advancing medical research. These technologies are being used in numerous facets of healthcare, including patient monitoring, therapy, diagnostics, and management. Diagnostics is one of the major fields where Industry 4.0 is having a substantial impact. To enable more precise and effective diagnoses, medical imaging data is being analyzed using AI and ML algorithms. Advanced analytics and predictive modeling are also assisting in pattern recognition and disease progression prediction, enabling early intervention and individualized treatment strategies.

15.4.2 Integration of Technology in the Healthcare Sector

Provider access and management of healthcare services are all being revolutionized by the use of technology in this industry. The sector is undergoing a substantial shift because of the introduction of cutting-edge digital solutions, which will improve patient care, increase efficiency, and improve health outcomes. Numerous applications and advancements fall under the broad category of technology integration in healthcare. Healthcare practitioners can now securely store, access, and share patient information in real-time thanks to electronic health records (EHRs), which have taken the role of outdated paper-based systems. This not only increases the accuracy and effectiveness of managing medical records, but it also makes it easier for healthcare professionals working in various settings to collaborate easily. Telemedicine, which enables the provision of medical services remotely, is a critical component of technological integration. Patients can seek medical advice, receive diagnoses, and even undertake some treatments from the comfort of their homes, thanks to video consultations, remote monitoring devices, and secure communication platforms. Telemedicine has shown to be very helpful in lowering healthcare expenditures, increasing healthcare access for individuals in underserved or rural locations, and eliminating the need for unnecessarily frequent hospital visits [16]. Additionally, the use of wearable technology and mobile health apps has grown, enabling people to actively monitor their health and well-being. These tools may monitor physical activity, sleep patterns, vital signs, and even spot anomalies or potential health problems.

15.4.3 The Role of Industry 4.0 in Revolutionizing Healthcare

The Fourth Industrial Revolution, often known as Industry 4.0, is changing many industries, including healthcare. It is possible to revolutionize patient care, increase efficiency, and improve outcomes by integrating cutting-edge technologies and digital advances into healthcare. The present analysis explores the ways in which Industry 4.0 is fundamentally transforming the healthcare industry. With the help of these technologies, massive volumes of patient data may be collected, analyzed, and used, leading to diagnoses that are more precise, individualized therapies, and proactive preventive care.

Diagnostics is one of the major areas where Industry 4.0 is revolutionizing healthcare. AI-powered algorithms with remarkable speed and accuracy, enabling medical personnel in more quickly identifying and diagnosing problems. This innovation could speed up treatment processes, lessen diagnostic blunders, and eventually, save lives. In addition to diagnostics, Industry 4.0 is revolutionizing patient monitoring and healthcare delivery.

15.4.4 Advancements in Healthcare through Industry 4.0 Technologies

Healthcare organizations are utilizing Industry 4.0 technology in this age of digital transformation to increase patient outcomes, enhance healthcare delivery, and streamline operations. The whole healthcare industry is being transformed by these technologies, from hospitals and clinics to academic institutions and pharmaceutical firms. The use of electronic health records (EHRs) and health information exchange (HIE) systems is an important part of Industry 4.0 in healthcare [17]. Through the seamless sharing of patient data made possible by these digital platforms, various healthcare providers may now deliver coordinated, individualized care. Rapid and accurate diagnoses are guaranteed, medical errors are decreased, and care coordination is improved with real-time access to comprehensive patient data. Through Industry 4.0, artificial intelligence and machine learning algorithms are fundamental to the transformation of healthcare. Healthcare practitioners are helped in their analysis of enormous amounts of patient data, medical imaging, and genetic data by AI-powered diagnostic tools and decision support systems. These technologies can help with early disease detection, treatment result prediction, and the creation of individualized treatment regimens. Industry 4.0 technology also make it easier to manage healthcare resources effectively. Systems for automating inventory management make ensuring that medical supplies are always available and minimize waste. RPA streamlines administrative processes, giving healthcare personnel more time to focus on patient care. Forecasting healthcare resource requirements with the aid of predictive analytics models enables proactive planning and resource allocation. However, issues including data privacy, security, and interoperability must be addressed as healthcare organizations use Industry 4.0 technologies. To safeguard sensitive patient data and preserve public confidence in the healthcare system, effective cybersecurity measures and data governance frameworks are crucial.

15.4.5 Industry 4.0: Shaping the Future of Healthcare Delivery

The integration of advanced technologies and automation across various industries, known as Industry 4.0, is rapidly transforming the delivery of healthcare. This summary examines how Industry 4.0 is transforming patient care, improving overall healthcare results, and redefining the future of healthcare. By streamlining procedures, collecting and analyzing massive quantities of patient data, and making better-informed decisions in real time, these technologies are empowering healthcare professionals. The focus on personalized medicine and patient-centric care is one of the major components of Industry 4.0 in healthcare. Healthcare providers can use patient-specific data to create personalized treatment plans, forecast illness development, and spot potential dangers by integrating AI and big data analytics. This level of individualization and precision results in better patient outcomes and a more effective healthcare system. Through remote monitoring and telemedicine systems, Industry 4.0 is also improving the delivery of healthcare. Continuous patient monitoring outside of conventional healthcare facilities is made possible by IoT-enabled devices and wearable sensors, allowing for the early identification of health conditions and prompt action. Platforms for telemedicine make remote consultations possible, increasing access to healthcare services, particularly in underdeveloped areas.

The developments brought about by Industry 4.0 in healthcare also heavily include robotics and automation. Robotic surgery offers more precision, less invasiveness, and faster recovery times than traditional surgery. Healthcare personnel can concentrate on providing complex patient care by using autonomous robots to complete monotonous duties [18]. Additionally, administrative chores like appointment scheduling and medical record administration are streamlined by robotic process automation, increasing productivity and lowering errors.

15.4.6 Smart Healthcare Systems and Industry 4.0 Innovations

The delivery of healthcare is being revolutionized by Industry 4.0 developments and smart healthcare technologies, which are also enhancing patient outcomes and changing the entire healthcare ecosystem. These Industry 4.0 technologies are used in smart healthcare systems to build intelligent, interconnected networks that improve the effectiveness, accuracy, and accessibility of healthcare services. Healthcare practitioners may make educated decisions, optimize resource allocation, and provide patients with individualized treatment by utilizing real-time data gathering, analysis, and communication. Innovations in healthcare brought forth by Industry 4.0 heavily rely on robotic technology. Surgery, precision medicine, drug dispensing, and patient support are among fields where robots are used. They can enhance the ability of healthcare workers by performing complex surgeries with more accuracy, completing repetitive jobs quickly, and offering patients round-the-clock monitoring and support. Additionally, the incorporation of cloud computing into smart healthcare systems allows for the safe storage, accessibility, and sharing of medical data between various organizations and healthcare practitioners. This encourages fluid communication, makes it easier to make decisions based on the best available information, and improves patient-centered treatment.

15.4.7 Transforming Patient Care with Industry 4.0 Technologies

"Transforming Patient Care with Industry 4.0 Technologies" examines the tremendous effects of Industry 4.0 on the healthcare industry with a particular emphasis on how patient care is changing. This summary digs into how Industry 4.0 technologies are changing how health care is provided. It emphasizes the use of AI-driven diagnostics and predictive analytics to enhance disease early detection and diagnosis. The use of IoT devices for remote patient monitoring is also highlighted, enabling healthcare professionals to gather real-time data and deliver individualized care outside of conventional hospital settings. The discussion goes into detail about the developments in robotic-assisted surgery, which allow doctors to carry out minimally invasive treatments with more control and precision. It looks at how wearable technology and smart sensors can be used to monitor wellness and encourage preventive healthcare, giving people control over their own health. The description also talks about potential difficulties and issues that can come up when applying Industry 4.0 technology to the healthcare sector, like data security issues, privacy worries, and the requirement for qualified personnel to handle and comprehend the enormous amounts of created data.

15.5 INDUSTRY 4.0 IN MANUFACTURING

Due to the development of new technologies, the concept of Industry 4.0 has become widespread for adopting them in industrial processes like smart automation, smart production, etc. It can be used to integrate physical objects like sensors, machines used in industries through cyber components like software can be used to control the operation and maintenance of the device and the integration of them into the real-world environment. Figure 15.1, illustrates the concept of how the Internet of Things (IoT) can be used in smart manufacturing applications comprised of various technologies like cloud manufacturing, cyber physical systems (CPS) in manufacturing, fog manufacturing, and data analytics.

Manufacturing CPS Layer: Industries that want to control the operations of their respective physical device, like robotic arms, in case of the packaging industry, can be connected using this unit, which, in turn, facilitates remote operation of the device. Each industry that avails this facility should maintain a proper service policy across different or specific units of the concerned factory.

Fog Manufacturing Layer: The fog layer is used as an intermediate layer to enable smooth integration of manufacturing CPS services with cloud

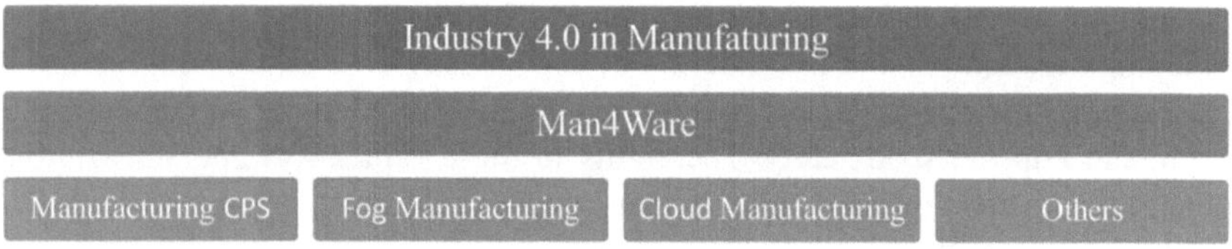

FIGURE 15.1 Connected Components: Harnessing Industry 4.0 for Advanced Manufacturing.

manufacturing. This layer remains close to those parts that are involved in manufacturing processes [18]. It can be used to enable services as digital twins. Using digital twins, the manufacturing process can be visualized virtually.

Cloud Manufacturing Layer: This layer provides a cloud computing facility for the manufacturing process in terms of performing complex computational tasks. It is equipped with reasonable storage capacity for providing instructions to remote equipment autonomously. They also have a knowledge database, which enables them to learn from the environment and act accordingly.

Blockchain: Blockchain technology includes methods that can be used to enable information sharing, automation, and negotiation to create agreement between collaborating enterprises. One such example of blockchain usage in smart factory environment is it integrates production and customer services. Based on the inputs from customers, the production of a particular good will be increased or decreased.

Service Middleware: As a service-oriented middleware, Man4Ware's functionality may be readily increased by including more core services and developing the capacity to serve more complex application services. Man4Ware has been expanded to include a crucial set of blockchain services that are increasingly beneficial for applications related to smart manufacturing.

15.6 INDUSTRY 4.0 IN TRANSPORTATION

The Fourth Industrial Revolution, sometimes known as "Industry 4.0," is changing several industries, including the transportation industry. "Industry 4.0" refers to the application of cutting-edge technology and digitalization to enhance efficiency, safety, and sustainability across a range of transportation-related domains. Here are some crucial areas in transportation where Industry 4.0 is having an impact: Intelligent Fleet Management: Industry 4.0 technologies support intelligent fleet management systems that maximize vehicle use, route planning, and maintenance scheduling by utilizing data analytics, IoT sensors, and connectivity. To increase fuel efficiency, save maintenance costs, and improve fleet performance overall, real-time data can be collected and evaluated on fuel consumption, vehicle health, and driver behavior. Connected and autonomous vehicles (CAVs): Industry 4.0 technologies are helping the creation and introduction of CAVs. CAVs use cutting-edge sensors, AI algorithms, and real-time connectivity to increase traffic efficiency, decrease congestion, and improve road safety. The ability of these vehicles to interact with one another and the infrastructure allows them to make decisions based on real-time data, improving traffic flow and lowering accident rates. Predictive Maintenance: Transportation systems can use predictive maintenance, thanks to Industry 4.0 technologies. By recognizing and addressing maintenance needs before they cause major disruptions, this proactive maintenance method helps to reduce maintenance costs, maximize operational efficiency, and limit downtime. Smart Logistics and Supply Chain Management: Industry 4.0 technologies are transforming supply chains and logistics in the transportation industry. To increase efficiency, cut costs, and improve overall supply chain visibility, innovative inventory management systems, real-time tracking and tracing of shipments, robots and automation-enhanced warehouse operations, and other measures are being put in place. Blockchain technology integration can improve supply chain operations' traceability, transparency, and security. Improved Safety and Security: Industry 4.0 solutions help to raise the level of security and safety in the transportation sector. With the aid of cameras, sensors, and AI algorithms, advanced driver assistance systems (ADAS) can identify possible hazards, alert drivers, and even start autonomous emergency maneuvers. Additionally, real-time monitoring of cars, cargo, and infrastructure by IoT-based security systems enables the early detection of security breaches, theft, or unauthorized entry. Sustainable Transportation: The development of more environmentally friendly transportation systems is being aided by Industry 4.0 technologies. With the help of intelligent charging infrastructure and renewable energy sources, fossil fuel dependence and greenhouse gas emissions are reduced by electric and hybrid automobiles. Additionally, route planning, load consolidation, and data analytics and optimization algorithms allow for the optimal use of resources, which reduces energy use and carbon impact. Advanced Mobility Services: Thanks to Industry 4.0, new mobility services like ride- and car-sharing as well as multimodal transportation platforms are now possible. These services make use of digital platforms, real-time data, and clever algorithms to enhance travel options, lessen traffic, and give customers more convenient and tailored travel experiences. As a result of utilizing cutting-edge technologies, data analytics, and networking, Industry 4.0 is changing the transportation sector. Industry 4.0 is advancing efficiency, safety, and sustainability in the transportation industry, ultimately influencing the future of mobility. This includes intelligent fleet management, networked cars, predictive maintenance, and sustainable transportation solutions. Smart Transportation enhances flexibility, equips according to market needs, and helps production units come closer to customer needs. This will result in proper delivery of goods, and production will be controlled based on customer needs [19]. This, in turn, helps the manufacturing units in procuring raw materials beforehand, thereby reducing the time required to deliver the goods to the customer. Figure 15.2 highlights the possibilities and advantages of incorporating Industry 4.0 technology into transportation networks while illustrating the main pillars of intelligent transportation systems. Some of the technological aspects smart transportation are:

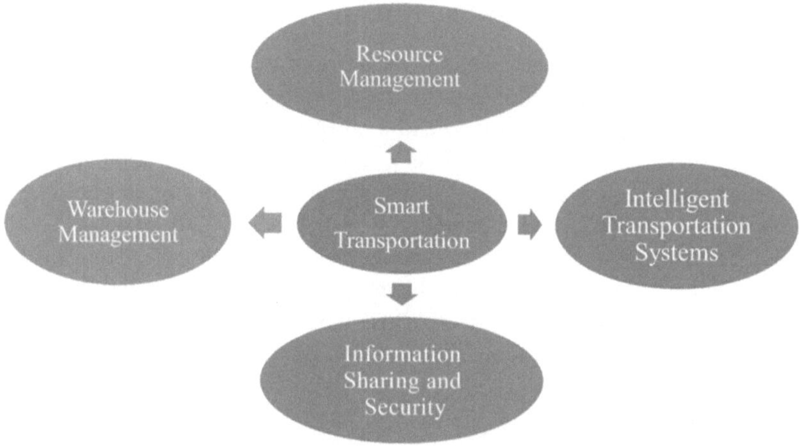

FIGURE 15.2 Intelligent Transportation Systems: Transforming Mobility with Industry 4.0 Technologies.

15.6.1 Resource Management

Through the use of resource management procedures, the adoption of the Industry 4.0 paradigm and the installation of cyber-physical systems (CPS) will boost overall productivity, adaptability, and response to changes that may occur in supply chains [5]. Enhanced levels of transparency and inclusivity, coupled with effective coordination and consolidation of the primary stakeholders in the supply chain, can facilitate an adequate projection of resources such as human capital, materials, and machinery. This approach would facilitate the synchronization of time to market, optimization of resources and processes, and utilization of additional assets.

15.6.2 Warehouse Management

The warehouse has traditionally served as the central point of the supply chain for the transportation of goods. In the present economic scenario, logistics providers must ensure that they embody a substantial basis of competitive advantage. The achievement of seamless coordination and alignment across all stages of the value chain requires integration among the various actors and stakeholders involved in the supply chain. Transporters have the capability to transmit their current whereabouts and anticipated time of arrival to the intelligent warehouse management system. This feature can be utilized to optimize the delivery process by ensuring that the goods are delivered precisely on time and in the correct sequence [20]. Subsequently, the system will possess the capability to select and arrange a suitable location for docking. Radio Frequency Identification (RFID) sensors will exhibit the delivered items and concurrently transmit track-and-trace information to the entire supply chain. The Warehouse Management System (WMS) will allocate storage space and initiate equipment requisition for the transportation of goods to the designated destination, based on the delivery information.

15.6.3 Intelligent Transportation Systems

A field that is still in its infancy, intelligent transportation systems (ITS), interact with a wide range of transportation system components, including management, control, infrastructure, operations, policy, and control methods. The ITS has adapted new technologies, such as data processing, sensor technology, communications, hardware for computing, positioning systems, and programming approaches. We live in a global society, and ITS is an essential part of that society [20]. The use of virtual technology in transportation is a fresh topic that needs to be discussed and has a big impact on finding solutions to world problems. ITS systems are crucial for enhancing security, reliability, traffic speeds, mobility, and reducing hazards, accident rates, carbon emissions, and air pollution.

15.6.4 Information Sharing and Security

To process the specific task and make decisions, this information must be shared with the various components. Verifying that the data was acquired from a trustworthy source is essential. As long as they can benefit from new technologies to a sufficient extent, users usually adopt them without considering any inherent security concerns. By fostering and advocating for a security culture, recognizing the inherent vulnerabilities of technological applications and systems, and acknowledging the human factor as the weakest link, organizations can attain satisfactory levels of security and progress towards their business goals.

15.7 CONCLUSION

It can be stated that Industry 4.0 is instigating a substantial transformation across various sectors such as manufacturing, communication, transportation, and healthcare. Contemporary technological advancements such as the Internet of Things (IoT), big data analytics, artificial intelligence (AI), and automation are being assimilated to transform traditional approaches and introduce novel opportunities for efficacy and ingenuity. The Fourth Industrial Revolution, commonly referred to as Industry 4.0, has facilitated the development of intelligent factories within the manufacturing industry. These factories leverage the Internet of Things (IoT), big data, and artificial intelligence (AI) to enhance production processes, minimize inefficiencies, and augment productivity. Collaborative robotics, additive manufacturing, and digital twin simulations, opening the door to greater efficiency and flexibility, are transforming manufacturing operations. Industry 4.0 is also transforming communication systems, as smart networks adapt to changing conditions and provide real-time data and insights. The evolution of 5G technology further empowers the development of smart cities and intelligent infrastructure. With the introduction of Industry 4.0, the transportation industry is undergoing a change. In the transportation sector, autonomous vehicles, intelligent traffic systems, and real-time data analytics are driving efficiency improvements, cost savings, and higher levels of safety. These developments are transforming supply chain management, passenger experiences, and logistics, which will lead to a future that is more sustainable and connected. More personalized and accurate medical treatments are now possible because of Industry 4.0. Big data analytics, IoT devices, and other technologies assist doctors in identifying and diagnosing diseases, while 3D printing enables the creation of individualized implants and prostheses. Telemedicine also improves accessibility and care quality by enabling remote patient monitoring and consultations. In its entirety, Industry 4.0 is significantly altering how businesses interact with their clients and conduct business in the fields of manufacturing, communication, transportation, and healthcare. We should expect even more groundbreaking uses of Industry 4.0 across a variety of industries as technology develops, which will further encourage creativity, efficiency, and better outcomes for both people and businesses.

REFERENCES

[1] M. Javaid, A. Haleem, R. P. Singh, R. Suman and E. S. Gonzalez, "Understanding the Adoption of Industry 4.0 Technologies in Improving Environmental Sustainability," Sustainable Operations and Computers, vol. 3, pp. 203–217, 2002, https://doi.org/10.1016/j.susoc.2022.01.008.

[2] A. K. Inkulu, M. V. A. R. Bahubalendruni, A. Dara and K. Sankaranarayana Samy, "Challenges and Opportunities in Human Robot Collaboration Context of Industry 4.0—a State of the Art Review," Industrial Robot, vol. 49, no. 2, pp. 226–239. 2022, https://doi.org/10.1108/IR-04-2021-0077.

[3] M. Karatas, L. Eriskin, M. Deveci, D. Pamucar and H. Garg, "Big Data for Healthcare Industry 4.0: Applications, Challenges and Future Perspectives," Expert Systems with Applications, vol. 200, p. 116912, 2022, https://doi.org/10.1016/j.eswa.2022.116912.

[4] M. Javaid, A. Haleem, R. P. Singh, S. Rab and R. Suman, "Exploring Impact and Features of Machine Vision for Progressive Industry 4.0 Culture," Sensors International, vol. 3, p. 100132, 2022, https://doi.org/10.1016/j.sintl.2021.100132.

[5] M., Mamoona, S. Habib, A. R. Javed, M. Rizwan, G. Srivastava, T. Reddy Gadekallu and J. C.-W. Lin, "Applications of Wireless Sensor Networks and Internet of Things Frameworks in the Industry Revolution 4.0: A Systematic Literature Review," Sensors, vol. 22, no. 6, p. 2087, 2022, https://doi.org/10.3390/s22062087.

[6] A. Souri, A. Hussien, M. Hoseyninezhad and M. Norouzi, "A Systematic Review of IoT Communication Strategies for an Efficient Smart Environment," Transactions on Emerging Telecommunications Technologies, vol. 33, p. e3736, 2022, https://doi.org/10.1002/ett.3736.

[7] A. Jamwal, R. Agrawal, M. Sharma and A. Giallanza, "Industry 4.0 Technologies for Manufacturing Sustainability: A Systematic Review and Future Research Directions," Applied Sciences, vol. 11, no. 12, p. 5725, 2021, https://doi.org/10.3390/app11125725

[8] A. G. Frank, L. S. Dalenogare and N. F. Ayala, "Industry 4.0 Technologies: Implementation Patterns in Manufacturing Companies," International Journal of Production Economics, vol. 210, pp. 15–26, 2019, https://doi.org/10.1016/j.ijpe.2019.01.004.

[9] D. Guo, M. Li, Z. Lyu, K. Kang, W. Wu, R. Y. Zhong and G. Q. Huang, "Synchroperation in Industry 4.0 Manufacturing," International Journal of Production Economics, vol. 238, p. 108171, 2021, https://doi.org/10.1016/j.ijpe.2021.108171.

[10] B. Bajic, A. Rikalovic, N. Suzic and V. Piuri, "Industry 4.0 Implementation Challenges and Opportunities: A Managerial Perspective," IEEE Systems Journal, vol. 15, no. 1, pp. 546–559, 2021, https://doi.org/10.1109/JSYST.2020.3023041.

[11] D. Bianco, M. Godinho Filho, L. Osiro and G. M. D. Ganga, "Unlocking the Relationship Between Lean Leadership Competencies and Industry 4.0 Leadership Competencies: An ISM/Fuzzy MICMAC Approach," IEEE Transactions on Engineering Management, vol. 70, no. 6, pp. 2268–2292, June 2023, https://doi.org/10.1109/TEM.2021.3069127.
[12] A. Sajjad, W. Ahmad, S. Hussain and R. M. Mehmood, "Development of Innovative Operational Flexibility Measurement Model for Smart Systems in Industry 4.0 Paradigm," EEE Access, vol. 10, pp. 6760–6774, 2022, https://doi.org/10.1109/ACCESS.2021.3139544.
[13] D. Adebanjo, T. Laosirihongthong, P. Samaranayake and P.-L. Teh, "Key Enablers of Industry 4.0 Development at Firm Level: Findings from an Emerging Economy," IEEE Transactions on Engineering Management, vol. 70, no. 2, pp. 400–416, 2023, https://doi.org/10.1109/TEM.2020.3046764.
[14] S. Qahtan, K. Yatim, A. A. Zaidan, H. A. Alsattar, O. S. Albahri, B. Bahaa, A. H. Zulzalil, M. H. Osman and R. T. S. Mohammed, "Novel Multi Security and Privacy Benchmarking Framework for Blockchain-Based IoT Healthcare Industry 4.0 Systems," IEEE Transactions on Industrial Informatics, vol. 18, no. 9, pp. 6415–6423, 2022, https://doi.org/10.1109/TII.2022.3143619.
[15] I. Ahmed, G. Jeon and F. Piccialli, "From Artificial Intelligence to Explainable Artificial Intelligence in Industry 4.0: A Survey on What, How, and Where," IEEE Transactions on Industrial Informatics, vol. 18, no. 8, pp. 5031–5042, 2022, https://doi.org/10.1109/TII.2022.3146552.
[16] D. Cordero, K. L. Altamirano, J. O. Parra and W. S. Espinoza, "Intention to Adopt Industry 4.0 by Organizations in Colombia, Ecuador, Mexico, Panama, and Peru," IEEE Access, vol. 11, pp. 8362–8386, 2023, https://doi.org/10.1109/ACCESS.2023.3238384.
[17] T.-A. Tran, T. Ruppert, G. Eigner and J. Abonyi, "Retrofitting-Based Development of Brownfield Industry 4.0 and Industry 5.0 Solutions," IEEE Access, vol. 10, pp. 64348–64374, 2022, https://doi.org/10.1109/ACCESS.2022.3182491.
[18] M. F. Prim, J. D. O. Gomes, H. Kohl, R. Orth, M. Will and G. B. Vargas, "Identifying the Dynamics of Intangible Resources for Industry 4.0 Adoption Process," IEEE Access, vol. 10, pp. 101029–101041, 2022, https://doi.org/10.1109/ACCESS.2022.3208250.
[19] M. Sverko, T. G. Grbac and M. Mikuc, "SCADA Systems with Focus on Continuous Manufacturing and Steel Industry: A Survey on Architectures, Standards, Challenges and Industry 5.0," IEEE Access, vol. 10, pp. 109395–109430, 2022, https://doi.org/10.1109/ACCESS.2022.3211288.
[20] M. Mishra, M. Biswal, R. C. Bansal, J. Nayak, A. Abraham and O. P. Malik, "Intelligent Computing in Electrical Utility Industry 4.0: Concept, Key Technologies, Applications and Future Directions," IEEE Access, vol. 10, pp. 100312–100336, 2022, https://doi.org/10.1109/ACCESS.2022.3205031.

16 Artificial Intelligence-Based Anomaly Detection for Industry 4.0

A Sustainable Approach

R. Ravinder Reddy, T. Sridevi, Ravi Uyyala, and Amit Kumar Tyagi

16.1 INTRODUCTION

With rapid developments in Artificial Intelligence and the IoT, every industry has become automated, and this makes the system more effective and improves production; it enables the system to function more efficiently. This generation of industry is termed Industry 4.0, which connects many devices to make the system function more effectively. Industry 4.0 became a reality with the exponential growth in cyber-physical systems; it makes for a more effective connection among things. Along with this, the integration of Artificial Intelligence and Machine Learning (AIML) has become the backbone of the Industry 4.0 revolution; it generates vast amounts of data. Analyzing this data necessitates intensive computational resources with new technologies like AI and ML. With the involvement of these new technologies the system becomes more efficient and robust. But the anomalies may cause the entire system, sometimes outcome may deviate, which we need to avoid in the early state for more sustainability of the system. With the advent of these new technologies, detecting anomalies at each stage is important. Devices can't stop functioning automatically while it deviates from normal behaviour. These machines can't detect abnormality like humans; they need a protective system which monitors and sets off an alarm to stop the abnormal activity. The involvement of Industry 4.0 makes the system more robust that can detect anomalies from automated industries to reduce the error rate and improve the sustainability of the model. As shown in the Figure 16.1 Industry 4.0 more efficiently manages things by adopting advanced technologies easily to make the system more sustainable.

Industry 4.0 is connected with a huge number of devices, which generates huge volumes of data. An abnormal or anomaly pattern is a rare event in the huge chunks of data; the big challenge, which is very difficult, is to identify the anomaly from these data. Searching for abnormalities from huge data is become a big challenge in these days. The latest developments in hardware and Artificial Intelligence improves the computation power with intelligent behaviour of the systems. Recent advancements in hardware components, along with AI and ML technologies, are employed to find anomaly patterns from huge data more precisely and accurately. These technologies

DOI: 10.1201/9781003473886-16

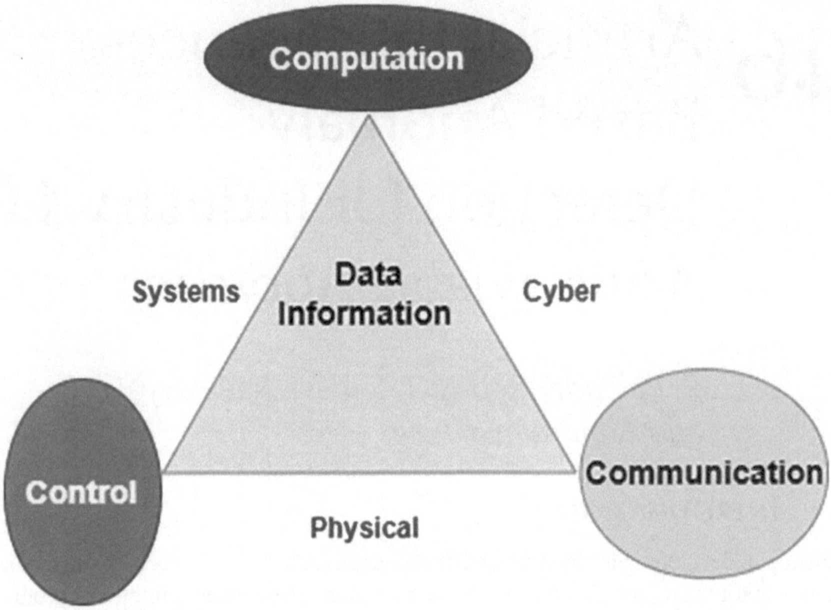

FIGURE 16.1 The Revolution of Industry 4.0 with CPS.

can help us to find abnormal patterns from the system. The combination of AI with Industry 4.0 can easily identity hidden, abnormal patterns from the data. Industry 4.0 is progressing at an unprecedented pace, surpassing the previous revolutions witnessed by humanity. Cloud computing, along with cyber-physical systems, serves as the underlying pillar of Industry 4.0. Communication and data-storage equipment have developed in the recent days, which makes Industry 4.0 evolve.

AI-ML systems are used to transform Industry 4.0 data into appropriate forms and analyze it and identify the appropriate patterns to detect the anomalies. The collaboration of AI-ML systems is vital for the advancement of Industry 4.0, and it can open up new avenues for research. Therefore, there is a growing demand to build and develop robust models based on AI-ML that can ensure the security of cyber-physical systems and devices. The overall system architecture is shown in Figure 16.1, which describes data-gathering units like sensors and IoT devices. The captured raw data can be processed for AI systems; it can be used to generate abnormal patterns.

The advent of Industry 4.0 has brought about a transformative impact on manufacturing practices, product enhancement, and distribution within companies. Smart factories play a crucial role in the collection and analysis of data by utilizing advanced sensors, embedded software, and robotics, thereby facilitating more informed decision-making processes. By seamlessly integrating production operation data with information from enterprise systems like supply chain management, ERP, and customer service, a new realm of visibility and understanding is achieved, surpassing the constraints imposed by isolated data sources.

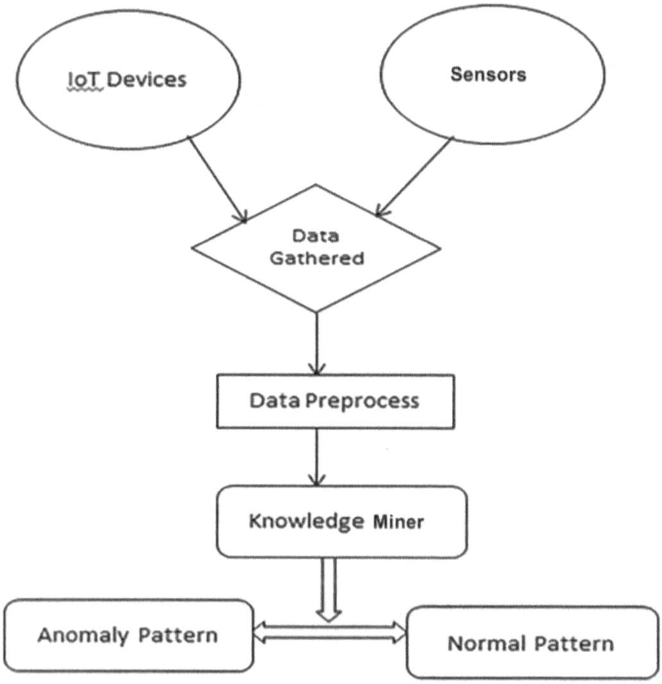

FIGURE 16.2 The Architecture of the System.

Emerging technologies like the Internet of Things, AIML, data analytics, and cloud computing are being actively incorporated into manufacturing facilities and operations. These technologies combine to lead Industry 4.0 to new heights and create the more user-friendly devices. The concept of Industry 4.0, delineated by Rubmann et al. [1], encompasses the convergence of diverse technologies. AIML involves employing automated methods to identify patterns in data and utilize these patterns for predicting future outcomes of interest. The Internet of Things (IoT) and cloud computing serve as key driving forces behind Industry 4.0. Through the digital transformation facilitated by Industry 4.0, manufacturers have gained the ability to create digital twins [2], which are computer simulations of real-world systems like production lines, manufacturing facilities, and distribution networks, which encompasses more reliable and high quality tasks.

16.2 RELATED WORK

Industry 4.0 is majorly focused on the atomization of industries, which reduces human interventions. For this purpose, it encompasses various technologies such as cloud computing, AIML, Cyber Physical Systems (CPSs), cognitive computing, and IoT. The intelligent industrial applications that make use of cutting-edge AI and ML technologies are being deployed with the express purpose of enhancing automation

[3], communication, and interaction of devices. The gathered data are used for further analysis and deploy advanced approaches. Collaboration among academia, research institutions, and industries from different domains is essential to design and implement modern smart applications that merge IoT, data engineering, AI, ML and cloud technologies. Therefore, the integration of these technologies is used to create reliable, smart applications. Consequently, one of the key challenges that Industry 4.0 must address is securing smart networks, particularly in the case of smart machines. The security aspect is crucial for providing the sustainable solutions. The malfunction of device identification is very important.

Research difficulties in the realm of cloud computing and its related fields, such as the Internet of Things, software-defined networking, and fog computing, are constantly expanding. Problems with network security and data availability are two examples of the difficulties presented by today's networking circumstances. For example, DDoS attacks on cloud networks raise brand-new problems and obstacles for the networking industry as a whole. Due, in large part, to the substantial non-linearity of real-world network traffic, the detection of DDoS attacks is an important research challenge in cloud security. DDoS attacks in cloud and IoT environments typically target vulnerable services, causing network sites to slow down or become unavailable until they are recovered. Also, it's crucially important to be able to tell the difference between normal network activity and malicious attacks. Finding such incidents in advance is beneficial to Industry 4.0 to provide sustainable solutions. The major role of AIML is to find such incidents by analyzing the previous incidents data; it will be help to avoid future attacks.

Despite numerous studies focusing on many distinct strategies, procedures, and tools for identifying attacks on Industry 4.0, none of the currently known strategies, procedures, or equipment has to effectively combat these attacks. Majorly, IoT devices are connected to cloud infrastructure; as a result, the cloud infrastructure and its users are under attack. The main reasons for the increase in such attacks is the lack of trust among various endpoints in a global network, making it difficult for global cooperation to enforce security. Socioeconomic factors also contribute to this issue. Furthermore, the nature of DDoS attacks prevents the deployment of a single-point solution that can provide the best defence against these attacks. These problems underline the importance of investigating and pinpointing the causes of the failure of current approaches [4], [5]. A DDoS attack is a sample example, but various kinds of attacks are coming to the picture to access the system. As the technology trends are increasing, the novelty of the attacks is also increasing. The compatibility among devices is a major challenge in threat analysis in Industry 4.0. Patterns among these device communications is also impacted.

In Industry 4.0, anomaly detection portrays a crucial part in ensuring the smooth operation of systems. To address this, an approach known as Artificial Intelligence of Things is employed to make every node intelligent. Real-time decision-making becomes essential, as machine failures can lead to significant losses. Real-time, intelligent devices for different machinery with remote monitoring are now available with the developments in information and internet-based technology [6]. Historically, anomaly detection has predominantly relied on a model-based approach, which involves leveraging statistical- and mathematical-based models to simulate various properties of the system, including mass, stiffness, damping matrices, and other characteristics of machines [7]. However, loading characteristics and an in-depth knowledge of the machines' internal structures are crucial to the success of such techniques. Dealing with complex mathematical expressions and managing multiple expressions for real industrial systems can be challenging. Additionally, this approach lacks the capability to adapt and update real-time detection with newly observed data. In recent decades, statistical and AIML methods supported by probabilistic models have gained prominence in fault detection [8]. However, these methods introduce additional computational complexity due to the involvement of various stochastic processes. By introducing the Machine Learning approaches, these tasks become simple but require a lot of data and computation power. The predictive accuracy of the attacks is more crucial in AIML approaches.

In recent times, data-centric approaches that incorporate signal processing and Artificial Intelligence (AI) have gained attraction for fault detection [4]. In the present era, there is considerable preference for the Industrial Internet of Things (IIoT) over traditional model-based and statistical methods in the field of anomaly detection. These approaches have the leverage of AIML Techniques [9]. Kumar Sharma et al. (2022) highlighted the effectiveness of using Temporal Convolutional Networks for Predictive Maintenance of Industrial Systems Based on Data [10]. The deep learning approaches improved significantly in the detection of anomaly accuracy.

AIML and Industry 4.0, together, created much of the hike in the industry toward automation. Industrial automation is where AI and manufacturing will be focusing their attention in the near future. However, the emergence of smart factories marks just the beginning of a revolution in search of increased productivity. Artificial Intelligence (AI) adds a new dimension to the manufacturing landscape by unlocking unprecedented opportunities for businesses. It encompasses various aspects of AI within the Industry 4.0 paradigm, which demonstrates how this powerful technology is already being leveraged by manufacturers to drive efficiency, enhance quality, and achieve better control over supply chains. Artificial Intelligence and Industry 4.0 can be called Industrial AI [11], and its impact on manufacturing artificial intelligence's effect on production can be prepared into five fundamental areas:

1 Predictive quality and yield
2 Human robotic collaboration
3 Generative layout
4 Supply chain for market change
5 Predictive maintenance

With the help of these, we will forecasting quality, decreasing production waste and increasing yield, manufacturing procedure in efficiently to enhance the sustainability. Experts have trouble explaining them, but industrial AI can use continuous multivariate analysis and machine learning algorithms that are uniquely skilled to thoroughly comprehend each manufacturing process to reveal the hidden causes of the perpetual manufacturing losses that producers face every day. This is accomplished by predictive quality and yield.

The International Federation of Robotics (IFR) reports that human-robot interaction is expected to grow in popularity this year. Millions of commercial robots had been in use across the world. AI will play a prime element in making sure of the protection of human employees, in addition to giving robots greater duty to make choices, which could similarly optimize techniques [12].

Market adaptation for supply chain management by using the Artificial Intelligence permeates the whole industry. AI algorithms optimize the delivery chain of producing operations and assist them with higher replies to and anticipating modifications inside the marketplace. This ground-breaking system aggregates market forecasts and establishes a rule set, categorizing them based on a variety of factors including date, geographical location, socioeconomic attributes, macroeconomic trends, political climate, weather patterns, and more. It gives the information to producers with valuable insights to optimize their inventory management strategies. Staffing strength consumption and uncooked materials make higher economic choices concerning the company's strategy.

Predictive analytics play a crucial role in supply chain management, demand predictions, and reaching the target appropriately. The system utilizes historical data and current availability to predict demand within the supply chain accurately, enabling the production of quality products tailored to meet the requirements of the intended users.

16.3 EVOLUTION OF INDUSTRY 4.0

Industry 4.0 is the next stage of the Industrial Revolution, which uses information and communication technologies to facilitate the intelligent networking of equipment and processes in the manufacturing sector. It bridges the gap between the real and the virtual. Factory-wide machine and system connectivity is the goal of Industry 4.0. AI allows machines to think and communicate with one another. Because of this, it can easily identify the faults in the system. The introduction of the AIML will enhance productivity and reduce the fault rate in the system to improve sustainability.

Digital technologies are part of our life in recent times. It used in all the routine jobs like teaching, grocery booking, ticket booking, daily activities, and health care apps. Like this every-day routine, it has become part of our life. With the tremendous usage, Industry 4.0 became the integral part of us. These devices will store the data from all the transactions, and it will be used for future analysis. The combination of IoT devices and the AIML technologies have changed the entire system to make it more intelligent.

Originally proposed by the German government in 2011, Industry 4.0 has now gained popularity because to the efforts of Klaus Schwab, Chairman of the World Economic Forum. By combining automation, cloud computing, the Internet of Things (IoT), and Artificial Intelligence (AI), it transforms business practices and efficiency on a massive scale [13]. Single-board computers, the Internet of Things (IoT), and machine learning (ML) are the backbone of Industry 4.0. To further meet the needs of remote monitoring and operational management, sensory detectors, controllers, and communication interfaces are implemented. The complex nature of today's industrial world makes it difficult to comprehend machinery and provide reliable interpretation and prognostication.

Industry 4.0 refers to the enhanced capability of industries in production aspects; because of including the smart devices, it reduces human interactions and improves production quality. The major characteristics of the industry are listed here:

16.3.1 Characteristics of a Smart Factory

The smart factory exhibits the certain properties which we can claim as the major services which are enabled by the Industry 4.0. The characteristics are listed as follows:

- Data analysis for better decisions
- IOT integration
- Customized production
- Aspects of the supply chain

The supply chain is the backbone for the majority of industrial operations. It requires a supply chain structure that is both open and effective and which, as a part of a comprehensive Industry 4.0 plan, must be included in manufacturing processes. IoT devices and communication networks, by adopting AIML, will become Industry 4.0 more powerful, which will be helpful for the further support of the supply-chain mechanism. The AI is more popular in the detection of anomaly in Industry 4.0.

16.3.2 Industry Revolution

Human evolution took millions of years, but the Industrial Revolution is very fast; it moved from one industry to another industry very rapidly. Here, as shown in Figure 16.2, the different industrial revolutions have taken places in history, starting from mechanical mass production to current cyber-physical systems. The term "Industrial Revolution" refers to a period of profound transformation brought about by the introduction of novel techniques and technology in the manufacturing and industrial sectors. A basic overview of the Industrial Revolution is presented here. As shown in the Figure 16.3, the different ages of the industries are shown. The different industry evolutions started in the year 1700 with steam engines, from there onwards, which shows the developments in the different industries. The developments of Industry 4.0 changed the entire phase of the industries.

In Industry 1.0, machines were powered by water and stream; this is an early phase of the Industrial Revolution, which was used to generate the mass production and speed of devices. Industry 2.0 is majorly focused on electrical engines and automobiles, which took place around 1870; the machines are run by the electricity. Industry 3.0 is focused on digital technology; more about automation and computers, it changes the entire industry. Industry 4.0 is focused on connectivity, where the actual world is linked to the digital one. This changes the entire thinking process by connecting; it changes the entire world with the fast communication updates and live streaming. The major advantages of the Industry 4.0 are listed as follows:

a Effectiveness and efficiency
b Cost minimization

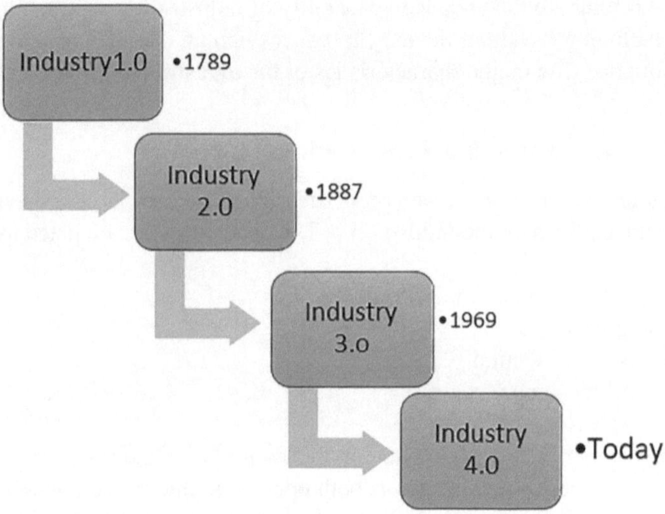

FIGURE 16.3 The Industrial Revolution.

c Emerging opportunities
d Customer satisfaction
e Safety
f Sustainability

Adopting the Industry 4.0 and Artificial Intelligence makes the industry change rapidly towards the development of things. Because of these changes, Industry 4.0 is more adopted and effective. The identification of a problem and detecting will improve the performance of the system. It will provide more sustainability in the industry.

16.3.3 Key Technologies

Machines, computers, and other electronic gadgets are all networked together in this Fourth Industrial Revolution. It takes all of our current digital technologies to the next level of optimization, hands-free. With this rapid revolution, the economic growth Industry 4.0 plays a crucial role; with this, every part of the world will be connected. The revolution in industry becomes more adaptable to users in all aspects. All these advantages come due to the integration of the AIML. The growth comes from data collection from different sources. The analysis and prediction part of AIML makes these devices more usable and user friendly. Different key technologies are involved in Industry 4.0, like IoT, AIML, networks, cloud, and big data. These key technologies make Industry 4.0 more usable and effective.

Usually referred to as the "Fourth Industrial Revolution", "Industry 4.0" signifies an ongoing changes in the production and processing industries driven by advanced digital technologies and the coordination of cyber-physical systems. This

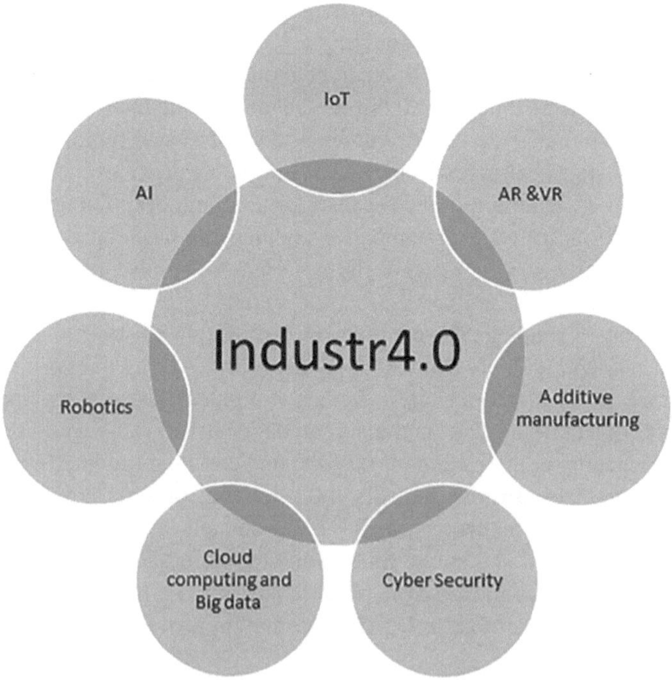

FIGURE 16.4 Key Technologies Associated with Industry 4.0.

revolutionary shift fundamentally alters how industries operate and encompasses the convergence of multiple latest and substantially upgraded technologies. The notion of Industry 4.0 places significant emphasis on the interconnectedness of machines, systems, and humans within a smart and autonomous manufacturing environment. Its primary objective is to establish highly flexible, efficient, and intelligent industrial systems capable of adapting to dynamic demands, optimizing production processes, and facilitating real-time data-dependent decision-making. The key characteristics and technologies linked with Industry 4.0 as shown in Figure 16.3. The key components used in Industry 4.0 are discussed as follows:

Connectivity: Communication is the basic requirement in Industry 4.0. Creating connections between elements can significantly boost productivity and improve the efficiency of the supply chain. Industry 4.0 makes use of IoT technology to link physical assets, machinery, and sensors, enabling the real-time collection and exchange of data. This seamless communication allows for better coordination, sharing of information, and control across different parts of the production system. The true power of Industry 4.0 lies in the accurate communication among interconnected devices, facilitated by cutting-edge protocols that ensure connectivity on a global scale.

Cyber-Physical Systems (CPS): CPS combines physical components (e.g., machines, robots) with digital technologies, allowing them to interact and cooperate with each other and with humans. The control and monitoring of the system becomes easy by adopting CPS. It helps the automation of the process, and human intervention can be reduced by further levels [14].

Big Data Analytics: Industry 4.0 utilizes advanced data analytics techniques to process and analyze large volumes of data produced by interconnected devices. Insights extracted from this data facilitate predictive and prescriptive analytics, enabling proactive decision-making, optimizing processes, and identifying patterns or anomalies. Enhanced power comes to Industry 4.0 from analytics, which makes the system more reliable and adaptable to all the users [15].

AIML: These technologies are employed in Industry 4.0 to enable systems to learn, adapt, and make autonomous decisions. AI-based algorithms can optimize production schedules, predict equipment failures, and perform quality control tasks. The anomalies can be identified by using these methods of learning of things and making the devices are more intelligent.

Robotics and Automation: Industry 4.0 relies on robotics and automation to enhance manufacturing processes. Automated vehicles, collaborative robots, automated assembly lines, lead to increased efficiency, precision, and safety. To increase the productivity without human intervention, robots will play a crucial role. The automation process helps to improve production.

IoTs: This enables us to connect every small device to networks, which are embedded into the electrical and mechanical machines to monitor the devices and record characteristics of the device.

16.3.3.1 Artificial Intelligence

Artificial Intelligence (AI) software creates robots that learn, plan, reason, share knowledge, and solve problems in human-like ways. The precise methods of AI and machine learning are employed to determine which set of rules is adept at uncovering the correct patterns, trends, and styles. This will enhance manufacturing processes and teams. Predictive maintenance/predictive renovation is one of the maximum fundamental and famous packages of industrial AI. Predictive renovation makes use of algorithms to expect the subsequent failure of a device; renovation approaches to save the failure. Predictive renovation structures rely upon machine learning strategies to formulate their predictions. The deliberate downtime, in lots of cases, by pre-empting a failure with a device, getting to know a set of rules structures, can keep featuring without useless interruptions.

Types of Learning: In the learning process, the devices can be updated by using different phases of learning in the system. In this chapter, machine learning algorithms are used to find the anomalies in the Industry 4.0. In machine learning, the types learning can be broadly classified as:

a Supervised learning
b Unsupervised learning
c Reinforcement learning

Machine learning algorithms are used to find the abnormal activities in the Industry 4.0. There are a variety of algorithms that are used to find the anomalies in the system. Each of the methods has its own advantage and disadvantage. The major goal of the early anomaly detection is to provide the sustainable solutions.

16.4 ANOMALY DETECTION

Identifying the anomalous patterns in the system is crucial, as Industry 4.0 generates a huge quantity of data, which contains normal patterns of data also, but very few records or devices malfunction. Identifying these patterns is crucial; we need to ensure which are the normal and abnormal patterns in the system. Anomaly behaviour deviates from normal activity. Suddenly, the change in devise function also can be treated as an anomaly.

The major anomaly behaviours in Industry 4.0 will be as follows:

i A module built for Industry 4.0 can identify problematic inputs and issue warnings
ii There are a number of things, including heat, vibration, and physical damage, that can cause machines to break down at any time
iii Predictions of the gadgets' remaining useful life
iv Interaction among the devices
v Communication protocol functionality

Any of the previously mentioned ones are deviated the Industry 4.0 functionality. The major role is to provide a sustainable solution for the system. The AI based approaches are used to detect these anomalies beforehand and provide sustainable solutions.

16.5 AI MODELS FOR ANOMALY DETECTION

Artificial intelligence (AI) has proven to be highly effective in anomaly detection across various domains. Anomaly detection involves identifying patterns or events that considerably depart from the expected norm or standard. AI-based anomaly detection systems leverage machine learning algorithms and statistical techniques to detect anomalies in data. There are several approaches to using AI for anomaly detection:

Supervised Learning: In this approach, anomalies are identified based on labelled data that includes both normal and anomalous examples. Machine learning models are trained on this data to learn the patterns of normal behaviour and then classify new instances as normal or anomalous. The major algorithms used for anomaly detection are Support Vector Machine (SVM), Random Forest, Decision Tree, and Naïve Bayes algorithms. These algorithms work only for the existing attacks, unable find the novel attacks. The risk of these algorithms is what kind of data is used to train the model will decide the attacks. It works well for known data and known attacks for limited training data.

Unsupervised Learning: It can identify the anomalies by analyzing labelled data containing both typical and abnormal instances. Machine learning models are trained on this dataset to recognize patterns of typical behaviour and subsequently classify new instances as either normal or abnormal. Common algorithms utilized for anomaly detection include SVM, Random Forest, Decision Tree, and Naïve Bayes. However, these algorithms are limited to detecting known attacks and may struggle to identify novel threats. The efficacy of these algorithms depends on the quality and variety of the training data; they perform well with familiar data and known attacks but may struggle with limited training data.

Semi-Supervised Learning: It works on combining both supervised and unsupervised learning. Initially, a model is trained on a labelled dataset of normal instances. The model then uses this knowledge to identify anomalies in unlabelled data. These kinds of algorithms are used in real time scenarios, trained with the existing data and trying to predict the real time attacks.

Deep Learning: To capture inside patterns for huge datasets, deep learning techniques are used, which are capable of handling large chunks of data; along with this data, the performance is also increased. Such models as neural networks can be applied to anomaly detection. Auto encoders, for example, are popular neural network architectures used for unsupervised anomaly detection. They are trained to learn the representation of normal data and are subsequently evaluated on how well they reconstruct new instances. Instances that are poorly reconstructed are considered anomalies. These algorithms become much more famous on anomaly detection for huge chunks of data.

Reinforcement Learning: In the reinforcement learning, agents are trained to take actions in the environment, when they encounter anomalous states or events that deviate from expected behaviour, they receive negative rewards. The agent learns to minimize these rewards in identifying the anomalies. Performance depends on the specific problem, available data, and the nature of anomalies being detected. Additionally, domain expertise is crucial for effective anomaly detection, as anomalies can vary widely across different domains.

Among these approaches, anomaly detection can vary based on the detection process. In Industry 4.0, devices are spread across the world. The point where the anomaly is detecting is important. Based on detection place the anomaly detection algorithms are further classified into the categories:

a Centralized anomaly detection
b Cloud-based anomaly detection
c Distributed anomaly detection

16.5.1 Anomaly Detection in Industry 4.0

In Industry 4.0, manufacturers generate huge volumes of data from machine sensors, product testers, and employees. Almost every aspect of manufacturing has a data stream. When analyzed, this data can provide a massive competitive advantage. If you spot an anomaly, an unexpected change in the usual data pattern, you might prevent a small issue from turning into a widespread, costly, time-consuming problem. It could be that machine settings have slipped or maintenance is needed. By knowing about it immediately and taking action, you can avoid degraded quality and potential profit loss. Early detection of issues can save time and money. Anomaly detection helps us to keep operations under control and efficiently managed, all without time-consuming manual processes. Many cost-saving opportunities are possible with technology. Global manufacturing leaders, including Hemlock Semiconductor, Siemens Mobility, and Brembo, all utilize TIPCO capabilities in anomaly detection use cases that help them make faster, smarter decisions, reduce costs, improve revenue, gain efficiencies, and reduce their impact on the environment.

With the tremendous growth and usage of the Artificial Intelligence and machine learning, we are able to identify anomalies patterns in Industry 4.0. With the help of a machine learning model using the neural processor, the system is able to recognize the anomaly. Anomaly detection is also quality inspection; it is not only about machine learning; it also comes with very high secure features that allow you to protect system and to provide sustainable solutions to the system.

Anomaly detection plays a crucial role in Industry 4.0, which focuses on the integration of advanced automation, data exchange, and real-time analytics in industrial settings. Anomaly detection in Industry 4.0 helps to identify abnormal behaviour, equipment malfunctions, process deviations, and other irregularities that can impact productivity, quality, and safety. Here are a few key areas where anomaly detection is applied in Industry 4.0:

Predictive Maintenance: To minimize the down time of the system, historical data is used to identify the anomalies by the using machine learning approaches. By analyzing sensor data, such as temperature, vibration, pressure, and energy consumption, anomalies can be detected and potential failures can be predicted. This allows for preventative maintenance, which cuts down on breakdowns and maximizes uptime.

Quality Control: Anomaly detection techniques are employed to detect deviations in production processes that may lead to defective products. By monitoring sensor data, machine vision systems, or other quality control measures, anomalies in product dimensions, surface defects, or other quality indicators are used to identify the anomalies in the system.

Supply Chain Management: Anomaly detection is used to monitor and optimize supply chain operations to promote the quality supply. By analyzing the data related to inventory levels, logistics, transportation, and demand patterns, anomalies such as delays, disruptions, or inefficiencies can be identified. This enables proactive management of supply chain risks and helps optimize the flow of materials and products.

Cyber Security: In Industry 4.0, where connectivity and data exchange are prominent, anomaly detection is critical for cyber security. Anomalies in network traffic, communication patterns, or system behaviour can indicate potential cyber threats, such as intrusions or attacks. By monitoring and analysing network data, anomalies can be detected, and appropriate security measures can be taken to mitigate risks.

Energy Management: Detecting anomalies in power consumption is crucial for optimizing energy usage and identifying abnormal patterns. By continuously monitoring energy data, anomalies like excessive consumption, equipment inefficiencies, or irregular usage patterns can be identified. This proactive approach enables organizations to improve energy efficiency, reduce costs, and support sustainability initiatives.

In each of these areas, AI models and algorithms, including those mentioned earlier (e.g., Isolation Forest, Auto encoders, LSTM), can be employed for anomaly detection. These models learn from historical data, sensor readings, or process parameters to identify deviations from normal behaviour, enabling proactive decision-making, improved operational efficiency, and enhanced productivity in Industry 4.0 environments. Anomaly detection makes use of a wide variety of artificial intelligence models and techniques. Just a few instances:

Forest Isolation: This method employs an unsupervised learning technique to detect outliers. By randomly selecting features and partitioning the data points into halves, it constructs a collection of isolation trees, each represented as a binary tree. Data points that require fewer splits to separate are classified as anomalies.

SVMs with Only One Class of Data: This model is trained exclusively on data from a single class, typically representing normal cases. It learns a boundary to encompass typical data points, while any outliers are identified for closer examination.

Deep learning techniques like auto encoders LSTM and GANS are also used to detect the anomaly patterns in the system. GANs have been used for anomaly detection by training the generator to capture the distribution of the normal data, and instances that the discriminator fails to differentiate as real or fake are considered anomalies. These are just a few examples of AI models and algorithms that can be employed for anomaly detection. The choice of model depends on the nature of the data, the availability of labelled or unlabelled data, and the specific requirements of the anomaly detection task.

16.5.2 The Advantages of AI

Using AI in Industry 4.0 anomaly detection has several advantages:

1 Enhanced capability for demand projection
2 Workplace safety through direct automation
3 Improved root cause analysis
4 Improved workforce intelligence
5 Increased yields
6 Improved workplace security

7 Reduced operational expenditures
8 Better detection of manufacturing flaws
9 Automated quality control
10 Higher supply chain efficiency
11 Enhanced productivity in the supply chain
12 Quick decision making

The usage of the AI in industry 4.0 for anomaly detection can be availed to improve the system performance to provide sustainable solutions to the environment [16, 17]. Quality control and anomaly are the major things in Industry 4.0.

16.5.3 Sustainability

Resource utilization and the abnormality of devices will play a crucial role in abnormality detection. The appropriate use of resources and productivity is the major goal of sustainable resources. Abnormal behaviour needs to be identified as early as the system is deviated to warn the control system to overcome the fault. Most industries have manual-level abnormality checking; this process is tedious and cumbersome. In Industry 4.0, with the automation of things and the availability of devices at cheaper cost, abnormality detection becomes an easier task. The sensors and IoT devices will track the performance and utilization of devices to detect the early abnormalities.

The major goal of the Industry 4.0 is to provide sustainable solutions. The adoption of AIML methods can track the anomaly from the IoT data to provide sustainability to the system [18].

16.6 CONCLUSION

The goal of Industry 4.0 is to transform traditional industries into intelligent, interconnected, and data-driven ecosystems, enabling increased productivity, flexibility, quality, and sustainability. The revolution in the manufacturing process done by Industry 4.0, supply chain management, product customization, and quality control are in a new era of industrial efficiency and innovation.

Industry 4.0 relies on an anomaly detection approach, enabling us to ensure the sustainability of devices within the Industry 4.0 framework. Artificial intelligence is used to make the production process more absorbed and robust to overcome the appropriate anomalies which can be used to produce the appropriate values. They can make production process more appropriate and sustainable. Anomaly detection using AIML approaches is more accurate; early detection of these anomalies can make the systems and environment more sustainable.

REFERENCES

[1] Rubmann, Michael, et al. "Industry 4.0: The future of productivity and growth in manufacturing industries." Boston Consulting Group 9.1 (2015): 54–89.
[2] Kashpruk, Nataliia, Cezary Piskor-Ignatowicz, and Jerzy Baranowski. "Time series prediction in Industry 4.0: A comprehensive review and prospects for future advancements." Applied Sciences 13.22 (2023): 12374.

[3] Javaid, Mohd, et al. "Artificial intelligence applications for Industry 4.0: A literature-based study." Journal of Industrial Integration and Management 7.1 (2022): 83–111.
[4] Sambangi, Swathi, Lakshmeeswari Gondi, and Shadi Aljawarneh. "A feature similarity machine learning model for DDoS attack detection in modern network environments for Industry 4.0." Computers and Electrical Engineering 100 (2022): 107955.
[5] Justus, Vivek, and G. R. Kanagachidambaresan. "Intelligent single-board computer for Industry 4.0: Efficient real-time monitoring system for anomaly detection in CNC machines." Microprocessors and Microsystems 93 (2022): 104629.
[6] Mian, Tauheed, et al. "Artificial intelligence of things based approach for anomaly detection in rotating machines." Computers and Electrical Engineering 109 (2023): 108760.
[7] Ullah, Waseem, et al. "Artificial intelligence of things-assisted two-stream neural network for anomaly detection in surveillance big video data." Future Generation Computer Systems 129 (2022): 286–297.
[8] Ahmad, Tanveer, et al. "Energetics systems and artificial intelligence: Applications of Industry 4.0." Energy Reports 8 (2022): 334–361.
[9] Alenizi, Farhan A., et al. "The artificial intelligence technologies in Industry 4.0: A taxonomy, approaches, and future directions." Computers & Industrial Engineering (2023): 109662.
[10] Keleko, Aurelien Teguede, et al. "Artificial intelligence and real-time predictive maintenance in Industry 4.0: A bibliometric analysis." AI and Ethics 2.4 (2022): 553–577.
[11] Ahmed, Imran, Gwanggil Jeon, and Francesco Piccialli. "From artificial intelligence to explainable artificial intelligence in Industry 4.0: A survey on what, how, ` where." IEEE Transactions on Industrial Informatics 18.8 (2022): 5031–5042.
[12] Angelopoulos, Angelos, et al. "Tackling faults in the Industry 4.0 era—a survey of machine-learning solutions and key aspects." Sensors 20.1 (2019): 109.
[13] Jan, Zohaib, et al. "Artificial intelligence for Industry 4.0: Systematic review of applications, challenges, and opportunities." Expert Systems with Applications (2022): 119456.
[14] Alohali, Manal Abdullah, et al. "Artificial intelligence enabled intrusion detection systems for cognitive cyber-physical systems in Industry 4.0 environment." Cognitive Neurodynamics 16.5 (2022): 1045–1057.
[15] Stojanovic, Ljiljana, et al. "Big-Data-Driven Anomaly Detection in Industry (4.0): An Approach and a Case Study." 2016 IEEE international conference on big data (big data). IEEE, 2016.
[16] Ahmad, Tanveer, et al. "Artificial intelligence in sustainable energy industry: Status quo, challenges and opportunities." Journal of Cleaner Production 289 (2021): 125834.
[17] Jan, Zohaib, et al. "Artificial intelligence for Industry 4.0: Systematic review of applications, challenges, and opportunities." Expert Systems with Applications 216 (2023): 119456.
[18] Mezair, Tinhinane, et al. "A sustainable deep learning framework for fault detection in 6G Industry 4.0 heterogeneous data environments." Computer Communications 187 (2022): 164–171.

17 Future of Industry 5.0 in Society 5.0

Human-Computer Interaction-Based Solutions for Next Generation

Amit Kumar Tyagi and Meghna Manoj Nair

17.1 INTRODUCTION

The global landscape of industries, manufacturing unities, and societies have been exposed to massive transformation in the recent few years with the rise of Industry 5.0. This phase is an upgrade from its previous stage, Industry 4.0, and focuses majorly on the effective utilization and integration of advanced science and technological strategies with a human-centric aspect. This is distinctly tied to the core concept of Society 5.0, which is a term used to describe the societal framework that harnesses tech developments to enrich the lifestyle and wellness of individuals and as community as a whole [1]. The Industrial Revolution has several characteristics, including technological, cultural, and socioeconomic ones. The very first industrial revolution, Industry 1.0, was born towards the end of the eighteenth century and continued for the next five decades. This phase was merely identified by the use of mechanized production, coal, steam power, and the establishment of some of the very first factories and manufacturing companies. Industry 2.0, which followed this period, was fuelled by the discovery of electricity, and it led to the mass production of various products in industries including those of automobiles and vehicles. While Industry 3.0 referred to a period that gradually adopted the use of electronics and the introduction of automation in industries, Industry 4.0 was all about the interconnectivity of machine equipment and devices through Internet of Things (IoT), Artificial Intelligence (AI), Big Data, etc. This is the phase that saw a spike in the rise of smart factories and intelligent equipment that drew attention to the need for reduced human interference. The stage that we're currently at represents the Fifth Industrial Revolution, which rests on the major pillars of sustainability, resilience, and human centricity. These features are further powered by the local environment and the stability of the country and its governance [2, 3]. Industry 5.0 is an inclination towards combining the strengths and benefits of machines as well as humans throughout the process of product formation and indicates a departure from the previous industrial revolution stage that sheerly focused on automation and data-driven approaches. This transition is due to the acceptance and acknowledgement of the limitations and

DOI: 10.1201/9781003473886-17

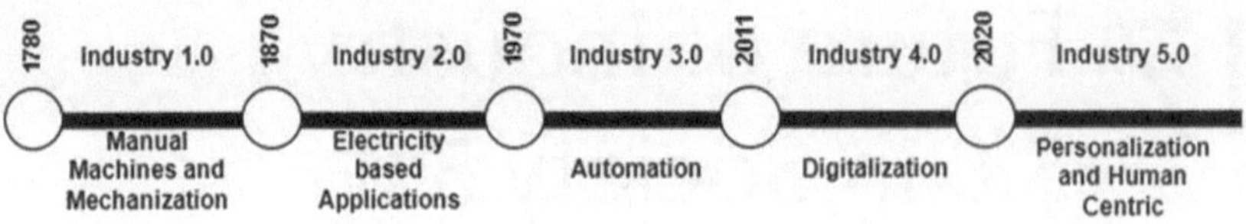

FIGURE 17.1 Timeline of Industrial Revolution.

drawbacks of highly autonomous devices. The current industrial phase envisions a future where the impact of technology isn't leveraged for efficiency alone but also for the well-being and cohesive adoption in the bigger society [4]. Figure 17.1 elucidates the previously described timeline of industrial revolution.

On a parallel note, Society 5.0 is a related and underlying concept that was put forth by the then Japanese prime minister Shinzo Abe to signify an evolution that was strongly sustained by technological advancements. It aims to ease in the integration of cyberspace and physical space through AI, IoT, and big data so as to curate a highly connected and intelligent society. This concept is an umbrella term that overarches the framework within which Industry 5.0 co-exists, highlighting the alignment of tech progress with human goals and welfare. In this dynamic topography where Industry 5.0 and Society 5.0 go hand-in-hand with each other, the importance of Human Computer Interaction (HCI) is deep rooted in this domain. Right from the initial days of punch cards all the way to modern day touch screens, HCI has also undergone a highly transformative journey by adapting to the growing needs and wants of humans. This concept adopts a deeper importance as it takes up the role of the medium through which humans engage and interact with rampantly growing technologies. The interfaces used in Industry 5.0 are not just devised to be user-friendly but they also act as the conduit to support humans in navigating through complex and sophisticated systems easily. It extends way beyond conventional interfaces, tapping into immersive technologies such as virtual/augmented reality, voice recognition, gesture recognition, etc., and showcases how HCI is a cornerstone for ensuring enhanced human experiences than overshadowing narratives [5]. This chapter discusses the various aspect of Industry 5.0, Society 5.0, HCI, and how they're strongly interlinked to each other while paving way for futuristic developments.

Hence, this chapter is divided into various sections, as follows. Section 2 focuses on the various existing researches and studies in this field. Sections 3 and 4 highlight the origin, importance, and core values of Industry 5.0 and Society 5.0, respectively, while section 5 sheds light on how HCI integrates to Society 5.0 and takes a core shape there. Section 6 discusses the possible future trends in this field, and section 7 concludes the work.

17.2 LITERATURE REVIEW

The domain of industrial and societal revolution has found its roots in a global landscape showcasing a paradigm-based shift that is transformative and constantly focused on how to improve the evolving lifestyles of humans by harmoniously integrating advanced technologies and devices. The conjunction of Industry 5.0 at the very core of Society 5.0 yearns for the integrity provided by HCI, and this is a domain that has garnered quite the attention from the academia and industry leaders. Some of the relevant researches, studies, and surveys are discussed in this section. Y. Kambayashi et al. gives an in-depth overview and important information on the Society 5.0 framework while focusing on the effortless integration of the cyberspace and physical space [6]. They've argued that such integrations of advanced technologies not only accelerate the industrial process, but it also complements the well-being of humans in a broader society. One of the other research projects digs deeper into the intricacies of HCI and how it relates to the domain of Industry 5.0 [7].

The authors of this work debate that HCI is important in accommodating impactful and efficient interactions between humans and machines with the aim of achieving certain goals. The major emphasis was on the pressing need for intuitive and easy-to-use interfaces that ultimately enhance the experience and how this aspect plays a major role in resonating with the key principles of Industry 5.0. Another lens through which this domain can be assessed is through that of ethical issues and implications. R. Lee et. al. surveys and examines the possible societal consequences and impacts of human-machine coordination [8]. The authors of this work towards establishing a holistic perspective to HCI by including the privacy, security, and ethical implications in a socio-cultural environment. In [9], the researchers have discussed the human-centric design approaches with respect to Industry 5.0 with a key concentration on how HCI shapes user experiences. The research sheds light on various design methodologies and conventional user interface design guidelines that prioritize things from the users' point of view, which, in turn, ensures that Industry 5.0 upskills more towards enhancement of technologies than replacement of the same. This indicates how HCI is a bridge that fosters a reciprocal relation between humans and technology. One of the other interesting works in this domain is a study carried out by L. Wang et al., which forages into the concept of cognitive ergonomics and its role in Industry 5.0 and Society 5.0 [10]. This is an important aspect when conducting research about industrial and societal revolution because this infusion helps invigorate and optimize human chores and doings. Their research searches for the various ways through which HCI can actively contribute towards increased efficiency and reduced cognitive load within the Industry 5.0 context.

The next major aspect to be addressed in this field is that of sustainability and growth, as highlighted by J. Park et al. through their work that examines the intuitive role of HCI in pursuing a sustainable approach towards the fifth revolutionary phase of industries and societies in common [11]. Their research delves into how HCI can effectively contribute to resource efficiency, reduction in waste and excessive production, and encourage the art of conscious decision making in smart manufacturing. N. Gupta et al. has put forward insightful approaches on the various challenges and possible opportunities associated with HCI in the revolutionary phase of both industries and societies and how it ties both together [12]. The authors have analyzed the existing HCI situation and have recognized the need to bridge the skill gap while taking into account the ethical issues and the potential landscape for social impact. The study gives a highly nuanced and detailed breakdown of the complexities involved in extracting the full potential of HCI within the industrial and societal context. One of the other interesting aspects highlighted by the researchers in [13] is the cultural dimension involved. Exploring the dynamic cultural dimension helps provide how HCI converges with the human norms and regulations by being culturally sensitive to guarantee inclusivity and acceptance. The research study put forth by H. Li et al. gives a fine overview of the human engineering factors that correlate with Industry 5.0 and Society 5.0, respectively [12]. The authors appreciate the contribution of these driving factors in accelerating and optimizing the collaboration between humans and developing technologies and elaborate on how these elements can upscale productivity, safety, and overall user experience. The increasing amount of research on the complex connections between Industry 5.0, Society 5.0, and HCI is aided by these extra evaluations. A more thorough knowledge of how human-computer interactions are not just technological but also intricately entwined with cultural settings and ergonomic issues may be gained by looking at cultural implications and the function of Human Factors Engineering.

17.3 INDUSTRY 5.0—EMERGENCE, ROLE OF ETHICS IN INDUSTRY 5.0, HUMAN-CENTRIC SMART MACHINE (HSM APPROACH)

The start of Industry 5.0 has marked a turning point in the industrial revolution landscape with a strong emphasis on its departure from the highly automated systems during the preceding stage of the Fourth Industrial Revolution to a framework that inclines way beyond technological creativity and innovation to bring in a harmonious collaboration between humans and machines. The core value here lies in a human-centric approach, and this section delves deeper into the evolution of Industry 5.0 and certain important aspects that contribute to it. The emergence of Industry 5.0 may be attributed to a widespread recognition within the international industrial community that fully automated systems, although very effective, do not possess the flexibility, originality, and sophisticated judgement that characterize human intelligence [14]. The term was originally coined about a decade ago to represent the concept of a collective acknowledgement within the industrial society that majorly worked towards automating systems, is highly efficient, and goes hand in hand with the adaptability, inventive thinking, and nuanced decisive skills from human intelligence at a global horizon. The evolution of industrial manufacturing in its fifth cycle is not just about rampant technological developments; its key driving factor is to curate a space that's inclusive, adaptive, and, most importantly, a human-centric industrial environment [15].

One of the salient features of Industry 5.0 is its commitment and consideration for ethical aspects. The integration of AI, IoT, big data, and other growing tech strategies within the industrial sector often raise a number of ethical questions especially in terms of privacy, security, transparency, and the societal impact of utilizing these tools and tech gizmos. However, there's a conventional ethical disclosure followed by the regulatory frameworks within the industrial community that calls for a proactive, sustainable, and anticipatory tactics to ensure that the tech advancements resonate with the values of humans while simultaneously contributing to the overall well-being of the society. As a result of organizations' growing recognition of the need for responsible innovation, ethical issues are being integrated into all phases of the technology life cycle, from design and development to implementation and continuous usage. This ethical framework serves not only to limit possible risks and liabilities but also to build public trust and acceptance of Industry 5.0 projects, guaranteeing a more sustainable and harmonious integration of technology and society [16]. The core perspectives of Industry 5.0, which revolve around people centricity, sustainability, and resilience, are illustrated in Figure 17.2.

The Human-Centric Smart Machine (HSM) approach is the core ideology of Industry 5.0, a paradigm that sets this period apart from its predecessors. Deeply contrasted with the ideology of Industry 4.0, HSM is all about a collaborative and fusion-based model that is built to augment human capabilities rather than outgrow/replace them. Through advanced sensory gadgets, communication and information gathering tools, and decision-making capabilities, HSMs tend to develop a mutual and complementary relation between humans and machines. In terms of priorities, HSMs tend to prefer user experience and the broader socio-cultural context within which the technology functions. The interfaces used are designed to be intuitive, flexible, and responsive to human needs. The point to be highlighted here is that the design philosophy followed here extends beyond that of mere functionality as it

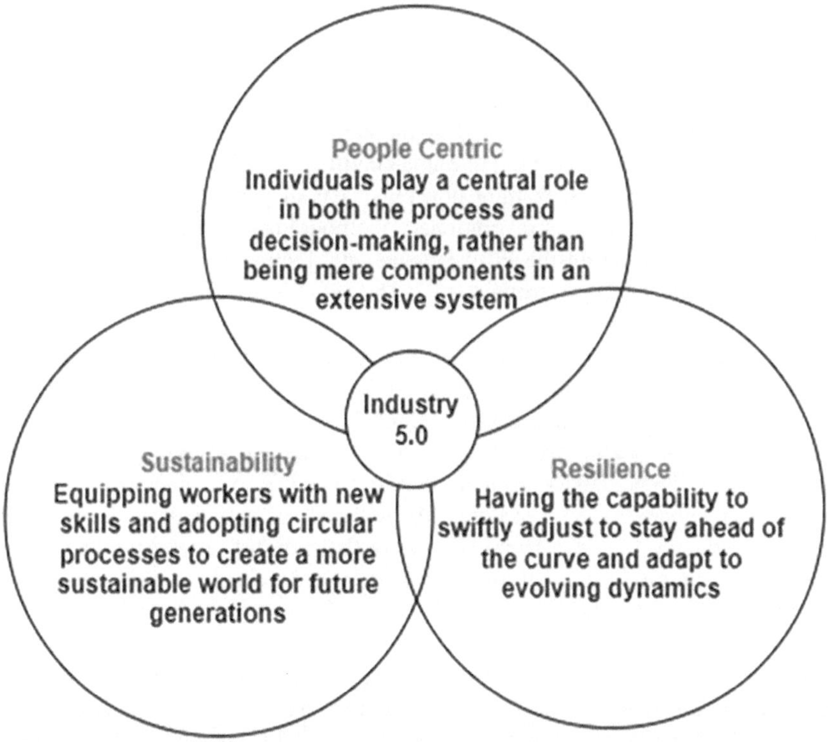

FIGURE 17.2 Core Principles of Industry 5.0.

envisions to enhance the industrial efficiency while empowering associated individuals by augmenting their cognitive abilities and creativity into the production system. In simpler words, HSMs tend to take up the role of companions throughout the industrial journey and cease to be just mere tools [17]. This contributes to creating an atmosphere in which human-machine cooperation is not only productive but also fulfilling, raising the standard of work in the end and creating a sense of shared accountability.

In summary, Industry 5.0 represents a turning point in the development of the industrial sector by stressing human-centeredness, ethical issues, and teamwork. Industry 5.0 represents a conscious shift in the industrial environment towards inclusive and responsible behaviours, rather than just a technical breakthrough. This idea is demonstrated by the integration of Human-Centric Smart Machines, which show a commitment to improving human skills while resolving the ethical conundrums posed by cutting edge technologies. The continuous evolution of Industry 5.0 emphasizes how important it is to carefully balance the advancement of technology with the timeless values that characterize mankind as a whole. This complex industrial paradigm is a reflection of a dedication to constructing a future where human-machine cooperation results in good advances for society as a whole [18].

17.4 SOCIETY 5.0

The latest paradigm changes in human society's evolution, known as Society 5.0, is the result of the integration of cutting-edge technologies like big data, artificial intelligence, and the Internet of Things (IoT). This idea, which originated in Japan in reaction to the difficulties presented by the digital age, aims to balance technical development with the improvement of human civilization. This section will examine the formation of Society 5.0, examine how ethics has shaped its course, and examine the primary objectives of this revolutionary methodology. The term "Society 5.0" was coined in Japan during the Fifth Science and Technology Basic Plan summit initiated by the Japanese government in 2016 [19]. It's a concept that envisions a society where technology and innovation is effortlessly integrated into human life, surpassing the drawbacks and limitations of the preceding industrial revolutions. The base foundation and core values of Society 5.0 builds off the components of Industry 4.0, which emphasize a human-centric approach to technology. A defining feature of Society 5.0 is the confluence of many cutting-edge technologies. Process automation is a key component of artificial intelligence (AI), which increases the intelligence and responsiveness of systems. As IoT devices proliferate, commonplace things become more linked, forming a network that supports effective data sharing. Large datasets may be used with big data analytics to extract valuable information and guide decision-making.

Society 5.0 also has an ethical aspect to it, which becomes increasingly important as it consists of issues related to privacy, security, transparency, and the equitable distribution of advantages [20]. It's essential to note that the development and implementation of technologies call for a succinct balance between creativity and defending basic human rights that are fundamental to each individual. When it comes to the collection of personal data and information, privacy issues pave their way through.

One of the key ethical challenges is to strike a balance between the convenience and comfort put forth by interconnected devices and safeguarding one's privacy. Security issues gain heightened importance when important infrastructure tends to depend more on interconnected digital devices, as there's more chance of data leakages and interceptions, which calls for robust cybersecurity strategies. One of the other imperatives for ethical issue is the transparency involved in algorithmic decision-making process. This is because AI systems often make decisions at a scale that impacts both individuals or large communities. The element of transparency is essential to ensure that the advantages offered by Society 5.0 are equally accessible to different segments of a society from an obligatory manner [21]. Figure 17.3 illustrates the various aspects of a society that are significantly impacted when considering the principles of the fifth societal revolution such as caregiving and healthcare, industrial automation, sustainable energy resource utilization, and streamlining agricultural activities for optimized yields.

The main objective of Society 5.0 is to establish a human-centred society that uses technology to improve everyone's quality of life. This includes societal peace, environmental sustainability, human well-being, and economic advancement. Through the smooth integration of digital technology into different facets of life, Society 5.0 aims to create a society where innovation promotes equitable growth and tackles societal issues. The notion highlights the amalgamation of the tangible and virtual domains, culminating in the development of intelligent transportation networks, smart cities, and advancements in healthcare [22]. Society 5.0 seeks to address difficult problems, including resource depletion, ageing populations, and climate change

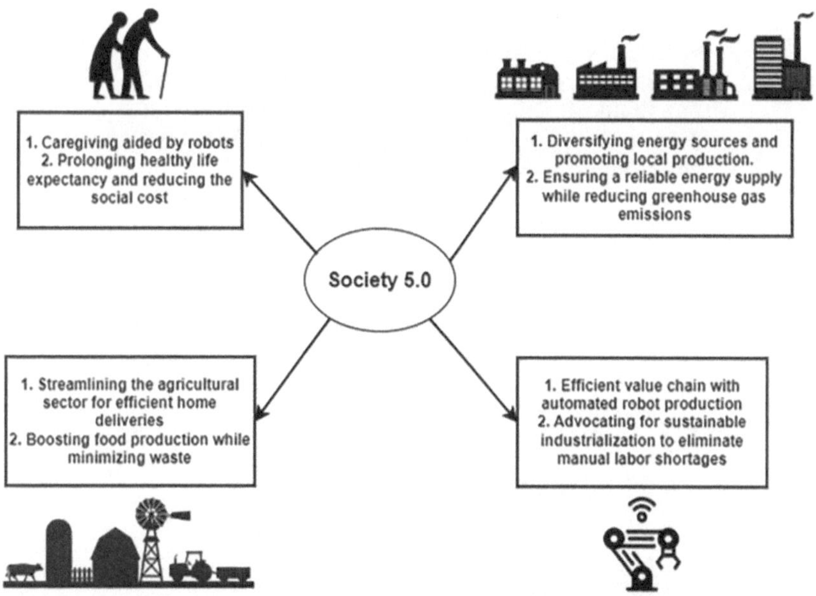

FIGURE 17.3 Major Components of Society 5.0.

using cutting edge technologies. Society 5.0 is a paradigm shift that combines technical breakthroughs with ethical issues, going beyond the boundaries of the merely technological sphere. Ensuring the conscientious and moral advancement of Society 5.0 is important as we manoeuvre through the intricacies of this revolutionary period. Society 5.0 aims to create a future in which technology is used as a potent instrument for societal progress rather than as a cause of conflict by placing a high priority on human well-being, inclusion, and sustainability.

17.5 SOCIETY 5.0, INDUSTRY 5.0, AND HCI INTEGRATION

As the point of intersection between digital systems and human demands, human-computer interaction (HCI) is at the forefront of technological integration. HCI is a key component of Society 5.0 and Industry 5.0, facilitating the smooth integration of cutting-edge technology with human-centred design concepts. This essay examines how HCI fits into Society 5.0 and Industry 5.0, emphasizing important components and offering illustrations of how HCI advances these paradigm-shifting ideas [23]. Figure 17.4 elucidates some of the major pillars and aspects of HCI that are closely considered during its integration with Society 5.0 and Industry 5.0, such as information security, ethnography, human factors, social psychology, design and layout, etc.

As elaborated in the previous sections, Society 5.0 is an umbrella term for the development of a society that has seamless integrations of technology to ease human

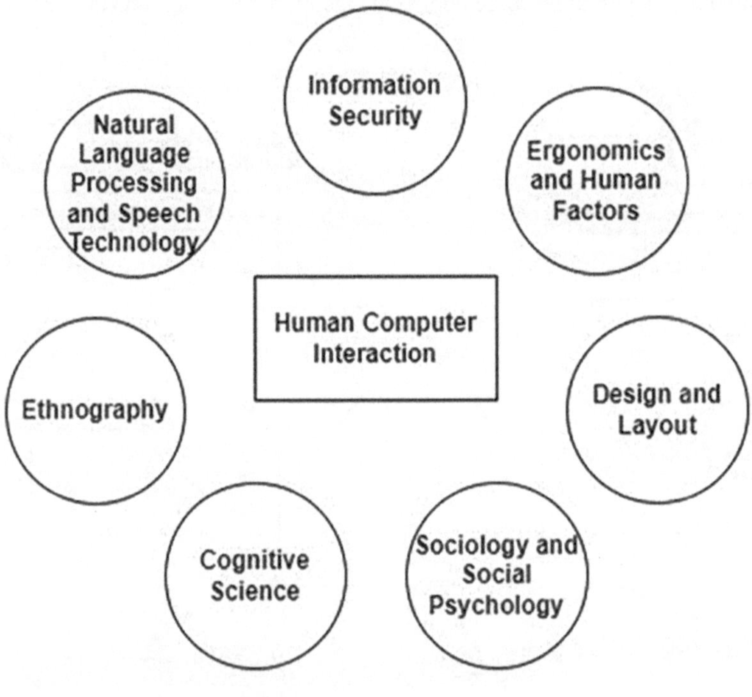

FIGURE 17.4 Core Pillars of Human-Computer Interaction.

life. It's all about curating an intelligent society where technology works hand in hand with human needs. HCI plays a significant role in supporting the vision of Society 5.0, as it helps lay down the basic principles that help integrate technology that is user-friendly, accessible, convenient, and, most importantly, aligns with social norms. HCI consists of various aspects and components which cover a wide range of tech integrations. Augmented Reality (AR) and Virtual Reality (VR) technologies are used by HCI in Society 5.0 to create interactive and immersive experiences. For instance, residents in smart cities may get real-time information on public transit, environmental conditions, and services by using AR applications on their smartphones [24]. VR may be used in healthcare, education, and entertainment to improve user experiences and encourage participation. One of the other aspects is Natural Language Processing (NLP) and voice interaction [25]. This is a key strategy that helps facilitate effortless communication and interaction between the devices and humans. Voice instructions are recognized by virtual assistants and smart home appliances, increasing the accessibility of technology for a wider variety of consumers. This improves the way people interact with smart homes, workplaces, and public areas. The use of wearable technology and biometric sensors has grown over the years and forms a key part of how HCI contributes to the growth of Society 5.0. The integration of such tech gizmos enhances personalization and health monitoring, be it for tracking individual's health metrics, providing customized dietary or physical activity recommendations, etc. The well-being of people in a technologically evolved society may be enhanced by HCI, as demonstrated by this integration [26].

The cooperative connection between people and machines in the production industry is the main emphasis of Industry 5.0. In order to create an industrial environment that is more responsive and efficient, human workers must be able to engage with intelligent equipment in a seamless manner. This is where HCI comes into play. Similar to how HCI plays a key role in Society 5.0, it plays a significant role in Industry 5.0 as well. Robots and automated machines take up central roles in the Fifth Industrial Revolution. On the production floor, HCI makes it easier for people and robots to interact naturally. With sophisticated sensors and user-friendly interfaces, collaborative robots, or co-bots, may operate alongside human workers. A collaborative and flexible industrial environment is encouraged by the design of user-friendly interfaces, which allow workers to operate and programme these robots without the need for specialized technical expertise [27]. Intelligent decision support systems is another way that HCI can be used for integrating decision support systems in smart industrial units that complement human operators in making well informed decisions. Employees can examine large, complicated data sets rapidly, thanks to sophisticated user interfaces and data visualisation tools. Operators in smart manufacturing, for instance, may monitor production processes, spot abnormalities, and streamline workflows with the help of user-friendly interfaces. Another aspect is that of digital twins combined with AR, and this transforms the manufacturing landscape to a great extent [28]. Through the use of digital representations of actual items, HCI improves the design and prototyping processes for employees. Employees may view and collaborate with digital twins of items using augmented reality interfaces, which improves teamwork in manufacturing and design.

The concepts of Industry 5.0 and Society 5.0 are tightly knit, and it paves a path that focuses on the shifting perspectives of humans with respect to their interaction and use of technology. At the core of this symbiosis is one of the key aspects of HCI, which plays an instrumental role. In terms of collaborative decision making, HCI plays an important part that blends the boundaries between societal and industrial developments to a certain extend. Through interactive platforms, individuals in Society 5.0 are enabled to voice their views and participate in the creation of policies. In Industry 5.0, HCI makes sure that intelligent machines and human operators work together to make decisions. It does this by using user-friendly interfaces to analyze data, spot trends, and improve industrial processes. Thus, the incorporation of HCI produces a dynamic feedback loop in which industrial decisions are influenced by social requirements, and societal well-being is enhanced by industrial achievements. One of the other key features made possible by HCI is the effortless collaboration between humans and robots. A fundamental aspect of human-robot interaction (HCI) is the development of smooth human-robot collaboration. This kind of cooperation is seen in Industry 5.0 on the production floor, where robots and humans collaborate together. By enabling workers to interact and manage robots in a natural way, HCI-driven interfaces promote an environment that optimizes the capabilities of both human and robotic capacities. This cooperative strategy transcends the industrial sector as, in Society 5.0, HCI allows citizens to engage with intelligent transportation systems, automated public services, and smart infrastructure. With respect to Society 5.0 and Industry 5.0, HCI also contributes significantly to the curation of personalized and tailored experiences and services. AI, big data, and data analytics can be easily integrated into various tech gadgets via HCI, and this ensures a customized interaction for users that caters to individual preferences and requirements. For example, HCI-driven interfaces in smart cities provide people tailored transit, healthcare, and entertainment recommendations. HCI makes sure that employees in Industry 5.0 receive individualized assistance and feedback, maximizing their output and sense of fulfilment at work [29–31].

17.6 FUTURE TRENDS

As we delve into the future of technological evolution, the symbiotic relationship between Industry 5.0 and Society 5.0 becomes increasingly apparent. The convergence of these paradigms lays the groundwork for a future where human well-being and industrial innovation are seamlessly intertwined. In this chapter, we provide the anticipated trends of Industry 5.0 within the context of Society 5.0, with a focus on Human-Computer Interaction (HCI) as a cornerstone for next-generation solutions.

- Advanced Collaborative Robotics: The development of collaborative robots is a major trend in Industry 5.0, which is framed within the context of Society 5.0. The creation of user-friendly interfaces that provide smooth communication between sophisticated robotic systems and human workers will be greatly aided by HCI. Collaborative robots with improved sensory capacities and natural language processing will be available in the future, creating a setting where humans and machines may work together easily in smart homes, public areas, and manufacturing floors [32].

- Enhanced Human-Machine Decision Making: Industry 5.0's use of HCI will improve how humans and machines make decisions together. Employees that have access to intelligent decision support systems with intuitive user interfaces will be better equipped to analyze complicated data, spot trends, and reach well-informed conclusions. HCI-driven platforms will help people in Society 5.0 by facilitating participatory decision-making and giving them the opportunity to influence governmental policies and services [33–40].
- Customizable Smart Environments: Future trends in Industry 5.0 will see the creation of personalized smart environments where HCI tailors interactions to individual preferences. Smart factories will utilize HCI to customize workstations, optimize workflows, and enhance the overall work experience for employees. Similarly, in Society 5.0, HCI will contribute to the development of smart cities that provide citizens with personalized services, from transportation to healthcare, based on their preferences and needs [34].
- Ethical and Inclusive issues: Emphasizing inclusive and ethical design is a key future trend in HCI for Industry 5.0 in Society 5.0. Privacy, security, and accessibility will be given top priority in HCI solutions as technology becomes more extensively incorporated into industry and society. Ensuring that the advantages of Industry 5.0 are equitable and available to all segments of society will require designing interfaces that take into account a wide range of user demands and values [35].

The future trends of Industry 5.0 within the landscape of Society 5.0 are intrinsically tied to the evolution of Human-Computer Interaction. As collaborative robotics advance, decision-making processes become more intelligent, environments become personalized, AR and VR are integrated, and ethical design principles are emphasized, HCI will be the driving force shaping the next generation of solutions. The synergy between Industry 5.0 and Society 5.0, facilitated by HCI, promises a future where technological advancements prioritize human well-being, inclusivity, and continuous innovation.

17.7 CONCLUSION

In summary, this study has examined the potential course of Industry 5.0 in the framework of Society 5.0, emphasizing the important role that Human-Computer Interaction (HCI) will play in forming the solutions of the future. The symbiotic link between business and society is becoming more and more apparent as we look to the future, with HCI emerging as the key that unites technology progress with human needs. The trends covered highlight the transformative potential of HCI in fostering a future where Industry 5.0 not only drives industrial innovation but also contributes to a human-centric and inclusive societal landscape. These trends range from personalized smart environments and ethical design principles to advanced collaborative robotics. This work shows the way to a future where HCI is seamlessly integrated with modern technology and empathy.

REFERENCES

[1] Huang S, Wang B, Li X, Zheng P, Mourtzis D, Wang L. Industry 5.0 and Society 5.0—comparison, complementation and co-evolution. Journal of Manufacturing Systems. 2022 Jul 1;64:424–428.

[2] Carayannis EG, Morawska-Jancelewicz J. The futures of Europe: Society 5.0 and Industry 5.0 as driving forces of future universities. Journal of the Knowledge Economy. 2022 Dec;13(4):3445–3471.

[3] Tyagi AK, Lakshmi Priya R, Mishra AK, Balamurugan G. Industry 5.0: Potentials, issues, opportunities, and challenges for Society 5.0. Privacy Preservation of Genomic and Medical Data. 2023 Oct 17:409–432.

[4] Rane N. ChatGPT and similar generative artificial intelligence (AI) for smart industry: Role, challenges and opportunities for Industry 4.0, Industry 5.0 and Society 5.0. Challenges and Opportunities for Industry. 2023 May 31;4.

[5] Paschek D, Luminosu CT, Ocakci E. Industry 5.0 challenges and perspectives for manufacturing systems in the Society 5.0. Sustainability and Innovation in Manufacturing Enterprises: Indicators, Models and Assessment for Industry 5.0. 2022:17–63.

[6] Kambayashi Y, Terada T, Motohashi K. Society 5.0: Japan's vision for a super smart society. Journal of Science Policy & Research Management. 2017;1(1):1–5.

[7] Johnson S, Wang J. The role of human-computer interaction in Industry 5.0: A review. International Journal of Human-Computer Interaction. 2019;35(12):1041–1054.

[8] Lee R, Kim Y, Choi J. Ethical considerations in human-computer interaction for Industry 5.0. Journal of Business Ethics. 2021;170(2):253–268.

[9] Chen M, Zhang Y, Hu Q, Ma Y. Human-centric design principles for Industry 5.0: A review. Frontiers in Mechanical Engineering. 2018;4:18.

[10] Wang L, Guo Y, Chen G. Cognitive ergonomics in Industry 5.0: A Comprehensive review. Human Factors and Ergonomics in Manufacturing & Service Industries. 2019;29(3):176–191.

[11] Park J, Kim J, Kim S. Human-computer interaction for sustainable development in Industry 5.0. Sustainability. 2020;12(9):3751.

[12] Gupta, N, Singh, P, Jain, P. Human-computer interaction in Industry 5.0: Challenges and opportunities. Computers, Materials & Continua. 2021;67(1):1165–1182.

[13] Rahman, A, Zhang, Q, Chua, T. Cultural implications of human-computer interaction in Industry 5.0. International Journal of Human-Computer Studies. 2022;151:102612.

[14] Brückner A, Hein P, Hein-Pensel F, Mayan J, Wölke M. Human-centered HCI practices leading the path to Industry 5.0: A systematic literature review. In International Conference on Human-Computer Interaction (pp. 3–15). Cham: Springer Nature, 2023.

[15] Garcia A, Quartulli M, Olaizola IG, Barandiaran I. Relevance of visualization and interaction technologies for Industry 5.0. Proceedings. http://ceur-ws. org. 2022;1613:0073.

[16] Passalacqua M, Pellerin R, Doyon-Poulin P, Del-Aguila L, Boasen J, Léger PM. Human-centred AI in the age of Industry 5.0: A systematic review protocol. In International Conference on Human-Computer Interaction (pp. 483–492). Cham: Springer Nature, 2022.

[17] Kaasinen E, Anttila AH, Heikkilä P, Laarni J, Koskinen H, Väätänen A. Smooth and resilient human–machine teamwork as an Industry 5.0 design challenge. Sustainability. 2022 Feb 26;14(5):2773.

[18] Tallat R, Hawbani A, Wang X, Al-Dubai A, Zhao L, Liu Z, Min G, Zomaya AY, Alsamhi SH. Navigating Industry 5.0: A survey of key enabling technologies, trends, challenges, and opportunities. IEEE Communications Surveys & Tutorials. 2023 Nov 2.

[19] Wahyuningtyas R, Disastra G, Rismayani R. Toward cooperative competitiveness for community development in economic Society 5.0. Journal of Enterprising Communities: People and Places in the Global Economy. 2023 Apr 28;17(3):594–620.

[20] Nair MM, Tyagi AK, Sreenath N. The future with Industry 4.0 at the core of Society 5.0: Open issues, future opportunities and challenges. In 2021 international conference on computer communication and informatics (ICCCI) (pp. 1–7). IEEE, 2021.

[21] Deguchi A, Hirai C, Matsuoka H, Nakano T, Oshima K, Tai M, Tani S. What is Society 5.0. 2020 May 29;5(0):1–24.

[22] Kasinathan P, Pugazhendhi R, Elavarasan RM, Ramachandaramurthy VK, Ramanathan V, Subramanian S, Kumar S, Nandhagopal K, Raghavan RR, Rangasamy S, Devendiran R. Realization of sustainable development goals with disruptive technologies by integrating Industry 5.0, Society 5.0, smart cities and villages. Sustainability. 2022 Nov 17;14(22):15258.

[23] Mourtzis D, Angelopoulos J, Panopoulos N. The future of the human–machine interface (HMI) in Society 5.0. Future Internet. 2023 Apr 27;15(5):162.

[24] Stryzhak O. Features of the relationship between human capital development and digital technologies in the context of Society 5.0 formation. Agricultural and Resource Economics: International Scientific E-Journal. 2022 Sep 20;8(3):224–243.

[25] Bungin B, Teguh M, Dafa M. Cyber community towards Society 5.0 and the future of social reality. International Journal of Computer and Information System (IJCIS). 2021 Aug 27;2(3):73–79.

[26] Coronado E, Kiyokawa T, Ricardez GA, Ramirez-Alpizar IG, Venture G, Yamanobe N. Evaluating quality in human-robot interaction: A systematic search and classification of performance and human-centered factors, measures and metrics towards an Industry 5.0. Journal of Manufacturing Systems. 2022 Apr 1;63:392–410.

[27] Murphy C, Carew PJ, Stapleton L. A human-centred systems manifesto for smart digital immersion in Industry 5.0: A case study of cultural heritage. AI & Society. 2023 May;30:1–6.

[28] Barata J, Kayser I. Industry 5.0–past, present, and near future. Procedia Computer Science. 2023 Jan 1;219:778–788.

[29] Stryzhak O. Features of the relationship between human capital development and digital technologies in the context of Society 5.0 formation. Agricultural and Resource Economics: International Scientific E-Journal. 2022 Sep 20;8(3):224–243.

[30] Longo F, Padovano A, Umbrello S. Value-oriented and ethical technology engineering in Industry 5.0: A human-centric perspective for the design of the factory of the future. Applied Sciences. 2020 Jun 18;10(12):4182.

[31] Murphy C, Carew PJ, Stapleton L. A human-centred systems manifesto for smart digital immersion in Industry 5.0: A case study of cultural heritage. AI & Society. 2023 May;30:1–6.

[32] Pizoń J, Gola A. Human–machine relationship—perspective and future roadmap for Industry 5.0 solutions. Machines. 2023 Feb 1;11(2):203.

[33] Rundo L, Pirrone R, Vitabile S, Sala E, Gambino O. Recent advances of HCI in decision-making tasks for optimized clinical workflows and precision medicine. Journal of biomedical informatics. 2020 Aug 1;108:103479.

[34] Rundo L, Pirrone R, Vitabile S, Sala E, Gambino O. Recent advances of HCI in decision-making tasks for optimized clinical workflows and precision medicine. Journal of Biomedical Informatics. 2020 Aug 1;108:103479.

[35] Rane, N. Transformers in Industry 4.0, Industry 5.0, and Society 5.0: Roles and Challenges. 2023.

[36] Singh, R, Tyagi, AK, Arumugam, SK. Imagining the sustainable future with Industry 6.0: A smarter pathway for modern society and manufacturing industries. In Machine Learning Algorithms Using Scikit and Tensor Flow Environments. IGI Global, 2024. https://doi.org/10.4018/978-1-6684-8531-6.ch016

[37] Deshmukh, A, Patil, DS, Soni, G, Tyagi, AK. Cyber security: New realities for Industry 4.0 and Society 5.0. In A. Tyagi (Ed.), Handbook of Research on Quantum Computing for Smart Environments (pp. 299–325). IGI Global, 2023. https://doi.org/10.4018/978-1-6684-6697-1.ch017

[38] Tyagi AK, Dananjayan S, Agarwal D, Thariq Ahmed HF. Blockchain—internet of things applications: Opportunities and challenges for Industry 4.0 and Society 5.0. Sensors. 2023;23(2):947. https://doi.org/10.3390/s23020947

[39] Tyagi AK, Fernandez TF, Mishra S, Kumari S. Intelligent automation systems at the core of Industry 4.0. In A. Abraham, V. Piuri, N. Gandhi, P. Siarry, A. Kaklauskas, A. Madureira (Eds.), Intelligent Systems Design and Applications. ISDA 2020. Advances in Intelligent Systems and Computing (vol. 1351). Cham: Springer, 2021. https://doi.org/10.1007/978-3-030-71187-0

[40] L. Gomathi, AK. Mishra and AK. Tyagi, Industry 5.0 for healthcare 5.0: Opportunities, challenges and future research possibilities. In 2023 7th International Conference on Trends in Electronics and Informatics (ICOEI) (pp. 204–213). Tirunelveli, 2023. https://doi.org/10.1109/ICOEI56765.2023.10125660

18 The Future of Manufacturing and Artificial Intelligence
Industry 6.0 and Beyond

Priyanga Subbiah, Amit Kumar Tyagi, and Bireshwar Dass Mazumdar

18.1 INTRODUCTION

As we live in this day and age, the increasing incorporation of AI into the complex web of production processes is announcing the arrival of a revolutionary era that has been dubbed Industry 6.0 [1]. In this chapter, we will begin an investigation into the limitless potential for change and development that is inherent in this emerging paradigm. The foundation is comprised of a careful investigation of the ever-changing landscape of contemporary manufacturing, with a purposeful emphasis placed on clarifying the mutually beneficial link that exists between artificial intelligence and industrial processes. This chapter presents a thorough overview of contemporary uses of artificial intelligence in manufacturing, giving information about the enormous influence that these applications have had on various aspects of production. Notable areas of effect include the promotion of adaptive manufacturing processes, the development of production efficiency, the optimisation of quality, the deployment of predictive maintenance strategies, and the optimisation of quality. The research casts its eyes into the future, imagining the directions that Industry 6.0 may take in the near future and pondering the regions that lie beyond. An array of prospective technical improvements, paradigmatic upheavals, and the integration of autonomous systems under the banner of artificial intelligence, edge computing, and human-robot interaction are among the things that are anticipated [2]. Taking centre stage is the central topic of responsible artificial intelligence (AI) utilisation in manufacturing, which addresses ethical issues, tackles implementation obstacles, and investigates the socio-economic ramifications that follow this revolutionary epoch. In order to facilitate the process of navigating the ever-changing environment of Industry 6.0, the purpose of this study is to provide a thorough roadmap. In order to provide stakeholders, lawmakers, and academics with important information about the most effective utilisation of artificial intelligence within industrial ecosystems, this strategy roadmap has been developed here. In order to encourage innovation, promote sustainability, and strengthen global competitiveness, the plan expands its scope of responsibility [3]. By shedding light on the complimentary nature of artificial

DOI: 10.1201/9781003473886-18

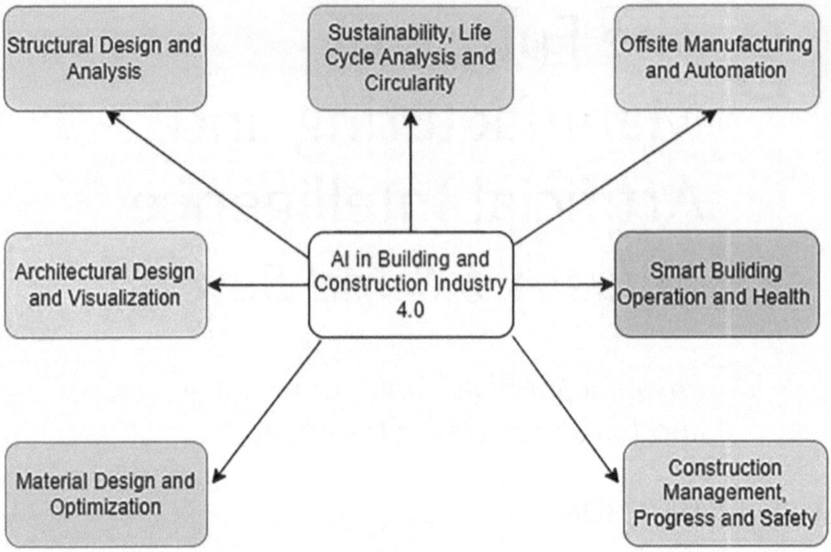

FIGURE 18.1 AI Application in Industry 4.0 Manufacturing.

intelligence and manufacturing, the paper highlights the possibility for these two fields to work together as future architects in the process of developing production systems that are both sophisticated and efficient. Figure 18.1 illustrates the applications of AI in the manufacturing of Industry 4.0 [4].

18.2 LITERATURE REVIEW

There is information in the provided document on the evaluation strategy for integrating AI into the Industry 6.0 metaverse [5]. Beyond Industry 6.0, this paper doesn't go into much detail on the AI and industrial future. We will go over the Key Enabling Technologies (KETs) that have been created from Industry 1.0 all the way up to Industry 4.0. It investigates real-world applications of artificial intelligence (AI) in robotics. This research paints a picture of how Industry 6.0 incorporates the metaverse.

There is the potential for increased efficiency, productivity, and sustainability in the manufacturing industry with the application of AI and ML [6]. Artificial intelligence has shown promise in a number of areas, including predictive maintenance, quality control, and process optimisation. There is the potential for increased efficiency, productivity, and sustainability in the manufacturing industry with the application of AI and ML. Artificial intelligence has shown promise in a number of areas, including predictive maintenance, quality control, and process optimisation.

This paper discusses about the pros and cons of AI integration into production processes, but it doesn't touch on "Industry 6.0" or the manufacturing of the future—topics that are outside the article's purview [7]. Increases in efficiency, output, and return on investment could result from the fast development of AI in industrial systems. Traditional mass production will give way to customised manufacturing in the not-too-distant future. The effectiveness, reliability, and financial gain of industrial systems may be enhanced with the use of artificial intelligence. Manufacturing systems of the future will have the brainpower to endure disruptions.

This study investigates the process of establishing interaction within the context of the metaverse as a framework [8]. It provides suggestions for the creation of things such as immersive virtual worlds, interactions that include several senses, and other things. Investigates the capacity of artificial intelligence technologies to facilitate communication in the metaverse. Recommendations might be useful in a number of different areas, including information connecting, multi-sensory interaction, time/space hopping, and immersive virtual space.

The purpose of this paper is to present a comprehensive summary of the achievements that artificial intelligence has accomplished in the industrial sector [9]. The topic of discussion in this article is the incorporation of artificial intelligence into high-precision engineering and production. This article aims to bring more attention to the various applications of artificial intelligence that might be beneficial to various industrial fields. The issue of artificial intelligence (AI) and its applications in the improvement of industrial processes and products is investigated in great detail during the course of this research.

In this work, a complete and methodical evaluation of the existing literature on the subject of artificial intelligence in manufacturing is presented [10]. The paper investigates the several applications of artificial intelligence in manufacturing and discusses the benefits and drawbacks of each use. This article presents a comprehensive and organised literature review on the subject of artificial intelligence and its influence on manufacturing. An explanation of the smart manufacturing processes, broken down into their individual components and the categories that they belong to.

The article is available for download and includes some thoughts on how smart manufacturing may benefit from AI [11]. A few examples of these technologies are intelligent manufacturing facility technology, generative product design, smart workbench, and anomaly detection. Excluding AI-related topics, the term "Industry 6.0" is not mentioned, and the article fails to address any other potential future of manufacturing. Here, we take a look at the many uses of artificial intelligence technology found in smart factories. Innovations in intelligent manufacturing facility technologies, smart workbench technologies, anomaly detection technologies, and generative product design will be covered. Productivity and efficiency in manufacturing are both enhanced by artificial intelligence technology. Engineering, quality control, autonomous production, and product design are some of the areas that use AI.

This article provides a comprehensive overview of Industrial Artificial Intelligence (AI) and its applications in real-world industrial settings, including a discussion of the challenges and opportunities associated with AI in Industry 4.0 production settings [12]. This paper provides a thorough review of the current state of investigation into industrial AI. The paper lays out the major challenges and opportunities that upcoming research projects will encounter. The current literature on industrial AI in manufacturing contexts is reviewed extensively in this article. The process of determining the most pressing research-related opportunities and challenges for the future.

The role of AI in smart manufacturing systems in industry is covered in a recently published paper [13]. The report omits any mention of what comes after Industry 6.0 in terms of manufacturing and AI. With this essay, we hope to present a synopsis of the key enabling technologies of industrial AI and their impact on the manufacturing industry. By systematically implementing these technologies, new opportunities for value creation may be created and problems can be avoided. This article provides an in-depth analysis of the important function of key enabling technologies of industrial AI within the framework of the manufacturing industry. It presents an analysis of how these technologies, when used extensively, will pave the way for new opportunities to create value and help us avoid problems.

This article will focus on the aspects of manufacturing that include the incorporation and application of information technology (IT) [14]. Additionally, the research emphasises the significance of data science technologies and intelligent system support for the implementation of smart manufacturing concepts. Next-generation information technology integration in the manufacturing industry is one of the national policies that are being implemented in many countries to facilitate the development of advanced manufacturing practices.

18.3 EVOLUTION OF INDUSTRY REVOLUTION

A historical journey that is characterised by transformative alterations in the way societies organise and carry out industrial activity is represented by the evolution of industry revolutions. Since the beginning of the First Industrial Revolution, which was driven by steam power and mechanisation, each successive phase has resulted in the development of new organisational paradigms and technological advancements. This chapter intends to give a contextual examination of the growth of industrial revolutions, charting the trajectory from the early days of mechanised production to the current period of Industry 6.0. Specifically, the chapter will focus on the evolution of the manufacturing industry. This background is situated within the larger historical narrative of industrial revolutions, which places an emphasis on the essential role that technical advances play in transforming economic systems, societal norms, and global contacts. Different socioeconomic shifts were brought about by each successive phase of the industrial revolution, beginning with the mechanisation and mass production of Industry 1.0 and culminating in the digitization and automation of Industry 4.0. At a time when we are on the verge of entering Industry 6.0, which will be characterised by the incorporation of artificial intelligence, the internet of things, and advanced data analytics, it is essential to have a solid grasp of the evolutionary route. The backdrop that has been constructed in this chapter not only reflects on the technological improvements that distinguished each industrial revolution, but it also illustrates the tremendous influence that these developments had on labour markets, business models, and the general fabric of society.

The purpose of this investigation is to serve as a guide for contextualising the contemporary age within the continuum of industrial revolutions. It does so by giving information about the driving factors, problems, and possibilities that have created and continue to influence the industrial environment. In doing so, it establishes the framework for grasping the ongoing progress and prepares the groundwork for anticipating the future trajectory of Industry 6.0 and beyond. Insightful and comprehensive, "Evolution of Industry Revolution" examines the past and present transformations of industrial landscapes as a result of technological advances. The present digital era, Industry 4.0, is the culmination of this comprehensive study that begins with the mechanization of Industry 1.0 and continues through the many stages of industrial revolutions. The rapid integration of AI into manufacturing processes, which has resulted in the rise of Industry 6.0, is an important part of this evolutionary trajectory. As events unfold, the novel emphasizes the complex connection between AI and industrial processes against the backdrop of a dynamic, modern manufacturing setting [15]. This study takes a close look at current AI applications in manufacturing and shows how these have changed many things, including production efficiency, quality optimization, adaptive manufacturing, predictive maintenance, and artificial intelligence overall. The chapter looks ahead and speculates on possible future trajectories, including but not limited to the arrival of Industry 6.0. Technological innovations, paradigm shifts, and the smooth integration of autonomous systems powered by AI, edge computing, and human-robot interaction are among the projected advancements. Finding the right way to use AI in manufacturing is a major focus of the aforementioned study. It discusses the bigger socioeconomic implications of this moment of change as well as ethical issue and implementation difficulties. In order to provide stakeholders, legislators, and academics with actionable advice in the midst of this dynamic environment, the study aims to provide an all-encompassing road map. This road map aims to facilitate industries in their pursuit of innovation, sustainability, and global competitiveness by providing guidance on how to make the most of artificial intelligence within their ecosystems. The article draws attention to the possibilities for cooperation between AI and manufacturing in an effort to put light on the common trajectory propelling the Evolution of Industry Revolution.

The term "Evolution of Industry Revolution" is a notion that embodies the dynamic and transformational journey that industries have taken via successive waves of technical developments. It is a story that takes place over the course of several different eras, each of which is characterised by a unique scientific advance that drastically alters the landscape of industrial operations. This investigation follows the progression of industrial revolutions throughout history, analysing the transitions from one phase to the next as well as the enormous effects that these revolutions had on society, the economy, and societal advancement. The First Industrial Revolution, which was characterised by mechanisation and the harnessing of steam power, led to the mechanised manufacturing of commodities. This revolution is considered to be the beginning of the progression of industrial revolution. Subsequent phases, such as the Second Industrial Revolution, which was characterised by electrification and mass production, the Third Industrial Revolution, which was characterised by the introduction of computers and automation, and the Fourth Industrial Revolution, which was fuelled by digital technologies and connectivity, have all brought about paradigm shifts in manufacturing and production.

Comprehending the evolution requires first acknowledging the ongoing pursuit of efficiency, innovation, and growth within industries. This provides the background for comprehending the evolution. The accomplishments of the revolution that came before it are built upon by each succeeding revolution, which brings up new opportunities and problems. A narrative is presented that investigates the ways in which many sectors have successfully navigated through these waves of transformation, adjusting to the disruptions brought about by technological breakthroughs and using these advancements to redefine their processes, products, and relationships with stakeholders. The current setting of the progress of the industry revolution is characterised by the integration of cutting-edge technologies such as artificial intelligence, data analytics, the Internet of Things (IoT), and other technologies. It is a reflection of the current status of industries that are on the verge of entering a new era, which is commonly referred to as Industry 6.0 [16]. This new age is characterised by the redefining of classic manufacturing paradigms by intelligent systems and automation. This investigation provides as a contextual framework for comprehending the historical continuity of industrial revolutions, recognising the influence that these revolutions have had on constructing the present industrial environment, and forecasting the possibilities that lay ahead in the continued evolution of industries.

18.4 NECESSITY OF SMART MANUFACTURING

Using intelligent manufacturing processes is now important for companies to thrive in today's economic world. With the help of state-of-the-art technology, this innovative method of manufacturing processes may boost efficiency, flexibility, and resilience. An assortment of factors highlights the important nature of implementing smart manufacturing processes:

Enhanced Efficiency: The Internet of Things (IoT), artificial intelligence (AI), and automation are examples of contemporary technologies that are being incorporated into smart manufacturing systems in order to make production processes more effective and simplified using these technologies. For this reason, there is an increase in total production, an increase in operational efficiency, and a decrease in downtime.

Data-Driven Decision-Making: The use of data analytics and monitoring in real time are the fundamental components of smart manufacturing, which is designed to deliver knowledge that can be put into action. The capacity to gather, assess, and act upon data makes it easier for manufacturers to make educated decisions, which, in turn, enables them to promptly adapt to changing market needs and operational issues.

Cost Reduction: There is a correlation between the automation and process optimisation that smart manufacturing provides and cost reductions. In the event that firms are able to reduce the amount of human error, energy consumption, and waste, they will be able to reduce their operating expenses.

Quality Improvement: Monitoring of product quality may be carried out in a continuous manner throughout the entirety of the production process, thanks to the utilisation of sensors and data analytics. Smart manufacturing provides for more accuracy, consistency, and adherence to quality standards, which ultimately results in an improvement in product quality over the course of the production process.

Supply Chain Resilience: Supply chain management also benefits from smart manufacturing, since it provides real-time visibility throughout the whole supply chain. This is one of the reasons why smart manufacturing is so beneficial. Businesses have the potential to increase their resilience by increasing their level of transparency [17]. This may allow them to react more quickly to disruptions, optimise their logistics, and engage with their suppliers in a more collaborative manner.

Adaptability and Customization: In today's fast-paced marketplaces, the ability to swiftly adapt to changing customer preferences and customise products is very necessary. Through the use of intelligent manufacturing methods, production procedures have the potential to become more adaptable and responsive, so enabling businesses to better satisfy the varied requirements of their customers.

Competitive Advantage: One of the most important factors in determining a company's competitive advantage is the broad use of smart manufacturing, which is rapidly becoming an essential component. In terms of creativity, efficiency, and responsiveness, businesses who invest in and make use of these technologies get a strategic advantage over their competitors. This advantage allows them to stay ahead of the competition.

Sustainability: Smart manufacturing is ecologically benign and consistent with sustainable practices, since it has a low impact on the environment, makes efficient use of resources, and produces less waste than traditional production methods. This commitment to sustainability will not only be recognised by individuals who are concerned about the environment, but it will also be beneficial to the community globally.

18.5 BENEFITS AND LIMITATIONS TOWARDS AUTOMATED/HUMAN MACHINE COLLABORATION TOWARDS INDUSTRY AUTOMATION

18.5.1 Benefits

Increased Efficiency: It is possible to improve the overall efficiency of an organization's operations by automating monotonous and repetitive jobs. This allows for the tasks to be completed in a timely manner while maintaining accuracy. Activities that require creative thinking, the analysis of complicated data, and the decision of difficult choices are activities that are especially suited to the human brain.

Enhanced Safety: Automation makes workplaces safer for everyone by taking over jobs that might be potentially harmful. This also minimises the likelihood of accidents occurring. Because of their ability to continually monitor and respond to potential safety issues, automated systems make it possible to prevent accidents before they occur. This is a significant advantage.

Improved Quality: There is a possibility that automated procedures will maintain quality standards, which implies that deviations in jobs that are handled by humans will be less likely to occur. Manufacturing and other processes can benefit from process automation because it guarantees the precise execution of tasks, which ultimately results in higher-quality outputs.

Cost Savings: By lowering the likelihood of human mistakes, removing procedures that aren't essential, and making more efficient use of the resources that are available, automating some processes might potentially save money. Because of their ability to operate continuously, automated systems frequently boost production and productivity without necessitating regular human supervision. This is because automated systems are able to function 24/7.

24/7 Operations: In contrast to processes that are dependent on humans, automated systems have a longer operating period, which guarantees continuous output and reduces the amount of time available for turnaround.

Data Analytics and Decision Support: With the large volume of data that are produced by automation, required information may be obtained, which can then be employed for the purpose of improving processes and making decisions. Through the use of data analytics, people are able to combine their own judgement with that of machine learning in order to arrive at conclusions that are informed.

18.5.2 Limitations

Initial Implementation Costs: It is possible that the installation of automation systems may be excessively expensive due to the large initial expenses connected with the infrastructure, training, and technology that are required.

Limited Adaptability: Automated systems may have difficulties when they are confronted with unexpected changes or variations in their tasks, which may necessitate reprogramming in order to accept the new conditions. Due to the fact that it lacks the cognitive skills and adaptability of humans, automation has difficulties while attempting to acquire autonomy in undertakings that are not customary or complicated.

Job Displacement and Workforce Transition: There is an issue that automation may result in the loss of jobs in particular industries, which may then lead to an increase in the number of people without jobs and the opportunity to retrain workers. The incorporation of automation may need the retraining and augmentation of skills of the existing personnel, which presents difficulties for the process of workforce transfer.

Dependency on Technology: Whenever an organisation places an excessive amount of reliance on automation, it exposes itself to the possibility of experiencing disruptions in its operations as a result of breaches in cybersecurity, system malfunctions, or technological failures. There is a possibility that an excessive dependence on automated technology might result in a loss in human abilities related with actual physical work.

Ethical and Legal Issues: When employing automated approaches, there are some issues that occur with regard to the assessment of culpability, which creates both ethical and legal difficulties. Issues have been raised regarding the integrity of the use of personal information and the protection of individuals' right to privacy in relation to automated systems that are responsible for data collection.

Resistance to Change: It is possible that some individuals are hesitant to accept automation for a variety of reasons, including a lack of familiarity with the technology, a reluctance to adapt to change, or worries about the stability of their employment.

Complex Maintenance: The maintenance of automated systems requires a specialist approach in order to function properly. Additionally, conquering technological problems can be challenging, which is why it is essential to have the experience of specialists so that you can overcome these challenges.

18.6 OPEN ISSUES AND CHALLENGES TOWARDS INDUSTRY 6.0 AND BEYOND

Ensuring seamless interoperability across many technologies is a challenging task. This includes ensuring that IoT devices, AI systems, and edge computing can communicate with one another. One thing that prevents the different parts from working together effectively is the lack of established protocols and frameworks for communication.

Security Issues: There has been an increase in the number of cybersecurity risks, including data breaches, ransomware attacks, and acts of sabotage, as a result of the increased connectivity that Industry 6.0 has brought about. Networked systems are responsible for the generation of sensitive data, which must be safeguarded through the implementation of stringent privacy measures and the maintenance of compliance with rules.

Data Management and Analytics: When trying to handle the large amounts of data that are created by Internet of Things devices and automated systems, several challenges arise in the areas of storage, processing, and the extraction of insights. In order to enhance our ability to make better decisions in a shorter amount of time, we will always have a need to develop our abilities in real-time analytics.

Reskilling the Workforce: In order to successfully transition to Industry 6.0, it is imperative to have a workforce that possesses high-level technical capabilities. This necessitates the implementation of comprehensive retraining programmes. Because they want to make the transition into the new position easier for their employees and soothe their anxieties about the possibility of job loss due to automation.

Ethical and Regulatory Frameworks: The work of defining ethical rules for the proper use of artificial intelligence and other sophisticated technologies is now being carried out by Industry 6.0. Because of this, it is necessary to be current with the ever-changing regulations that deal to data, privacy, and advances in technology.

Infrastructure Readiness: It is necessary to construct communication networks that are not only dependable but also speedy in order to fulfil the data-intensive needs that are imposed by Industry 6.0. Build capabilities for edge computing, which bring processing closer to the site where the data is being created, in order to obtain a decrease in latency. This is achievable in order to meet the goal of achieving a reduction in latency.

Environmental Impact: There is a possibility that the negative impacts that Industry 6.0 will have on the environment can be mitigated by addressing the issues surrounding energy consumption that are associated with the continuous operation of networked devices and systems. The achievement of this goal may be facilitated by locating a compromise between the development of new technologies and the adoption of ecologically responsible and sustainable practices.

Adaptability and Scalability: Systems that are a part of Industry 6.0 need to be able to adapt to new technologies and the ever-changing requirements of organisations. This means that designs need to be flexible in order to accommodate these requirements. It is important to make certain that the architecture and technology of the system can successfully scale in order to effectively serve the ever-increasing requirements of linked ecosystems. This is the case since it is the only way to properly satisfy these criteria.

Global Collaboration: To reach the goal of worldwide standardisation of the technology and practice of Industry 6.0 standards, it is important to expand the amount of international collaboration that is taking place. Data governance and the transportation of data across international borders are two of the challenges that are faced by networked industrial systems.

Social Acceptance: Within the context of resolving issues regarding privacy, employment stability, and the impact on society, it is of the utmost importance that the general people have faith in and support for the technologies that are a part of Industry 6.0. In order to meet the standards of this programme, each and every community must be able to experience a considerable improvement in their standard of living and be able to take advantage of the benefits that Industry 6.0 has to provide.

Cultural Shift and Leadership: In order to truly embrace digital-first thinking, collaboration, and innovation, it is necessary to give direction on how to effectively traverse the cultural transformation that is taking place within companies. The enhancement of one's capacity to lead others in order to guarantee that enterprises will be able to endure the unavoidable transformations that will be brought about by the implementation of technology that is part of the Industry 6.0.

18.7 FUTURE RESEARCH OPPORTUNITIES TOWARDS INDUSTRY 6.0 AND BEYOND

Autonomous Systems and AI Integration: In order to accomplish the objective of autonomously making decisions in industrial processes that need a significant amount of human labour, research is now being conducted to develop algorithms that are founded on artificial intelligence. This is being done with the intention of reaching the aim of autonomously making decisions. In order for Industry 6.0 to be successful, it is very necessary to do research on a variety of methods that make it feasible for people and autonomous systems to collaborate in a manner that is more effective.

Blockchain Applications: Conduct study on the potential of blockchain technology to create safe and decentralised systems for the administration of data, transactions, and transactions including smart contracts. This research should be conducted within the framework of Industry 6.0. This study aims to get a better knowledge of the ways in which blockchain technology has the potential to make global supply chains more transparent and simpler to monitor. A better understanding of these possible benefits is the goal of this study.

Quantum Computing: The conduct of research to identify whether or not quantum computing may enhance data processing with Industry 6.0 and develop answers to complex optimisation difficulties is an essential step that has to be performed. This study must be carried out. It is of the utmost importance to perform research on cryptographic algorithms that are resistant to quantum attacks in order to guarantee the security of data transfers that take place within networked industrial systems.

Explainable AI and Trustworthiness: Artificial intelligence models that contain reasoning that is clear and easy to understand are of the highest relevance in industrial situations. The creation and refining of these models are becoming increasingly important. When investigating the many methods that may be utilised to ensure the dependability of automated systems and artificial intelligence, it is important to take into consideration problems like equality, transparency, and responsibility.

Human-Centric Design: The fundamental purpose of this research is to improve the user experience for end-users as well as operators through the design of applications. This will be accomplished through the creation of systems for Industry 6.0. It is of the highest significance to do research on the impact of technology on human well-being, including work satisfaction and mental health, within the context of smart manufacturing. This study falls under the category of "smart manufacturing."

Edge Computing Optimization: The exploration of optimisation strategies for edge computing may result in an improvement in real-time processing capabilities, a reduction in latency, and an overall gain in system efficiency. These outcomes are all possible outcomes. For the purpose of facilitating edge computing in large-scale industrial settings, researchers are currently concentrating their efforts on the development of intelligence models that are distributed and decentralised simultaneously.

Green and Sustainable Manufacturing: In the context of smart manufacturing processes, it is of the highest significance to discuss and investigate technologies and methods that give increases in energy efficiency and sustainability. This is because these processes are becoming increasingly important. Determine the ways in which the concepts of a circular economy may be implemented by Industry 6.0 in order to reduce the quantity of waste that is generated and to enhance the effectiveness with which resources are utilised.

Human-Robot Collaboration: The research that is being done on smart manufacturing should prioritise the development of enhanced safety procedures and processes that will make it simpler for people and robots to collaborate with one another. In order to improve the cognitive capacities of humans, it is recommended that research be carried out to study the potential applications of wearable technology, augmented reality, and artificial intelligence (AI) in the industrial sector.

Experiential Learning Systems: It is recommended that, as part of the workforce training for Industry 6.0, virtual reality (VR) and augmented reality (AR) be utilised in order to develop training simulations that are not only realistic but also immersive. Taking into consideration the fact that the smart manufacturing industry is a dynamic sector, it is of the highest significance to research solutions that enable workers to acquire new skills while they are working.

Adaptive Cybersecurity: One of the areas of study that is now being conducted is the investigation of how to construct adaptive cybersecurity solutions that are able to recognise and respond to developing cyber threats in networked industrial systems. Consider the possibility of using zero-trust architectures in order to improve the level of security that is present in Industry 6.0 ecosystems.

Resilience Engineering: Conduct research on a wide variety of methods that may be employed to construct manufacturing systems that are robust. These systems ought to be able to adjust to interruptions, recover rapidly, and continue to function without interruption. By utilising technology-driven solutions and risk mitigation tactics, we are investigating the various potential ways that may be taken to enhance global supply chains.

Cross-Disciplinary Collaboration: In order to find solutions to the several issues that Industry 6.0 presents, professionals in the fields of engineering, social science, economics, and environmental science ought to collaborate more closely. We need to take a look at the ways in which sectors such as biotechnology, the Internet of Things, and artificial intelligence are coming together in order to develop novel solutions for smart manufacturing.

Regulatory Frameworks and Policy Research: If you want to overcome regulatory obstacles in the context of linked industrial systems, you should experiment with different sorts of global governance and ways of working together. Investigate the most effective approaches to establishing regulatory frameworks and rules in order to get further knowledge on the responsible implementation of Industry 6.0.

Human-Centric AI Governance: When it comes to the process of developing, implementing, and managing artificial intelligence in manufacturing, doing research on governance models that give ethical issues priority is absolutely necessary. Take into consideration the many different ways in which a varied collection of individuals, including as workers and communities, might be involved in the decision-making process about Industry 6.0 artificial intelligence.

Quantifiable Impact Assessment: In order to analyse the consequences that the rollouts of Industry 6.0 will have on society, the economy, and the environment, there are some parameters that may be quantified. In order to anticipate and minimise the consequences of unforeseen events and long-term implications that may be brought about by the widespread adoption of Industry 6.0, we are doing study into a number of different ways that are now accessible.

18.8 A FUTURE WITH EMERGING TECHNOLOGIES FOR EFFECTIVE AND SUSTAINABLE INDUSTRY 6.0-BASED ENVIRONMENT

The implementation of new technology is propelling Industry 6.0 in the direction of a future that will be characterised by capabilities that are ground-breaking, sustainability, and efficiency on a scale that has never been found before. A future in which intelligent and environmentally friendly manufacturing takes place in an environment where cutting-edge technologies coexist is the target of this study. The purpose of this research is to conceptualise such a future. The focus of this study is on the ways in which artificial intelligence, the Internet of Things, blockchain technology, and sophisticated robots all interact with one another to produce the desired results. The potential of artificial intelligence-driven decision-making, the function of blockchain in assuring transaction transparency and security, and the role of the Internet of Things (IoT) in enabling connections are all topics that are investigated. Recent advancements in robotic systems that allow for human-machine collaboration are also investigated. The necessity of sustainability in Industry 6.0 is emphasised as a result of the far-reaching repercussions that technology advancement has on ecosystems as well as social and economic institutions. The development of production systems that are both durable and adaptable, the utilisation of the potential of edge computing, and the guaranteeing of the deployment of technological applications in an ethical and responsible manner are all essential components. The study also dives into problems such as cybersecurity, workforce adaption, and international regulatory frameworks, all of which are essential to the successful deployment of Industry 6.0. In order for academics, policymakers, and industry stakeholders to be able to assist in bringing Industry 6.0 to its full potential, the purpose of this research is to create a future in which a variety of technologies will function together in harmony. The future that is envisioned involves the utilisation of technology to improve operational efficiency and to contribute to the establishment of an industrial environment that is sustainable, inclusive, and ethically controlled.

Industry 6.0 is an important turning point in the development of industrial processes, which is being pushed by the seamless integration of fast-growing technology. This turning moment is represented by the launch of Industry 6.0. To analyse the landscape of this revolutionary age and to research the potential of cutting-edge technologies to establish a foundation that is both effective and sustainable for Industry

6.0, the objective of this study is to investigate both of these aspects. The context becomes clearer when seen against the backdrop of rapid technological advancements and the huge repercussions that these advances have for the ecology of the industrial sector. In this day and age, industries are on the verge of undergoing a technological revolution, which is marked by the convergence of a number of diverse creative technologies that are rewriting the paradigms of conventional production. This revolution is expected to take place in the near future. In addition to the attention that is given on enhancing operational efficiency, there is also an emphasis made on lobbying for environmentally responsible practices in industrial settings. The mix of artificial intelligence, the Internet of Things (IoT), edge computing, and other emerging technologies is what distinguishes this environment. These novel technologies together steer the industry in the direction of a future that is more intelligent and connected. In order to achieve the goal of this study, which is to navigate through the complex interaction of these new technologies, the purpose of this study is to attempt to anticipate the role that these technologies will play in the establishment of an environment that is not only efficient in terms of production but also sustainable and environmentally conscious. Furthermore, it studies the synergies that exist between technological advancements and their capacity to handle challenges that are linked with the utilisation of resources, the effect on the environment, and the well-being of society. Specifically, all of these issues are addressed.

These new technologies are becoming increasingly important as industries strive to become more robust and flexible in the face of changeable global difficulties. The future is dependent on the strategic deployment of these new technologies, which is becoming increasingly important more and more. This chapter serves as a guide to understanding the trajectory towards an environment based on Industry 6.0 that is not only technologically advanced but also aware of its effect on the greater socio-economic and environmental landscape. This chapter is intended to serve as a foundation for comprehending the trajectory. The ambiance that is described in this article can be used as a point of reference.

18.9 CONCLUSION

When artificial intelligence is incorporated into industrial processes, a new period of tremendous change and advancement will begin. This new era is known as Industry 6.0. In order to bring attention to the significant influence that the integration of artificial intelligence and industrial processes has had on modern production, the purpose of this chapter is to demonstrate how these two technologies may work together to improve the manufacturing process. A comprehensive examination of the applications of artificial intelligence that are now being used in manufacturing has revealed significant gains in several areas, including production efficiency, quality optimisation, predictive maintenance, and adaptive manufacturing. The opposite of this is that this research does not stop at the here and now; rather, it looks ahead to the future and attempts to visualise how Industry 6.0 will develop beyond the paradigms that are now in place. Because it anticipates technological developments, paradigm shifts, and the inclusion of autonomous systems, the study emphasises the importance of making responsible use of artificial intelligence. These autonomous systems may include, among other things, artificial intelligence, edge computing, and human-robot interaction. In light of the fact that we are addressing ethical issues, implementation problems, and the socio-economic repercussions of this important juncture, it is imperative that we take a calm and thorough approach. In order to provide academics, researchers, etc., with a comprehensive road map that they may follow as they navigate the constantly shifting terrain of the industry 6.0 environment, the purpose of this project is to collect data. Not only does this road map outline the best practices for integrating artificial intelligence into industrial ecosystems, but it also stresses innovation, sustainability, and international competitiveness as two of the most important criteria that will determine success. This chapter provides light on the prospects of manufacturing and artificial intelligence working together, which leads to a future in which these two sectors will combine to transform the basic character of production.

REFERENCES

[1] Deshmukh A, Patil DS, Soni G, Tyagi AK (2023) Cyber Security: New Realities for Industry 4.0 and Society 5.0. In A Tyagi (Ed.), Handbook of Research on Quantum Computing for Smart Environments (pp. 299–325). IGI Global, USA. doi: 10.4018/978-1-6684-6697-1.ch017

[2] Eun Seo L, Heechul B, Hyun Jong K, Hyonyoung H, Lee Yong K, Ji Yeon S (2020) Trends in AI Technology for Smart Manufacturing in the Future. doi: 10.22648/ETRI.2020.J.350106

[3] Gomathi L, Mishra AK, Tyagi AK (2023) Industry 5.0 for Healthcare 5.0: Opportunities, Challenges and Future Research Possibilities. In 2023 7th International Conference on Trends in Electronics and Informatics (ICOEI) (pp. 204–213). Tirunelveli, IEEE. doi: 10.1109/ICOEI56765.2023.10125660.

[4] Heilala J, Singh K (2023) Evaluation Planning for Artificial Intelligence-based Industry 6.0, Metaverse Integration. AHFE International. doi: 10.54941/ahfe1002892

[5] Kim SW, Kong JH, Lee SW, Lee S-C, Lee S-C (2021) Recent Advances of Artificial Intelligence in Manufacturing Industrial Sectors: A Review. International Journal of Precision Engineering and Manufacturing. doi: 10.1007/S12541-021-00600-3

[6] Kovalenko I, Barton K, Moyne J, Tilbury DM (2023). Opportunities and Challenges to Integrate Artificial Intelligence into Manufacturing Systems: Thoughts from a Panel Discussion. IEEE Robotics & Automation Magazine. doi: 10.1109/MRA.2023.3262464

[7] Lee J, Singh J, Azamfar M, Pandhare V (2020) Industrial AI and Predictive Analytics for Smart Manufacturing Systems. doi: 10.1016/B978-0-12-820027-8.00008-3

[8] Li Y, Zhang L (2023) Research on Interaction Design Based on Artificial Intelligence Technology in a Metaverse Environment. Lecture Notes in Computer Science. doi: 10.1007/978-3-031-35699-515

[9] Mu-Yen C, Edwin L, Erol E (2022) Deep Learning and Intelligent System Towards Smart Manufacturing. Enterprise Information Systems. doi: 10.1080/17517575.2021.1898050

[10] Nair MM, Tyagi AK, Sreenath N (2021) The Future with Industry 4.0 at the Core of Society 5.0: Open Issues, Future Opportunities and Challenges. In 2021 International Conference on Computer Communication and Informatics (ICCCI) (pp. 1–7). IEEE, New York.

[11] Plathottam SJ, Iloeje CO (2023). A Review of Artificial Intelligence Applications in Manufacturing Operations. Journal of Advanced Manufacturing and Processing. doi: 10.1002/amp2.10159

[12] Ricardo S, Peres X, Jia J, Lee K, Sun A, Walter C, Jose B (2020) Industrial Artificial Intelligence in Industry 4.0—Systematic Review, Challenges and Outlook. IEEE Access. doi: 10.1109/ACCESS.2020.3042874

[13] Shreyas BS, Haripriya D (2023) Applications of Artificial Intelligence in Manufacturing: A Review. International Journal of Trendy Research in Engineering and Technology. doi: 10.54473/ijtret.2023.7105

[14] Singh R, Tyagi AK, Arumugam SK (2024) Imagining the Sustainable Future with Industry 6.0: A Smarter Pathway for Modern Society and Manufacturing Industries. In Machine Learning Algorithms Using Scikit and TensorFlow Environments. IGI Global. doi: 10.4018/978-1-6684-8531-6.ch016

[15] Tyagi AK, Dananjayan S, Agarwal D, Ahmed THF (2023) Blockchain—Internet of Things Applications: Opportunities and Challenges for Industry 4.0 and Society 5.0. Sensors. 2023; 23(2):947. doi: 10.3390/s23020947

[16] Tyagi AK, Fernandez TF, Mishra S, Kumari S (2021) Intelligent Automation Systems at the Core of Industry 4.0. In A Abraham, V Piuri, N Gandhi, P Siarry, A Kaklauskas, A Madureira (Eds.), Intelligent Systems Design and Applications. ISDA 2020. Advances in Intelligent Systems and Computing (vol 1351). Springer, Cham. doi: 1007/978-3-030-71187-0

[17] Tyagi AK, Lakshmi Priya R, Mishra AK, Balamurugan G (2023). Industry 5.0: Potentials, Issues, Opportunities, and Challenges for Society 5.0. In Privacy Preservation of Genomic and Medical Data (pp. 409–432). Wiley.

Index